FOSSILS AND EVOLUTION

Wenlock limestone from Dudley with bryozoa, corals, brachiopods, bivalves, trilobites, and crinoids.

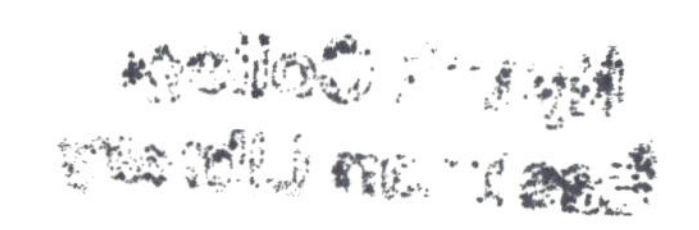

# FOSSILS AND EVOLUTION

■

T. S. Kemp
*University of Oxford*

OXFORD
UNIVERSITY PRESS

OXFORD
UNIVERSITY PRESS

Great Clarendon Street, Oxford OX2 6DP

Oxford New York
Athens Auckland Bangkok Bogotá Buenos Aires Calcutta
Cape Town Chennai Dar es Salaam Delhi Florence Hong Kong Istanbul
Karachi Kuala Lumpur Madrid Melbourne Mexico City Mumbai
Nairobi Paris São Paulo Singapore Taipei Tokyo Toronto Warsaw

and associated companies in
Berlin Ibadan

Oxford is a registered trade mark of Oxford University Press

Published in the United States
by Oxford University Press, Inc., New York



British Library Cataloguing in Publication Data
(Data available)

Library of Congress Cataloging in Publication Data

1 3 5 7 9 10 8 6 4 2

ISBN 0 19 850345 8 (Hbk)
0 19 850424 1 (Pbk)

Typeset by Hewer Text Ltd, Edinburgh
Printed in Great Britain on acid free paper by
Redwood Books, Trowbridge, Wilts

# Contents


**1** Introduction 1

**Part I PRINCIPLES**

**2** Some fundamental ideas 11
Pattern and process 11
Cladograms, trees, and scenarios 18
The epistemological gap 21

**3** Evolutionary theory: analysing process 24
The new evolutionary synthesis 26
Contemporary arguments 27

**4** Taxonomy: analysing pattern 48
Cladistics and genealogy 51
Phenetics and disparity 66

**5** Incompleteness and what to do about it 71
Organismic incompleteness 72
Stratigraphic incompleteness 85
Ecological incompleteness 94
Biogeographical incompleteness 100

**Part II PRACTICES**

**6** Fossils and phylogeny: if only we had more fossils 107
Fossils and phylogenetic analysis 107
Fossils and molecules 118
Fossils and formal classification 121

**7** Speciation: gradual, punctuated, or what? 129
The neontological perspective 129
The palaeontological perspective 133
The debate and the synthesis 138

**8** Rules and laws of taxonomic turnover: are there any? 156
The units of taxonomic turnover 156
The total diversity of the Earth's biota 159
Diversity changes within taxa 171
Clade interactions 179

**9** Mass extinctions: resetting the evolutionary clock 188
The epistemology of mass extinctions 189
The global Phanerozoic pattern 201
The Big Seven bio-events 202
Overview of the causation of mass extinctions 214

**10** The origin of new higher taxa: the ultimate question 217
Questions and speculations 218
Case histories 226
Generalizations 246

**11** Epilogue: where next? 251
The known 251
The knowable 253
A conclusion 255

*References* 257

*Index* 277

# 1
# Introduction

What are fossils good for? These remains of organisms that once lived represent the only direct evidence of what has been happening to life on Earth during the last 3000 million (3 billion) years of preserved palaeontological history. But, of course, they do not tell us much at all that is actually 'direct', because for the most part they consist only of a few hard bits of a handful of specimens representing a minute fraction of all the species that must have lived during this time. Indeed, as D.B. Kitts put it in his splendid essay of a quarter of a century ago, 'fossils by themselves tell us nothing; not even that they are fossils' (Kitts 1974). We must infer it for ourselves with the help of other kinds of information (Fig. 1.1).

It must be made absolutely clear from the start of this account that common statements such as 'the number of species has increased over the last 600 million years', 'the birds evolved from dinosaurian ancestors', 'ammonites illustrate the phenomenon of parallel evolution'; and so on, are not 'direct' observations of the fossil record at all but are complex interpretations derived from a multitude of sources. The fossilized remains, of course, play their part, but we also require assorted beliefs about genetics, ecological processes, taxonomic theory, chronostratigraphy, sedimentology, and so on.

The essential question really comes to this: should there be theories of evolution and ecology that are based on knowledge of the nature of living organisms, and that can then be applied to the fossil record as a way of interpreting these assorted bits of dead organisms? Or can the fossil record itself actually be the source of these kinds of theories? Responses to this question have resulted in two rather extreme myths about the usefulness of the fossil record, and particular palaeontologists can often be placed at some point on the intervening scale. The first, and older one might be termed the Stakhanovite Myth, which says that if enough palaeontologists were to work hard enough for long enough, then a good enough fossil record would be discovered to tell us practically everything about life in the past and how it evolved. The second one is the Phenomenalist Myth, which claims that the fossil record, even in principle, could never tell us anything beyond a bit about the anatomy of a few kinds of organisms that once lived.

Naturally the truth lies somewhere in between these two extremes and a main purpose of the chapters that follow is to show how rational, scientifically acceptable hypotheses can be created with contributions from fossils and from the biology of modern organisms combined: how the contemporary

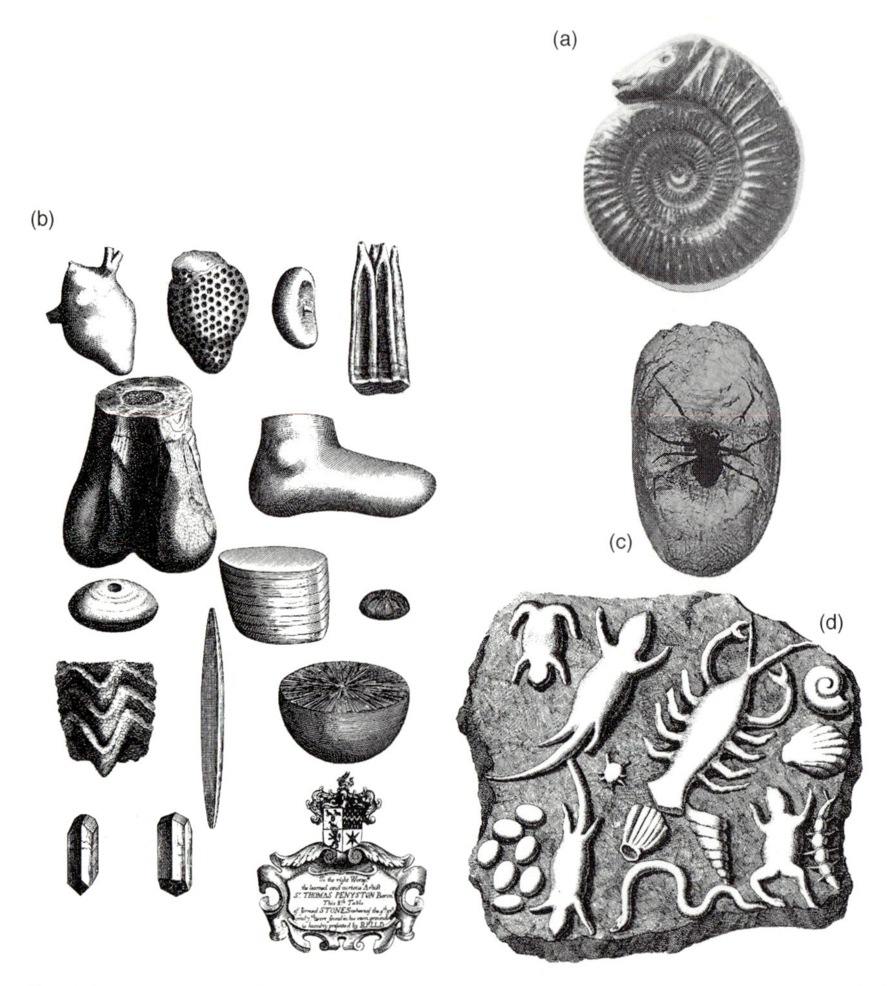


**Fig. 1.1** Misinterpreted fossils. (a) A 'snakestone' (an ammonite believed to be the petrified body of a snake). (b) Various geological artefacts believed by Robert Plot to be fossilized parts of the human body. (c) A spider enclosed in Kauri Gum in order to fake a fossil in amber. (d) Example of the 'Beringer Forgeries', a crudely carved representations of various living creatures designed to test the credulity of Dr Beringer, professor at Wurtzburg (they worked very successfully). ((a), Specimen in Oxford University Museum; (b) from Plot 1677; (c) from Grimaldi *et al.* 1994; (d) from Beringer 1726.)

partnership that supposedly exists between palaeontological and neontological studies invoked by Gould (1983) actually works and produces results.

Three general areas of interest are illuminated by the fossil record to the extent that they represent the three respective focuses of much palaeontological work. In fact, together, they form the straightforward answer to the question posed at the start: what are fossils good for? The first area concerns the discovery of kinds of organisms that are no longer found on Earth, such as trilobites, dinosaurs, and the earliest land plants. There is great intrinsic interest in the one-time existence of completely unfamiliar and often quite bizarre, unexpected organisms (Fig. 1.2), an interest shared by large numbers of non-palaeontologists, as a visit to any large natural history museum will soon attest. Reflected here is simply the deep human curiosity about the world we live in.

There is also a more formalized and scientific version of this interest in extinct kinds of organisms—the study of the nature of morphological form. Despite their great variety, all organisms are actually constructed from a rather small number of biological materials. Their form—that is, the way in which these materials are assembled—is subjected to various laws of physics, such as those that govern the need to be strong enough to resist external stresses, to have adequate surface areas for exchanging materials with the environment, and to be able to perform lots of different functions within a relatively small volume. There is also a demand imposed by the need to grow gradually from a fertilized egg by successive cell divisions, movements, and differentiations whilst remaining viable all

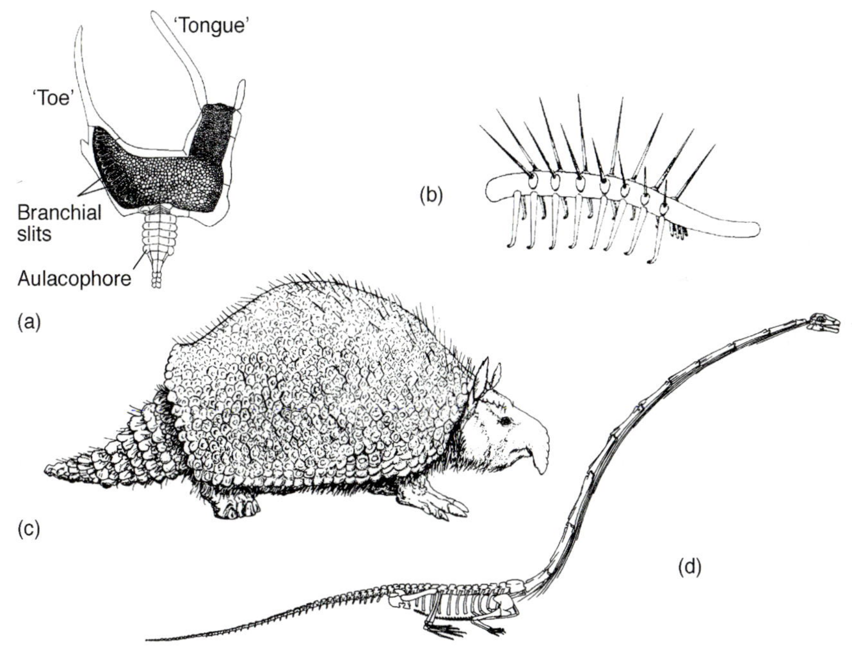


**Fig. 1.2** Weird organisms of the past. (a) *Cothurnocystis*, a member of a Palaeozoic group of organisms usually interpeted as echinoderms, but interpreted by Jefferies (1986) as possessing characters of vertebrates such as gill slits, brain, and postanal tail. (b) *Hallucigenia*, a spiny, long-legged onychophoran from the Middle Cambrian Burgess Shale of Canada. (c) *Glyptodon*, a 2-metre long edentate mammal completely enclosed in heavy armour, from the Pleistocene of South America. (d) *Tanystropheus*, a Triassic reptile, perhaps using its elongated neck as a fishing rod. ((a), (b) After Clarkson 1993; (c), (d) after Carroll 1988.)

along, and the requirement to make a living by appropriate adaptations to various environmental features.

All these considerations lead to the general idea that there are limits, or constraints, to what kinds of morphology can theoretically exist. One way to study this is to look at what kinds of organisms actually exist; fossils significantly extend the range of examples of morphologies that must be able to exist in theory, because they undoubtedly did in practice. A nice simple example is the question of what is the maximum possible size of a terrestrial organism. Without a fossil record, there would be a case for saying that elephants, weighing in at 7–8 tonnes, must be around the limit, otherwise bigger animals would be found. The discovery of sauropod dinosaurs disabused biologists of any such idea. In fact, sauropod dinosaurs offer the further moral that one should never be too confident that any theoretical limit is ever reached. The world record for body weight amongst sauropods continues to be regularly broken. *Diplodocus* amazed the nineteenth-century world with its estimated body weight of some 15–20 tonnes, but in due course *Brachiosaurus* quite dwarfed it with a weight variously estimated at between 50 and 80 tonnes. This certainly seemed that it must be at any conceivable limit. But in 1985, Jim Jensen announced his discovery of giant sauropod bones of such a size that they led to an estimate of up to 90 tonnes for *Supersaurus*. In the same year, David Gillette's 'Sam' was also announced. This partial skeleton, now known as *Seismosaurus*, could well indicate a body weight in excess of 100 tonnes and a body length of perhaps 50 metres. Each of these dinosaurs in turn could have been taken as indicating the limit to body mass for a land animal, so how confident may one really be that *Seismosaurus* is at this limit?

The second area of general interest in palaeontology is the way in which fossils illustrate the astonishing fact that today's organisms taken together are only the latest, momentary representation of an enormously complex pattern of turnover of different kinds of organisms through time. Species and higher taxonomic groups alike appeared and disappeared, replaced one another, invaded whole new habitats and geographical areas, suffered occasional massive synchronized extinctions and explosive radiations, and so on in a great never-repeating kaleidoscope of three billion years of history. Without a fossil record there would no doubt still be a theory of evolution. But there would be no inkling that it had followed such a tortuous and complicated course. The phylogenetic tree of life based solely on those species alive today would be a very modest plant compared with the evident truth (Fig. 1.3). (And humans would probably take the existence of their own species even more seriously than they do as it is.) The scientific study of this aspect of the fossil record is focused on recognizing the different kinds of organisms and looking for their taxonomic relationships to one another. The outcome is in the form of classifications and phylogenies, illustrating the nature of the diversity that exists amongst the organisms, extinct and extant.

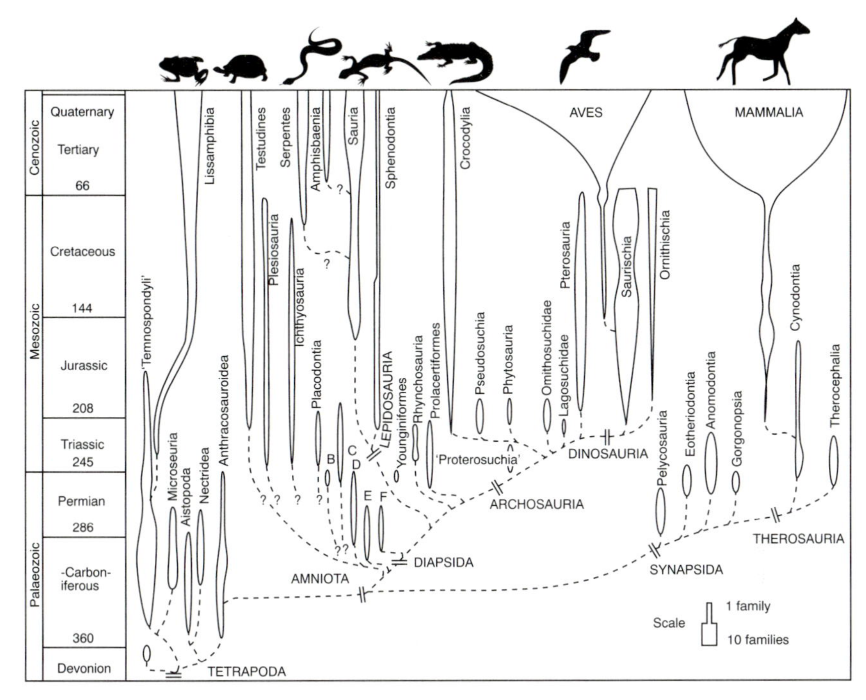


**Fig. 1.3** Phylogenetic tree of the tetrapods to show the complexity of the phylogeny underlying the emergence of modern groups. (After Benton 1989.)

The third area of palaeontological endeavour concerns the use of the fossil record to throw light upon the mechanisms of evolutionary change as they manifest themselves over very long periods of time. It is easy in principle to look at populations of living organisms and see what kinds of changes occur, generation by generation, when subject to particular environmental factors, natural or artificial. Thus, for example, a lot is known about the evolution of industrial melanism in insects such as the peppered moth, and about how fruit fly populations change when subjected to selection in the laboratory for practically everything imaginable, from the number of leg bristles to mating preferences. But it is a very long way indeed from this kind of knowledge to a confident belief that what has been discovered in the field and the laboratory offers a complete explanation of the kinds of evolutionary change that took millions, tens of millions, and even hundreds of millions of years to go to completion. So focusing on this aspect of palaeontology means looking at the taxonomic patterns of diversity evident in the fossil record and using them to infer what kinds of evolutionary processes could have been at work to produce them. This is to return to the concern of the opening paragraph of this introduction: how to use the fossil record to discover mechanisms of evolution if it appears to be necessary to know about the mechanisms of evolution in order to be able to describe coherently the fossil record in the first place.

The chapters that follow are arranged into two parts. Part I is an account of the principles of palaeobiological inference. In the last couple of decades or so, most active palaeobiologists have come to pay appropriate attention to the scientific principles that should govern their discipline. There has been a rapid spread of such commendable procedures as objectivity of observation (rather than an authoritarian choice of what to describe); use of simplicity or Occam's razor as the means for choosing between explanations (rather than defending a favourite hypothesis through the use of excessive extra assumptions); calculating statistical significance and confidence levels (rather than just eyeballing the data); a more open-minded approach to possible additional mechanisms of evolution (rather than relying solely on strict neo-Darwinism); awareness of the methodological and epistemological limitations of the fossil record (rather than assuming that what you expected or hoped to find is simply undiscovered or unpreserved due to ill-luck).

This new rigour has much to do with what has made palaeontology such an exciting field in recent years, but there is still a lot of disagreement about what actually are the correct procedures and acceptable kinds of explanations. Rather than confuse the discussion of the particular palaeontological questions by simultaneously expanding upon the underlying philosophy and methodology of the subject, it is easier to begin with an extended consideration of these matters of principle.

Chapter 2 is about the application to palaeobiology of the wider issues of philosophy of science, such as how should different kinds of evidence be combined to produce explanatory theories, and what kinds of questions can properly be asked.

Chapter 3 is a brief summary of those areas of evolutionary theory and the controversies surrounding them that are most relevant to long-term evolution and interpretation of the fossil record. Chapter 4 deals in a similarly brief fashion with taxonomic theory and how to recognise, analyse, and express the diversity of organisms. Chapter 5 is about the more practical matter of assessing the level of various kinds of incompleteness of the fossil record, pointing out the effect it has on palaeontological theories, and noting what can be done to allow for it. At any given level of incompleteness, and therefore of resolution, there are questions that can reasonably be asked of the data, and questions that cannot.

In Part II these principles are put into practice to tackle what are seen as the five central areas of research in modern palaeobiology. Obviously this is a somewhat arbitrary division of the whole field of endeavour, but at least roughly it is possible to envisage a scale of questions from the very fine-grained about classification of individual fossils, through the evolution of populations and species, to the course-grained ideas about long-term evolutionary events such as the large changes in diversity over millions of years, and the origin of new major kinds of organisms.

And, finally, if all the reader wants is a few reflections on what is happening now, and maybe will happen in the coming years in palaeontological research, then they might just read Chapter 11, the Epilogue.

# Part I

# PRINCIPLES

# 2
# Some fundamental ideas

This chapter is about the philosophy of science. The message is that the main reason why evolutionary biologists, including palaeobiologists, disagree amongst themselves so much is because they interpret their findings in the light of different beliefs about things that are chosen from a menu of options, none of which is known to be either true or untrue. Sometimes a second reason is that they are actually asking different questions from one another.

## Pattern and process

Some years ago now, Steven Stanley published a book entitled *Macroevolution: pattern and process* (1979) and Niles Eldredge and Joel Cracraft published one called *Phylogenetic patterns and the evolutionary process* (1980). These and other works stressed the distinction between two ways of looking at the diversity of organisms and their evolution, a distinction that has dominated thinking within palaeontology to this day. 'Pattern' refers to what is observed in the fossil record; 'process' refers to the mechanisms responsible for generating that pattern, usually evolutionary mechanisms, but they could equally well involve abiotic processes such as post-fossilizational events, climatic changes, or continental movements.

Fossils, and museum specimens of modern organisms, do not *do* anything, such as reproduce, evolve, or become extinct; they consist simply of dead organisms possessing collections of characters that can be described, and the specimens are then grouped in one way or another on the basis of which have what characters, which are similar to, and which different from one another, and so on. This exercise produces a classification, which is effectively a summary of the similarities and differences; in other words, the pattern of diversity. Conversely, populations of living organisms, whether artificially maintained in the laboratory or observed in their natural environment, certainly do *do* things. They reproduce, they eat or get eaten, their genes occasionally mutate, the frequencies of particular genes within the population alter from generation to generation, and so on. These are the mechanisms of change to which living organisms are subject; that is to say, the evolutionary processes.

Now where, one might ask, is the problem in such an obvious distinction as that? The problem lies in an apparently circular argument that connects taxonomic pattern with evolutionary process. Taking pattern first, there are many different possible ways of using the differences in characters between organisms to produce a classification. Should an organism only be allowed to be in either one group or in another, or should it be possible for it to be in two groups at once, as, for example, may a particular volume in a cross-referenced catalogue of library books? Should groups consist of those organisms with the most overall similarity, or of those organisms that have certain unique points of similarity even if in general appearance members of the group are rather different from one another? Can it be accepted that the classification produced represents the complete, true pattern, or should allowance be made for the possibility that there are parts of the true pattern missing because not all the different kinds of organisms that have ever lived are known?

All these questions indicate perfectly clearly that there is no self-evidently true or correct pattern of diversity to be discovered and expressed as a classification. What is needed in order to decide which particular kind of pattern to look for is some idea of the meaning of the pattern. For example, knowing how the diversity of organisms came about might indicate what form the description of the diversity should take. If, for example, it were believed, as it once was, that a Creator made all organisms in an ascending ladder of perfection, then the pattern to be looked for would be a linear sequence of forms that reflected this idea. If, as now accepted, divergent evolution generated the diversity, then the pattern must be a hierarchical arrangement of non-overlapping groups arranged within higher non-overlapping groups, and so on. So, a pattern of diversity that is claimed to be true, or real, can be discovered and represented only in the light of knowledge about the processes that brought the diversity about.

Fine, but how are these processes actually discovered in the first place? Intensive study of living organisms has resulted in a comprehensive theory of a genetical process of evolutionary change in populations of organisms. Organisms have an appropriate genetic mechanism, and appear to be subject to appropriate environmental conditions to promote differential survival of genes. There are, however, many different possible varieties of evolutionary mechanism. Plants, for example, often evolve by spontaneous polyploidy, and hybridization between species—methods largely unavailable to animals. Asexual organisms seem to do things rather differently from sexual ones, and so on. More importantly, evolutionary change as actually observed by the biologist never goes very far, barely to the level of new species at best, because it is a very slow process by the standard of the human lifetime. So how can the nature of longer-term evolutionary processes be discovered, such as how common is convergent evolution, which kinds of characters are most and which least susceptible to change, how rapidly does major evolutionary change occur, and so on? It is not enough simply to scale up the very short-term observations of evolution in action in modern populations to the long-term. This would automatically exclude any possibility of discovering evolutionary processes that are too rare, or too slow-acting to be seen in the field or laboratory, and the whole of evolutionary biology would be reduced to population genetics, to its great impoverishment. The answer, of course, is to inspect the results of long-term evolution by looking at the pattern of diversity of organisms produced by the processes.

So, the way to decide the correct kind of pattern of diversity to seek is to consider the processes that produce diversity; and the way to discover the nature of these processes is to look at the pattern of diversity that they generated. This is the apparent circularity of argument at the heart of the pattern/process distinction (Fig. 2.1a).

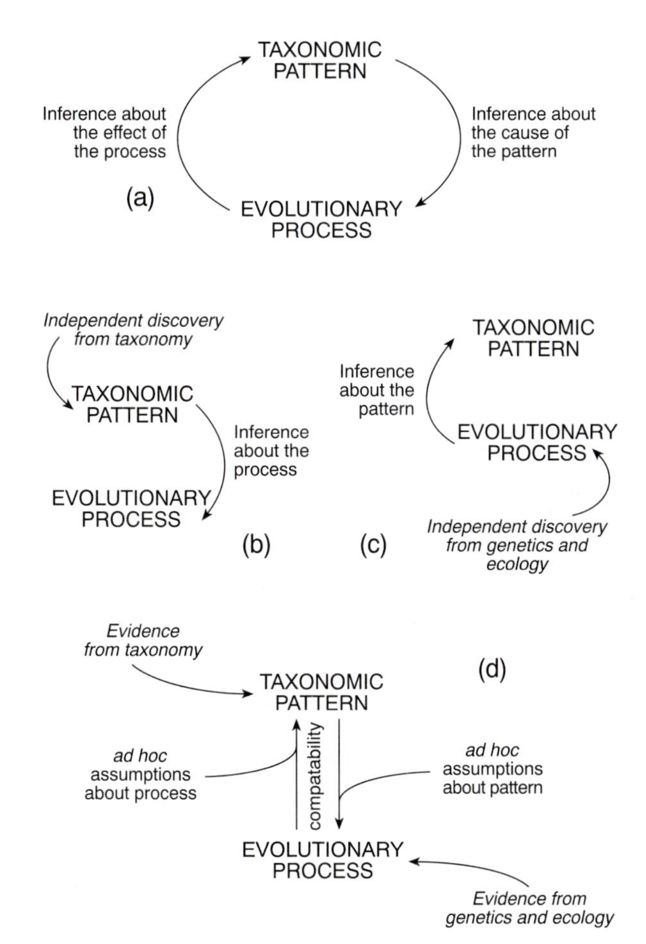


**Fig. 2.1** The relationship between taxonomic pattern and evolutionary process. (a) Pattern and process as apparently circular. (b) Deriving pattern independently and inferring process. (c) Deriving process independently and deriving pattern. (d) Creating compatibility between observed pattern and discovered process, using *ad hoc* assumptions.

There has been no shortage of palaeontologists prepared to ignore the circularity of argument and put their faith in the pattern of diversity as superficially seen by them in the fossil record. Henry Fairfield Osborn, a dominant figure in American palaeontology early in the twentieth century, described evolutionary trends and happily inferred from them a process of evolution based on an internal drive that he termed aristogenesis. In Germany, Otto Schindewolf was similarly attributing paramount status to the pattern of fossils he found in the record, and using the lack of intermediate forms to support his typostrophic theory that included the instant origin of major new kinds of organisms.

More recently, taxonomists known as transformed or pattern cladists have argued that because the processes that produced a particular taxonomic pattern cannot be independently discovered, then the pattern should be recognized and described without recourse to any assumptions about the processes involved. Such independently discovered patterns could in principle be used subsequently to indicate the kinds of processes that must have produced them, thereby avoiding the circularity (Fig. 2.1b). Unfortunately the difficulties are not so easily overcome. The conversion of the raw information about organisms and their characters into a taxonomic pattern, or classification, still requires certain rules. If evolutionary rules are not to be allowed, then rules based on general ways of handling information must be applied, and these come in different varieties. Decisions have to made about what constitutes homology, the meaningful similarity of characters; on whether a group should be based on overall similarity of its members or on possession of shared characters unique to the members; on whether the classification should be strictly hierarchical in structure. The pattern cladists's particular belief, that the natural taxonomic pattern is a hierarchical arrangement of taxa supported by as many characters as they can find, is just that—a belief and not a self-evident truth. The point is that it is not actually possible to recognize a taxonomic pattern without some preconception about the nature of the diversity, be it an evolutionary preconception or one based on precepts of general information theory. Classifications do not exist without a prior presumption about their cause or meaning.

The alternative escape from the pattern versus process circularity has been to treat as paramount some particular theory of evolutionary process that has been derived independently of any taxonomic pattern (Fig. 2.1c). The acceptable taxonomic pattern can then be discovered on the assumption that that process was the cause of the diversity of the particular organisms being studied. This is how the developers of the neo-Darwinian theory of evolution interpreted the fossil record. George Gaylord Simpson, in particular, accepted that gradual evolution by natural selection was the cause of diversity, and he interpreted the fossil record and the classification of fossils in the light of the kind of taxonomic pattern that natural selection would be expected to produce. This remains the normal approach to palaeobiology by neo-Darwinian scholars today. However, this idea of giving paramountcy to process also suffers from a difficulty. There is more than one possible evolutionary process that can be suggested in any particular case. Is evolution always by selection, or can chance play a part; is convergent evolution rare or common; is major change gradual or rapid? If one particular set of rules of evolution is assumed, then the possibility of discovering what actually happened if it was not in accordance with those rules is lost. An interpretation of the fossil record subjugated to preconceived rules will be unable to speak for itself about whether those rules are true.

Therefore it has to be concluded that the relationship between pattern and process is more complicated than is sometimes appreciated, and that they cannot be divorced from one another and treated independently. The very ability to perceive a taxonomic pattern demands some prior belief about its meaning; at the same time, the proposal of such an idea of meaning has to be constrained by the form of the taxonomic pattern. The true situation about pattern and process in palaeontology is a matter of the kind of compatibility that has to exist logically between any observed effect and its proposed cause. The taxonomic pattern into which the observations of the fossils are arranged (the effect) must be of such a kind as could have been generated by the processes hypothesized as having been at work (the cause); equally and inversely, the hypothesized cause must be of such a nature as to be capable of producing the perceived effect. What this means more simply is that palaeontology is like any other science. It creates theories, which are attempts to explain aspects of the natural world as the particular effects of certain proposed causes that are believed to operate. Evolution is a formulation of the cause of diversity; classification of the diversity is the description of the effect it produces.

Appreciating thus far that pattern and process cannot be separated as two distinct, alternative ways of interpreting the fossil record, the question now is how to combine them into a single explanatory theory. If all the information possessed about the fossil record readily formed the kind of pattern that could have been produced by the operation of the mechanisms of evolutionary change that are well known and understood from living organisms, there would be no problem; the perceived pattern would be immediately compatible with the proposed process. If, for example, all fossils fell into finely graded sequences exhibiting gradual changes in characters up the stratigraphic column, and if simple neo-Darwinian natural selection was the only known cause of evolutionary change, then the explanatory theory that would no doubt emerge would be that these various lineages of species form a series of named groups, within each of which the members are evolutionarily related to one another, having been produced by natural selection. The groups would form an evolutionary classification that offered an explanation for the existence and nature of those particular fossils.

Unfortunately the situation is not so simple, for the observed fossil pattern is invariably not compatible with a gradualistic evolutionary process. Fossils only extremely rarely come as lineages of finely graded intermediate forms connecting ancestors with descendants. It transpires that either the pattern as perceived or the processes as invoked (or indeed both) must be in some sense 'wrong'. A 'wrong' pattern is most likely to result from large numbers of kinds of organisms missing from the fossil record, which prevents the true pattern from being discovered. In a similar manner, a 'wrong' evolutionary process is likely to be due to the failure to discover, or to appreciate the extent of certain ways in which evolutionary change can occur in living organisms. The only way in which the essential compatibility between pattern and process can be achieved is by making some additional assumptions, called *ad hoc* hypotheses. These are neither known to be true nor known to be untrue, but by assuming them to be true, they make up for the missing knowledge. Undiscovered fossil forms can be proposed that modify the taxonomic pattern so that it becomes compatible with some particular evolutionary mechanism; unknown mechanisms of evolution can be proposed to create compatibility with the fossil record as it is.

There is an excellent illustration of this in the correspondence between Otto Schindewolf and George Gaylord Simpson in the 1950s. Both these palaeontologists claimed that the fossil record supported their respective theories of evolution; they were looking at the same fossil record, and yet their theories were utterly different from one another. How could this be? Simpson was involved in the establishment of the neo-Darwinian theory of evolution, or the modern synthesis (page 26), and believed that the major driving force was natural selection acting on interbreeding populations and producing gradual adaptive change. This process, however, is incompatible with a fossil record full of morphological discontinuities and sudden appearances of new kinds of organisms. So, like Darwin before him, he made the *ad hoc* assumption that the fossil record is very incomplete, and that the missing fossils must include long series of finely intergrading forms. On the other hand, Schindewolf created compatibility of process with his rather literal reading of the fossil record by the *ad hoc* assumption that there is an as yet undiscovered mechanism in living organisms for producing large instantaneous change.

Given this underlying incompatibility between the morphological discontinuity of the fossil record on the one hand and the gradualistic theories of evolution as derived from neontological study on the other, then *ad hoc* hypotheses cannot be avoided if an evolutionary explanation of the fossil record is to be offered at all (Fig. 2.1d); the question is, then, how should a choice be made between the two possibilities? Are Simpson's 'invented' fossils more or less acceptable than Schindewolf's 'invented' mechanism? Since the *ad hoc* hypotheses underpinning a theory are assumptions taken to be true, then the obvious way to assess the theory is to test whether they are indeed true or not. Many branches of science achieve exactly this effect by experiments, which is not a direct option in palaeontology. Here the nearest thing to experiment is to predict what kinds of fossils ought to exist at what geological times and places, and then to go and look. The continued absence of the postulated fossils, however, may be because they have not been preserved, not because the organisms never existed. In comparable vein, failure to discover by experiment alternative evolutionary mechanisms in living organisms could be because they are extremely rarely expressed and would only be apparent in observations over, say, tens of thousands of years.

Therefore the apparent impasse is reached whereby any proposed explanation of the fossil record can be defended by making whatever *ad hoc* assumptions are necessary. Anything *could* be true and therefore nothing can be proved to be false! The elephant *could* have acquired its trunk by having its nose pulled by a crocodile by the banks of the great green greasy Limpopo River. Why is the scientific community nevertheless inclined to reject this particular theory? It is because the 'Just-so story' theory requires an enormously large amount of *ad hoc* assumption compared with rival explanations such as natural selection. It would have to be true that at some time in the past crocodiles were in the habit of pulling elephant's noses, and that elephants at that time could inherit acquired characters by some completely unknown mechanism, unlike all organisms, including elephants, today. A possible alternative explanation is that of natural selection. Here the *ad hoc* assumption has to be made that proto-elephants had much the same genetical mechanisms and population structures as modern elephants, and that at some time they were subjected to a selection pressure such that an elongated, prehensile nose was advantageous in feeding. Both theories contain *ad hoc* assumptions; both could be true or false depending on whether these unknowns are true or false. The difference is that the *ad hoc* assumptions in the second case are very much simpler and less extensive than in the first case. This is the nub of the reason why the second theory is preferable.

The conclusion is that an explanatory theory in palaeontology on its own can be neither proved nor refuted. But two competing theories, both of which explain all the observations, can be compared in terms of how much additional *ad hoc* hypothesis has to be created. The one with the least *ad hoc* burden is the simpler, and therefore in accordance with the standard principle of science called simplicity, parsimony of explanation, or Occam's razor (page 60), that is the theory that must be preferred.

The reason for rather labouring this question of pattern and process is because it underpins implicitly, if not explicitly, all theorising in palaeobiology, yet is frequently ignored or misunderstood. First, pattern and process cannot be divorced from one another, but have the relationship to one another of effect and cause; any theory about the fossil record must embrace both. Second, in principle alternative theories or explanations in palaeobiology can be compared with one another with a view to preferring the one that requires the least amount of additional, *ad hoc*, assumptions on the grounds that it is simpler. Third, in practice there are often no easy criteria for measuring the *ad hoc* component of any explanation and therefore for comparing alternatives. This is why there remains so much dispute in practically every area of palaeobiology: different authors prefer different *ad hoc* assumptions and therefore are inclined to defend different theoretical explanations. While a choice between Simpson's gradualism and Schindewolf's typostrophism may be quite easy to make by applying this principle, there are many subtler cases to come where it is not at all clear which of the alternatives is really the simplest explanation. The warning is not to rely solely on authoritarian claims that certain things are universally true, when that has not at all been shown to be the case.

## Cladograms, trees, and scenarios

In 1979, Niles Eldredge wrote about a distinction between what he referred to respectively as 'cladograms, trees, and evolutionary scenarios'. Since then this trichotomy of terms has come to represent another cliché-like categorization of ways of approaching palaeobiology, generating comparable passion to that accompanying the relationship between 'pattern and process'. Cladograms are hierarchical diagrams showing the distribution of shared characters amongst the organisms and taxa being investigated. Trees are branching diagrams of the evolutionary relationships between these organisms, and between them and common ancestors. Scenarios are accounts of the causes of the particular evolutionary events undergone by the organisms. To Eldredge these three ideas represented increasingly complex theories that require increasing numbers of additional assumptions—that is, *ad hoc* hypotheses—for their creation. Nevertheless, he accepted that all three were valid goals of palaeobiological research, asking only that the assumptions made in each case should be explicitly recognized.

Since that time, some biologists have taken a much less liberal view. At the extreme, it is claimed that only the cladogam has scientific validity because it represents no more than the empirically discovered pattern of distribution of characters, which can be objectively and completely expressed. To proceed further and interpret a cladogram as an evolutionary tree requires some additional and completely unverifiable assumptions. For example, a tree might be postulated in which birds and crocodiles are related and descended as two separate lineages from an ancestral species. To accept this, it is necessary to believe at the very least that the characters shared by the members of these two groups were present in the common ancestor, and that the differences between the two arose by evolutionary changes at some particular times and places. Neither the ancestor itself nor the inferred character changes can be directly verified, and therefore they are actually *ad hoc* assumptions—necessary beliefs if the theory is to be accepted. Thus, the extreme view says that because these assumptions cannot be tested by direct evidence, then trees are not testable hypotheses and therefore have no business in scientific discourse.

How much more scathing about evolutionary scenarios, then, are those who hold such views? A scenario for the origin of birds might be that during the Late Jurassic there was a selection pressure favouring the adoption of increasingly arboreal habits acting on a group of small, lightly built bipedal dinosaurs. Arboreality increased their ability to escape predators and find new food sources. Subsequent selection forces promoted leaping, then gliding, and eventually powered flight from branch to branch and tree to tree. Absolutely none of these suppositions about the intermediate forms, the ecological conditions they lived in, or the selective forces to which they were subjected could be tested empirically. The outcome is the evolutionary scenario or, rather more pejoratively, the 'Just-so Story'.

This negative response to trees and scenarios highlights certain genuine difficulties, but misses a most important point. The three respective concepts—cladogram, tree, and scenario—actually represent the answers to three different respective questions about the diversity of organisms. A cladogram (Fig. 2.2a) is the answer to the question, What is the nature of the diversity of these organisms? A tree (Fig. 2.2b) is the answer to the question, How did this diversity come about historically? A scenario (Fig. 2.2c) is the answer to the question, What caused it to come about? So: birds and crocodiles share certain characters with one another, but differ in others (cladogram). These similarities and differences exist because birds and crocodiles respectively diverged from a common ancestor (tree). Birds evolved their differences from crocodiles because they were selected for flight, whereas crocodiles were selected for a semi-aquatic, piscivorous habitat (scenario).

There are some important conclusions to be drawn. One is that because the three respective ideas represent the answers to three different questions, they are not in competition with one another as explanations of the fossil record. A palaeobiologist is perfectly entitled to seek the best cladogram to express the pattern of characters found in the organisms. What is not logically allowable is to claim that a cladogram is a better or a worse explanation for the diversity than is a tree. Similarly, it is meaningless to claim that a tree is a superior or an inferior explanation compared to a scenario when accounting for some particular bit of the fossil record.

A second conclusion is that the cladogram–tree–scenario trio represents

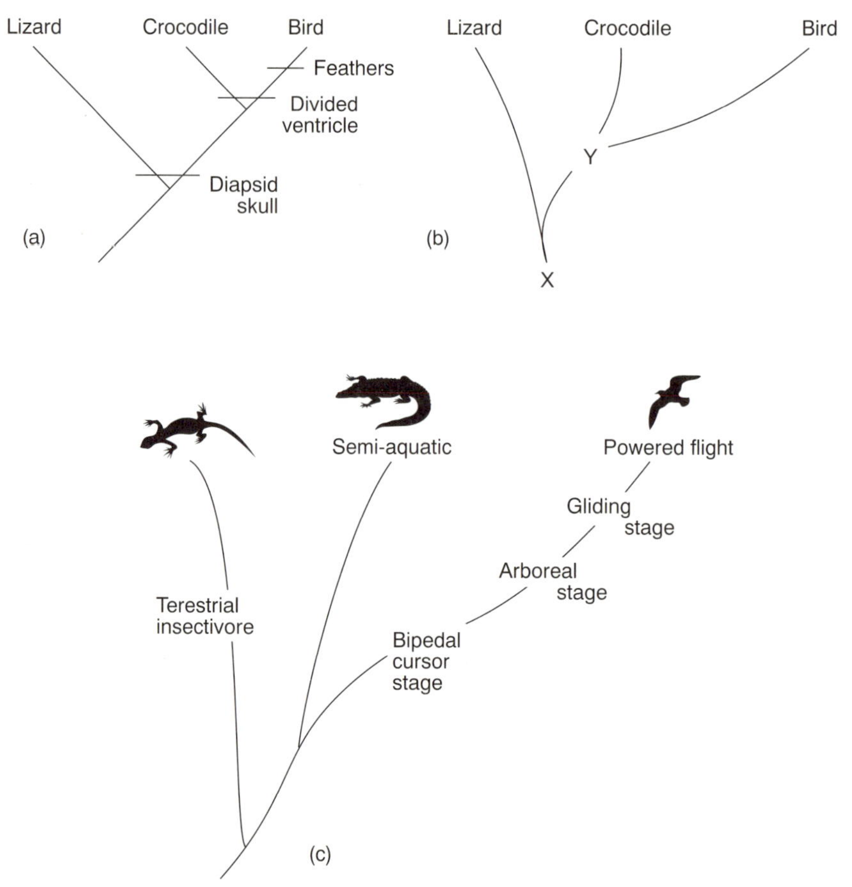


**Fig. 2.2** Theories about the relationships of squamates, crocodiles, and birds: (a) a cladogram; (b) a tree with X and Y successive common ancestors; (c) a scenario.

categories of theories offering answers to increasingly broad, bold questions about diversity, and indeed each one necessitates increasingly extensive *ad hoc* assumptions over the one before. Therefore they tend to inspire decreasing levels of confidence, and increasing difficulty in judging between theories. Any worker is entitled to express distaste for scenarios because of the very high *ad hoc* content, and to restrict their own activities to seeking the more confidently supported cladograms, or maybe trees. But no one should claim that trees or, even more so, scenarios are not scientific. They are theories purporting to explain phenomena; they contain both empirical observations and *ad hoc* hypotheses; in any particular case there will be competing versions that differ in the details of their *ad hoc* content. In principle, all three levels of question are rational and in scientific form, and subject to assessment between alternatives. The level of confidence placed in the three respective levels is not actually relevant.

# The epistemological gap

The astute reader will have appreciated that the two fundamental palaeobiological ideas dealt with have in common that they are both about causes and effects. The pattern-versus-process controversy is about the general relationship between effects and their causes. It was concluded that theories are explanations of particular, observable effects (patterns) by proposed evolutionary mechanisms (processes). The cladogram, tree and, scenario debate is really about chains of causes and effects. The scenario is a theory that proposes evolutionary and ecological causes for the growth of the phylogenetic tree. As well as being an effect, the phylogenetic tree in turn is the proposed cause of the distribution of characters in organisms that is expressed as an effect by the cladogram. A common way to say this is that the scenario is about the ultimate cause of the diversity, whereas the tree is its proximate cause.

These thoughts apply in principle to all science, not just to palaeontology. Yet palaeontologists, or at least those who think about these matters, have been inclined to make rather more of a fuss about them than other scientists. The reason has been called the problem of the epistemological gap. This rather whimsical term relates to the fact that there are two distinct sources of knowledge about evolution. First there is the evidence about the properties of living organisms in general. Populations of organisms can be studied and the circumstances and nature of heritable changes that take place can be observed from generation to generation. This evidence has very high resolution, in that successive parent–offspring generations, or for that matter changes within individual organisms, can be studied. But it is of very limited duration, for in practice a living population can be studied for at most only some tens of years. In contrast, the second kind of knowledge is that of fossils. These can be studied over enormous durations of time, tens and hundreds of millions of years. But the resolution of the fossil record is extremely poor and, with rare exceptions, even in the best of cases the relative dates of the individual specimens are accurate only within tens of thousands of years. The epistemological gap is an inevitable gap in knowledge concerning events that occur over certain time-scales. Anything that happens too slowly to be observable in living populations but too rapidly to be discernible in the fossil record falls into this category. Certain very important evolutionary events do exactly that. The most noteworthy is the process of speciation. As far as can be told from the indirect evidence available, the splitting of an ancestral species into two distinct daughter species typically takes a few thousand years. It cannot be followed by observing natural populations; yet neither can the succession of morphological forms connecting the descendant with the ancestral species be discriminated in time from one another in the fossil record.

A corollary of the epistemological gap is the recognition of two different time-scales—ecological time measured in decades and downwards and geological time measured in tens of thousands of years and upwards. The problem is that of how far it is justifiable to extrapolate from processes well understood at one time-scale to explain effects manifest only at the other. Is it reasonable to expect that ecological time-scale processes can just be scaled up in time and used to explain events occurring over geological time-scales? For example, can competition between living species be assumed to be the cause of extinction of a major taxon and its replacement by another one over a period of several million years? Can the same natural selection that affects such simple characteristics as the coloration of the peppered moth or the number of bands in snail shells be a completely adequate explanation for the emergence of mammals from their ancestors over the course of perhaps 100 million years? The pessimistic view of extrapolation is that one can never test whether an ecological time-scale process is or is not the cause of some geological time-scale event, so the attempt at any such explanations should be abandoned. The optimistic view is a form of reductionism that claims that if some particular ecological time-scale process *could* have caused the geological time-scale event, then because the existence of that process is known, one need look no further for an explanation.

The position best adopted is that spanning the epistemological gap by extrapolation from ecological to geological time-scales involves *ad hoc* assumptions. These should be at least made quite explicit, and at best should be compared to the *ad hoc* assumptions necessary for alternative explanations that do not entail extrapolation. Much thought will be devoted to this matter in the chapters to come. Meanwhile a brief review of how to analyse process, by the study of evolutionary mechanisms, and pattern by applying the principles of taxonomy is pursued in the next two chapters.

## Further reading

Attempts to divorce evolutionary process from taxonomic pattern became explicit at the hands of transformed (pattern) cladists such as Platnick (1979; Nelson and Platnick 1981) and Patterson (1982, 1988). Panchen (1992) accepts at heart the independence of taxonomy, but with careful thought. Grande and Rieppel (1994) continue to defend the idea, but remain unclear about how to solve the dilemma that they recognize; several authors such as Ridley (1986), and Mayr and Ashlock (1991) never quite grasp the problem.

The Simpson versus Schindewolf correspondence was discovered and discussed by Grene (1958) and used as an example by Kemp (1989). The classics by Simpson (1944, 1953) and Schindewolf (1950; English translation 1993) convey well their different views about the fossil record. Much of the recent philosophy of science has concerned itself with *ad hoc* assumptions; Lakatos (1970) remains a stimulating general study of it.

Those who have commented on pattern and process tend also to have added their views to the cladogram–tree–scenario issue, first discussed at length by Eldredge (1979).

# 3 Evolutionary theory: analysing process

The idea was mentioned in the last chapter that there is no simple, universally applicable theory of the cause of evolution that all agree explains all aspects of all cases. This must now be elaborated, though at the outset the problem may be attributed to the sheer complexity of living organisms and their relationships to the environment. If an organism really did resemble one of Richard Dawkins's charming computerized 'biomorphs', with its number of characters counted in single figures and its evolutionary choices restricted to one or two possibilities at each step, or if, indeed, real populations of organisms had but two alternative alleles at each of but two loci on the chromosomes, as population genetic models are apt to assume, then doubtless a simple deterministic (or at least probabilistic) theory would suffice to explain all evolutionary changes. But, of course, real organisms have many thousands of loci with alternative alleles available in the population as a whole for typically up to 30 per cent of them, and all of which can affect fitness. Worse still, the fitness bestowed by an allele also depends on the particular combination of alleles in which it occurs. Furthermore, the organisms live as heterogeneous populations in complex environments where selection forces differ from place to place with changing geography, and from time to time with the seasons.

There is no wonder, then, that not a single actual evolutionary event, however apparently simple, has ever been explained in full detail to everyone's unqualified approval. Karl Niklas (1995) has illustrated this very point by a simple but most expressive computer simulation (Fig. 3.1). He modelled a hypothetical ancestral land plant, based on the Silurian form *Cooksonia*, and imposed upon it simultaneous selection for three biological attributes. One was the area exposed to sunlight, which is enhanced by horizontally spreading branches. The second was mechanical stability, promoted by the branches being as vertically oriented or, if lateral, then as short as possible. The third attribute was spore-dispersal distance, which is most enhanced by branches holding the sporangia as high above the ground as possible. Given the conflict between some of these requirements, it is not perhaps surprising that the simulated selection process generated a wide variety of end-forms, presumably having the same overall fitness as one another but expressing it in different combinations of fitness levels for the three respective functions. Fascinatingly, the range

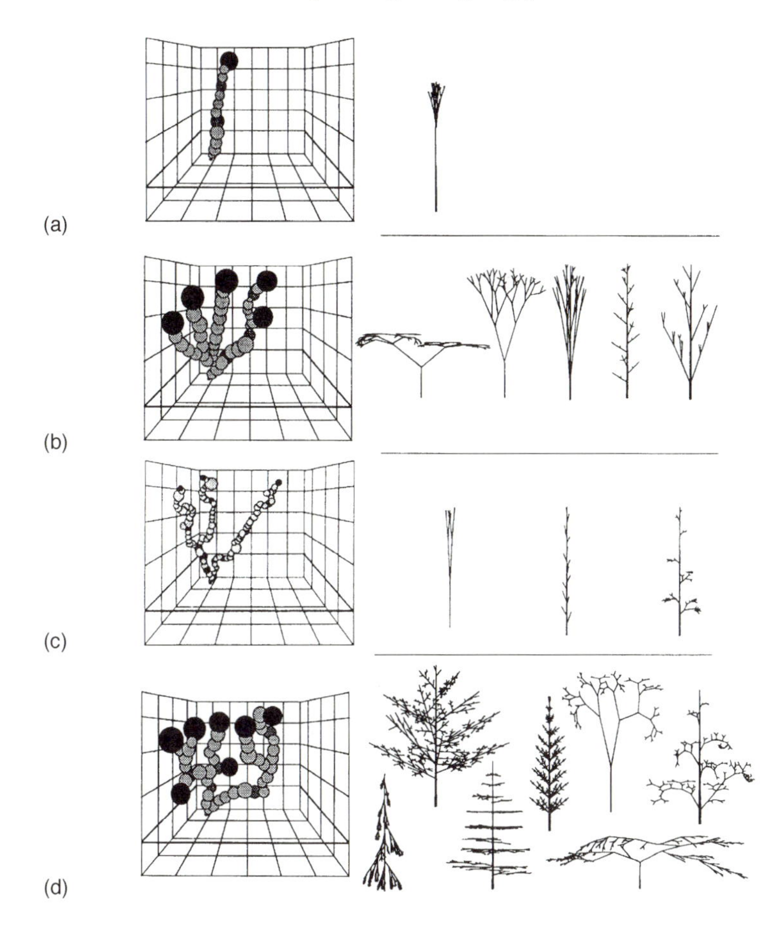


**Fig. 3.1** Computer simulation of simultaneous selection for combinations of attributes acting on a hypothetical primitive land plant. (a) Selection for mechanical stability and reproductive success; in this case a single best compromise exists. (b) Selection for mechanical stability and light interception; here there are several equally good compromise structures. (c) Selection for light interception and reproductive success; again there are several equally good compromises. (d) Selection for all three functions—mechanical stability, light interception, and reproductive success; in this case there is a surprisingly large number of compromises endowing the organism with the same overall fitness. (After Niklas 1995.)

of forms is extraordinarily reminiscent of the actual range of Devonian land plants found in the fossil record.

At this point a word of warning may be prudent. The serious arguments amongst contemporary biologists are not about whether evolution is the cause of the diversity of organisms, nor whether the phenomenon of natural selection is at least a major part of a satisfactory explanation for adaptation of organisms. They are about whether other processes in addition to natural selection are involved and need to be taken into account. In recent years, the intuitively obvious idea has been developed that some systems are too complex a set of interactions, involving too many variables, that can be measured with too little precision, to allow exact descriptions and predictions to be made. Evolution joins its sister science of ecology, along with meteorology, fluid dynamics, and so on as subject to chaotic behaviour. It is most unlikely that any single, simple evolutionary process could account for life's total diversity.

## The new evolutionary synthesis

The theory of evolution that is subscribed to by most biologists is the outcome of the well-known synthesis between Darwinian natural selection and Mendelian genetics, developed in the middle decades of the twentieth century and referred to more or less synonymously as neo-Darwinism or the modern synthesis. Actually mainstream evolutionary theory is by now well into a 'new evolutionary synthesis': in a manner analogous to the way in which Darwinism was made intelligible and acceptable by the addition of the Mendelian theory of inheritance, so Mendelism in its turn has been made intelligible by a physico-chemical theory of gene action derived from molecular genetics. This new evolutionary synthesis can be analysed as a series of interlinked theories about nature, which together purport to explain how and why the diversity of the organic world came about. It is convenient, if a little arbitrary, to recognize five such component theories:

1. *The theory of inheritance.* All heritable characters are passed on by the Mendelian inheritance of separate genes that consist only of lengths of DNA. The process of transfer of the coded genetic information within the developing and functioning organism is exclusively from DNA to protein and thence to the organism. Therefore there can be no inheritance of acquired characters, which would necessitate the transfer of information from the organism to DNA in the genome.
2. *The theory of genetic variation.* All the variation that is inherited by successive generations of a species ultimately originates from mistakes during the copying of the genetic material. This may be due to inaccurate replication of individual DNA molecules. It can also be a consequence of inaccurate division of the chromosomes, which alters the distribution of genetic material. These mistakes occur randomly as far as any selection forces or evolutionary trends in progress in the population are concerned, and therefore the character changes produced in the organisms by the modified genomes are similarly random. There can be no 'orthogenetic' directional evolution driven by forces from within the organism.
3. *The theory of sorting.* Evolutionary change within a population of organ-

isms occurs at least principally if not exclusively by means of Darwinian natural selection acting on individual organisms. Those kinds of organisms with heritable variation that increases the probability that they will produce offspring will tend to increase in proportion in the population. Those kinds that have a reduced probability of producing offspring will decrease in proportion. Therefore fitness, or adaptation, of the members of the population tends to increase over time. The possibility of accumulation or loss of genes within the population by the process of random genetic drift is allowed, but is regarded as generally of minor importance in the evolution of organisms.

4. *The theory of speciation.* The species is the evolutionary unit because members of different species by definition cannot successfully interbreed with one another and therefore speciation represents the moment in evolution at which an irreversible step has occurred. New species arise mainly as a result of the geographic isolation of parts of a once-continuous population, with different selection pressures causing differential changes in the physically separate populations. The possibility is accepted of parapatric speciation, whereby differentiation occurs between adjacent populations that still retain some degree of genetic exchange between their respective members. Sympatric speciation, whereby a new species differentiates within the existing range of the ancestral species, is regarded as very unlikely and only possible under exceptional circumstances. Asexual species are regarded as awkward aberrancies whose evolution is still believed to be primarily by natural selection. Mercifully they tend to be short-lived species and therefore of relatively little importance to evolutionary theory.
5. *The theory of macroevolution.* Long-term evolution that consists of whole series of species appearing, replacing one another, and going extinct over geological time results exclusively from the same mechanisms that operate within species; that is to say, Darwinian natural selection acting within interbreeding populations. Macroevolution results from the intrapopulational microevolutionary processes continuing for long periods of time.

## Contemporary arguments

Obviously there is no precisely delimited theory of evolution, but most biologists today who regard themselves as neo-Darwinists would accept the general ideas expressed above. Different individuals stress different aspects, as much perhaps because of their different foci of interest as because of different basic beliefs. But there are also numbers of thoughtful biologists, many of them are palaeobiologists, who have considerable difficulties with parts of the neo-Darwinian viewpoint. Thus, fundamental debates about aspects of evolutionary theory are going on, some of which have been around ever since the early days of Darwinism without resolution, others of which concern the interpretation of new information, from the molecular genetic to the palaeoenvironmental. Among these areas of concern are some that are particularly pertinent to the interpretation of the fossil record, and which will bear on much that follows. It is regrettable that evolutionary debates are not always as dispassionate or rational as science is supposed to be. On the one hand, there are those who dismiss any alternatives to straightforward neo-Darwinism as unnecessary troublemaking; this is to belittle the complexity of organisms and their histories. On the other hand, there are the iconoclasts, of the 'Darwinism is dead' mould, who in turn belittle the wonderful way in which natural selection offers a mechanistic explanation for the otherwise extreme improbability of adaptation.

The purpose of discussing here these areas of long-standing dispute is not to resolve them, but to draw attention to their existence, and potential relevance for understanding the fossil record. Three general areas of serious dispute, or at least lively argument, can be identified, which together cover most of the contemporary questions. First there is the matter of whether the genetic variation available is completely random or whether there are restrictions on the directions that evolution can take: this is the question of evolutionary constraints. Second, there is the matter of how much, if at all, random rather than selection processes are involved in the accumulation of evolutionary change: this is the question of non-adaptive evolution. Third, there is the matter of the possibility that a selection process could work on entities other than organisms, such as DNA molecules on the one hand and populations or species on the other: this is the question of units of selection.

## The question of constraints

Is any imaginable evolutionary change possible, given a selection pressure promoting it? Could any conceivable organism exist, provided only that there is a suitable niche for it to occupy? Or are there rules and laws about the way in which biological materials and processes can be assembled, which restrict the possible organisms to those in accordance with such laws? In contemporary style of expression, can the whole of multidimensional morphospace be occupied, or only certain specifiable regions within it? The standard neo-Darwinian response to these questions is that yes, of course, there are limits to the possible kinds of organisms that could exist, if only because they are always subject to the laws of physics. Pigs cannot evolve wings because they are too heavy to fly using bone for support and muscle for energy. All organisms must consist of cells that are neither too large for adequate diffusion exchange nor too small to contain the minimum structures needed for them to operate, and so on. Furthermore, what are referred to as historical or phylogenetic constraints limit possible evolutionary changes to those that consist of a modification of what went before. Thus tetrapod vertebrates have four legs because their ancestors happened to have two pairs of fins, for reasons that actually had nothing at all to do with walking on land. But, continues this argument, the restrictions are limited to such uncontroversial examples as these. A very wide range of morphological possibilities exists, including finely graded intermediates between the different kinds of organisms known, and therefore natural selection has a very wide range of variation to act upon. The role of constraints has no great importance for understanding evolutionary change compared to the role of natural selection.

Throughout the history of the study of evolution there has been a radical strand of thinking that argues that there is much more restriction on the possibilities of evolutionary change than neo-Darwinism generally allows. To understand any sequence of evolutionary changes, it is just as necessary to know the range of possible changes available as to know the nature of the selection forces acting. The desirable is not necessarily the same as the possible and any actual evolution comes about from the interplay between the two. There are two sources of evidence relevant here. One is the very nature of organisms. As extremely complex entities consisting of the integration of a large number of different parts, it is supposed that there must be laws dictating their assembly and maintenance, analogous perhaps to 'laws' that limit the ways that bricks, tiles, timber, glass, etc. can be assembled to produce a viable building. If these biological laws are very restrictive, then they will limit the ways in which organisms can theoretically exist, and therefore the number of possible different kinds of organisms. Evolution will not be able to produce 'forbidden' morphologies, perhaps including in some cases intermediate stages between allowable forms. The second source of evidence is the actual distribution of known morphologies amongst both the modern and the fossil biota. Is the almost invariable absence of intermediate forms between the known kinds of organisms because they have not yet been found, because they were not preserved, or because they could never have existed? Is the common occurrence of directional evolutionary trends in the fossil record a response to a selection pressure applying unchanged for tens of millions of years, or an indication of extreme restriction on potential evolutionary change to a single possibility? Does the repeated convergent appearance of particular kinds of organisms, such as uncoiled ammonites (Fig. 3.2a) or mammalian carnivores, reflect a limited suite of niches, or a limited set of theoretically possible morphologies?

Views on the role of constraints vary from the largely discredited extreme to the unexceptionably moderate. The most extreme expression is orthogenesis and its variations. Here the argument is that there are internal, and

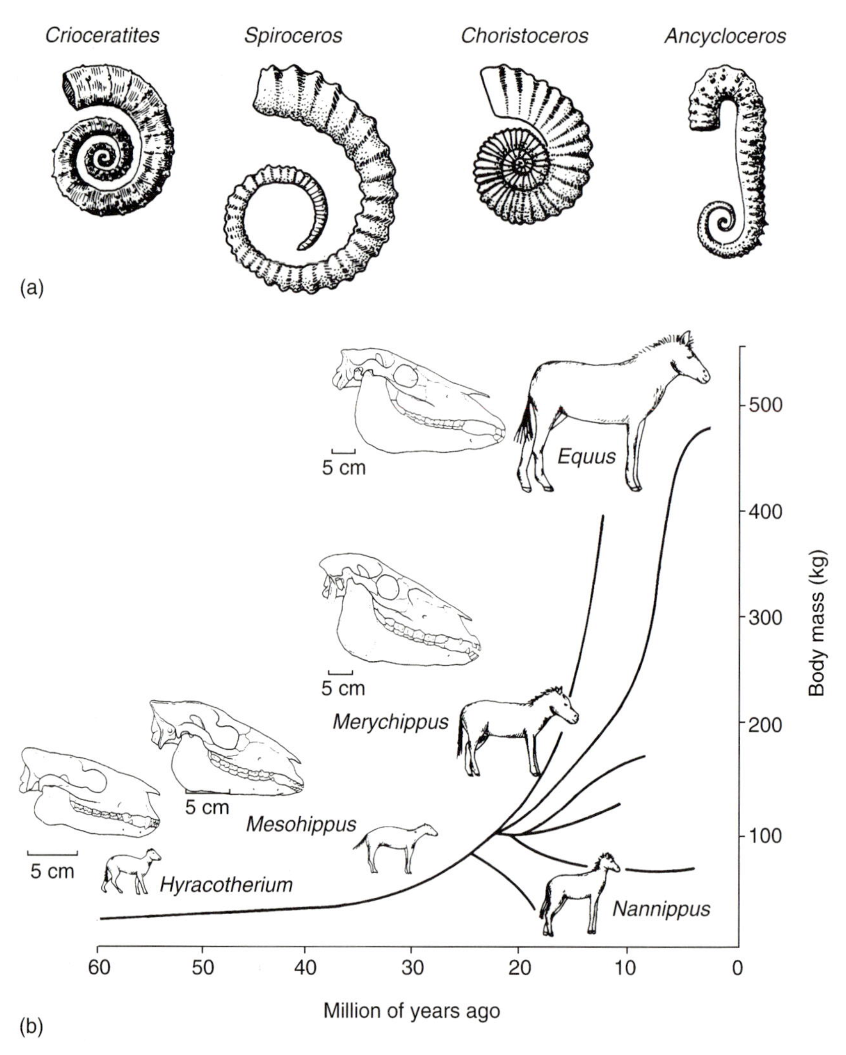


**Fig. 3.2** (a) Independently evolved uncoiled ammonites. (b) Orthogenetic pattern of evolution in horses, showing increase in body size and skull proportions such as preorbital length and size of jaw angle. ((a) After Clarkson 1993 and Kuhn-Schnyder and Reiber 1986; (b) after MacFadden 1992.)

as yet undiscovered forces that drive evolutionary change in a constant direction for long periods and without being affected by the environment. Such concepts include the aristogenesis of the leading pre-war American vertebrate palaeontologist Henry Fairfield Osborn, and the nomogenesis of L. S. Berg. Their origin lies in a rather simple explanation for the evolutionary trends found in the fossil record. Many cases have been documented of fairly gradual, if course-grained unidirectional change occurring over long periods of time, such as the classic study of the evolution of the horse

(Fig. 3.2b). These vitalistic ideas of orthogenesis are discredited nowadays because of the absence of any evidence for a mechanism that could cause it, and because the evolutionary trends seen in fossils can be explained by mechanisms that are known to occur.

At a somewhat less-extreme level, there are several theories of saltational evolution that link constraints to supposed laws of morphological organisation. Otto Schindewolf's typostrophism (page 16) is one of the most celebrated versions, and claims that there is a limited number of stable morphologies that could be constructed. Evolutionary change consists of spontaneous jumps, or saltations from one such stable morphology to another; intermediate stages cannot exist because they would be unstable states, which is why they are absent from the fossil record. In more recent years, Rupert Riedl has developed comparable ideas from his interpretation of the architecture of morphology. He argues that characteristics have different levels of what he terms 'burden', due to the hierarchical organization of morphology. Those features with high burden are deeply integrated into the overall structure and are very difficult to alter without destroying the integration of the organism. But characters carrying low burden have a superficial role in the structure and are correspondingly more able to be altered. This concept of burden has implications for the origin of major new kinds of organisms: a fundamental change in body plan could not evolve gradually as it would involve changes in characters of very high burden, and therefore the transition would require a quantum leap in morphology.

The most recent version of this mode of thought comes from Stuart Kauffman and his concept of 'self-organization' properties of complex systems. He argues that because a complex system like an organism has a tendency to organize itself spontaneously in certain ways dictated by physico-chemical laws, then the variation that is available for natural selection to work on is limited to the range of such self-organized possibilities. Kauffman does not claim that his views are anti-Darwinian in the way that Schindewolf quite explicitly did, but he does suggest that understanding evolution cannot result from understanding selection alone. There must also be an appreciation of the process of self-organization, because evolution is a consequence of the interplay between the two.

The argument against macromutational (or saltational) theories is the presumed improbability of a saltational jump proving successful. It is supposedly almost infinitely more likely that a large random change in morphology would actually produce a biological disaster rather than a new, stable, viable kind of organism. On the other hand, the probability is presumably not quite infinitely small, so that at extremely infrequent intervals such a saltation could be expected to work out. Just how infrequently is beyond realistic calculation, but the fossil record indicates that the actual origin of a radically new kind of organism is an exceedingly rare occurrence, whatever the mechanism. The distinction between 'impossible so never' and 'improbable so rare' cannot, in practice, be made.

In fact, the most moderate, and therefore most readily accepted versions of constraint theories are those claiming that morphology must be limited to what the developmental systems of organisms are capable of generating. This is the concept of developmental constraints, and along with the closely related concept of phylogenetic constraints, which stresses that evolutionary change must be restricted to what can be achieved by modification of the immediate ancestor's morphology, is relatively uncontroversial. Some biologists have claimed that these constraints are still sufficiently pervasive and extensive to limit significantly the evolutionary options available to organisms. Once again, it is argued that understanding actual evolutionary events must include a knowledge of the range of potential changes that could occur. An explanation based on natural selection alone would state that there was some selective force to which the organisms responded by adaptive change. If the fact is that the optimal adaptive response was impossible because it would have required the evolution of forbidden structure, then obviously natural selection alone has failed to provide an explanation of what actually happened. Some examples most clearly illustrate this point.

In a classic 1966 study, David Raup considered that the growth of spiral shells of molluscs was controlled by variation in a few growth parameters (Fig. 3.3a). The whole range of possible shell form could therefore be predicted, and proscribed shell form inferred. None of the proscribed forms has been discovered in either living or fossil groups, although neither have most of the permitted forms either. In 1983, Nigel Holder proposed a model of the mechanism for the development of tetrapod limbs that was unable to generate certain particular patterns of digits; for example, those with inner digits shorter than the flanking outer ones (Fig. 3.3b). Again, it is interesting that none of these proscribed types has been found to exist in any living or fossil species. From the perspective of constraints, both these examples suggest that even where there is a selection pressure favouring one such form, that form does not evolve because it is not possible to modify the developmental mechanism in such a way as to produce it.

Up to this point, even these moderate theories of constraints upon the potential for evolutionary change suffer one great disadvantage when applied in practice to explain particular evolutionary events, or lack of events. Failure to evolve in a certain way may be due to structural, developmental, or phylogenetic constraint. But it may also be due to the absence of particular selective pressures promoting that change. Perhaps none of Holder's forbidden limb morphologies would serve any useful purpose; perhaps the repeated evolution of certain forms of ammonites reflects only the continuity of the habitat appropriate to those forms. There is a need for criteria to tell whether the one or the other kind of explanation is better.

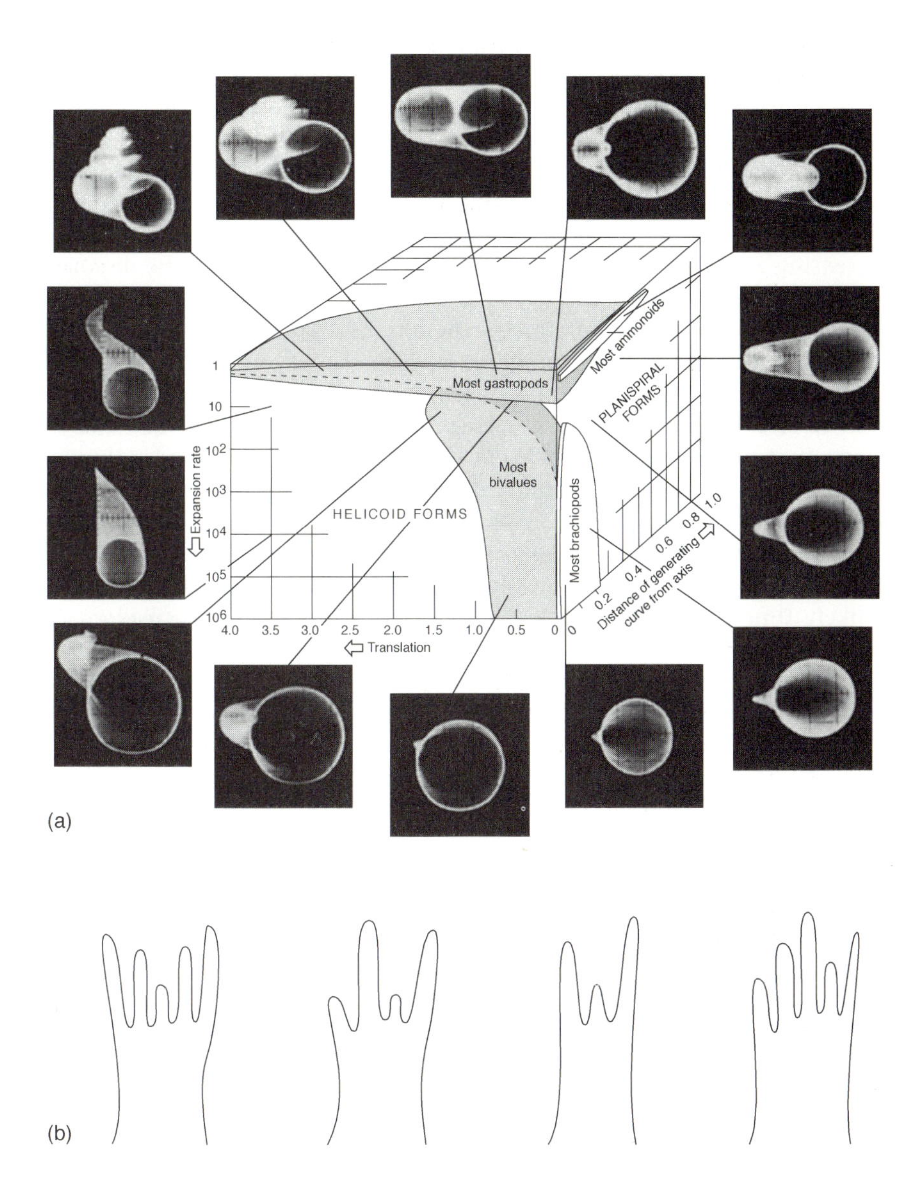


**Fig. 3.3** Constraints. (a) The theoretical morphospace occupied by mollusc and brachiopod shells in terms of the three parameters: translation, expansion rate, and distance of generating curve from axis. The parts of this space occupied by the main kinds of known mollusc and brachiopod shells are indicated. (b) Some impossible digit patterns of tetrapods, based on Holder's model, and which are not known to exist in nature. ((a) After Raup 1966; (b) after Holder 1983.)

An important discussion about this problem was published by Maynard Smith and eight colleagues in 1985. They proposed a number of tests, such as discovering the actual selection force in modern populations, or comparing the taxonomic distribution of particular characteristics with the respective genetic variability of that character within species. By their own admission, however, such tests are ambiguous, and certainly could not be applied in many cases. But the situation is rapidly changing with increasing knowledge of the molecular basis of development. The discovery of homeotic mutants in *Drosophila*, where single mutations cause major, but integrated modifications to the phenotype (Fig. 3.4a), has now led to the identification of homologous genes, the Hox genes, which are responsible for patterning development in a wide range of organisms. As more and more is discovered about how these and other regulator genes act, and the effects that mutant versions of them have during development, so the way in which morphology is organized will eventually be understood (Fig. 3.4b). For example (page 239), interesting evolutionary speculations about the ways in which segmented, arthropodan animals may evolve by quite simple mutational changes in regulator genes are being discussed. It is likely that true constraints on form will eventually be identified on the basis of molecular genetic studies, rather than on extrapolations from known and unknown form, as at present.

Meanwhile, assuming that there really are constraints on evolutionary possibilities, phylogeny should be understood as an interplay between what is adaptively desirable and what is structurally possible. Interpreters of the fossil record should bear this in mind, for the range of known morphologies to be found there will surely play an important part in framing questions about that relationship.

## The question of non-adaptive evolution

In principal several aspects of the evolution and diversity of organisms could be due to random change rather than to the strict determinism of natural selection. Genetic drift, for example, is the process where the proportion of the different genes in the offspring population can differ from the proportion in the parental population by chance, due to a form of sampling error. In a population with two allelic versions of some gene, one of the alleles can be lost and the other increase to complete representation with a simply calculable probability. Generally, the smaller the population, the greater the genetic drift effect. Neo-Darwinists have always to a greater extent (such as Sewall Wright) or lesser extent (notably R. A. Fisher) accepted this process at the level of population genetics. As with constraints, however, there has been a long tradition of the heterodox view that chance plays a much greater part in evolution than is usually allowed. Early expressions of the view were derived from incredulity that the small

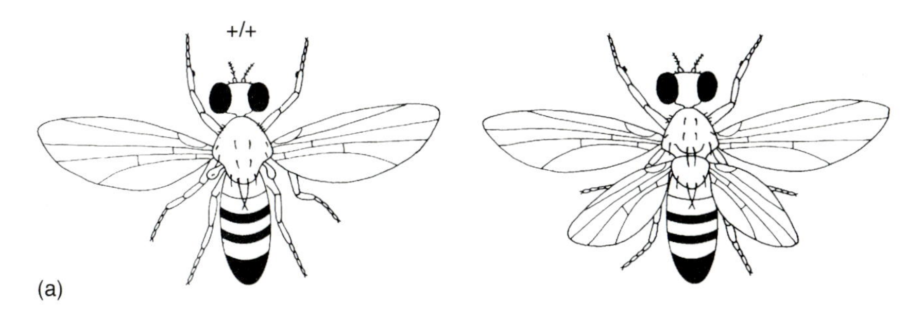


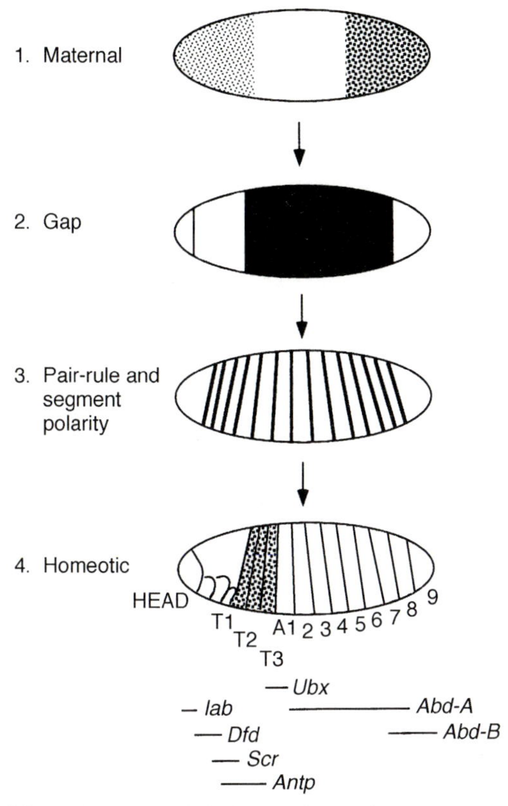


**Fig. 3.4** (a) Normal *Drosophila* on the left. On the right is the homeotic mutant *bithorax/post-bithorax* in which the third thoracic segment that normally carries the small haltere sense organs derived from wings now has a normal-sized pair of wings. (b) The hierarchical control of segment development in *Drosophila*: 1, the maternal RNA establishes the anterior/posterior axis; 2, the gap genes establish broad domains of segmentation; 3, the pair-rule genes establish the 14 parasegments; 4, the Homeotic (Hox) genes establish the precise identities of the respective segments. ((a) After Raff and Kauffman 1983; (b) after Raff 1996.)

degree of genetic variation in actual living populations could have any significant effect on the relative fitnesses of the individual organisms involved. Could a millimetre or two difference in bill length really affect feeding efficiency in sparrows, or the exact pattern of spots the degree of inconspicuousness of a leopard? This naive misconception has long since been refuted by innumerable studies of natural selection in action in the wild, showing that extraordinarily insignificant looking variation can indeed be subject to selection. A more sophisticated and serious version of non-adaptive evolution was found in Goldschmidt's much derided theory of hopeful monsters. His view was that a new species is formed initially by the occurrence of a chromosomal mutation that so modifies the pattern of gene expression in the individual that a significantly modified organism results immediately. Speciation is then a matter of chance and not natural selection.

In recent years, the theme of non-adaptive evolution has expanded from these historical views into four areas of evolutionary theory: molecular evolution, phenotypic evolution, speciation, and patterns of macroevolution, with a number of serious implications for palaeontology.

The neutral theory of molecular evolution attempts to account for numerous facts about the evolution of macromolecules, both proteins and nucleic acids. It claims that most of the changes that occur in the sequence of the amino acids of proteins or the nucleotides of nucleic acid are selectively neutral and have become fixed in the population by genetic drift. One of the main lines of arguments by its proponents comes from the very high levels of variation seen at the molecular level within modern populations. The argument that this variation is too excessive to be explicable by natural selection, however, requires assumptions about populations that are hard to verify, and there is a continuing controversy about whether neutralism or selection is the best explanation A second line of evidence for the neutral theory is the apparently constant rate of evolutionary change in particular macromolecules in different lineages (Fig. 3.5). If

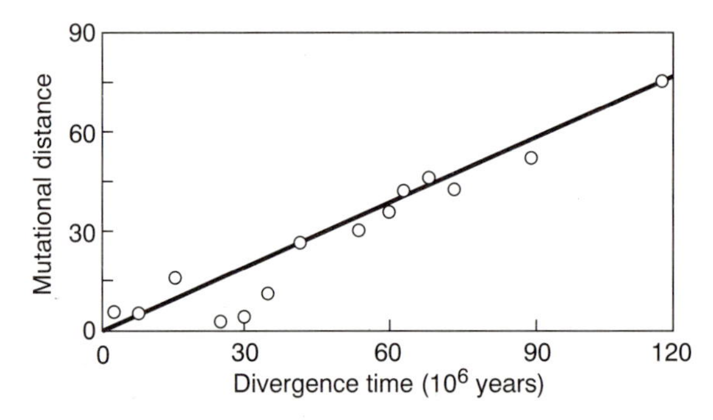


**Fig. 3.5** Accumulated molecular substitutions in a series of seven proteins of a selection of mammal species plotted against their divergence times based on the fossil record, showing the clock-like evolution of the molecules. (After Avise 1994.)

these rates of substitution are indeed constant, then this is most easily explained by the assumption that the rate of mutation is constant and that the mutations become fixed by chance drift rather than by selection. Other aspects of macromolecular evolution, particularly the different rates of evolution that are found in different molecules, and the different rates between different parts of the same molecules can also be explained by the neutral theory. The rate of evolution depends on the rate that neutral mutations occur, not on the rate of all mutations. If the proportion of mutations that is neutral rather than deleterious differs in different molecules, but only neutral mutations spread to the whole population, then the rate of accumulation of substitutions will differ in the different molecules. The same argument will apply between different parts of the same molecule. If a protein consists of a functional part whose structure is strongly constrained, most mutations in that part will be deleterious. But another part of the same protein, whose structure is less critical, will be less constrained; a mutation there is more likely to be neutral, so a greater rate of evolution will occur.

There are considerable computational difficulties to proving the neutral theory true. The current position is that most of the molecular differences can be accounted for at least equally well by natural selection, with one main exception. Rates of evolution of non-functional and non-transcribed lengths of DNA, such as introns, are best explained by neutral evolution. As far as the molecular clock and the implied constancy of neutral mutation rate are concerned, great caution is necessary. On the negative side, there is little doubt that rates of evolution of particular molecules do vary, both in different lineages, and from time to time in the same lineage. This is often demonstrated by different molecules indicating different relationships, or different divergent times for some particular group of organisms. Any apparent constancy of rate over a long period can actually be the average of highly varying short-term rates. On the positive side, there are certainly some convincing-looking cases of constancy, and molecules undoubtedly show a better relationship between time elapsed and degree of difference than does morphology. The direct implications for palaeontology of a possible molecular clock are obvious if ambiguous. On the one hand, the fossil record can be used in principle to calibrate a clock by indicating independently when the lineages actually diverged. This requires faith that the record is sufficiently complete, and that the phylogenetic relationships proposed are sufficiently well based. On the other hand, a molecular clock itself can be used to test palaeontological hypotheses about phylogenetic relationships. This requires prior acceptance of the validity of the clock.

At the level of organism evolution, within which the concept of selection is so embedded, there is a suggestion that phenotypic features may nevertheless result from non-adaptive processes too. Gould and Lewontin's evocatively entitled 1979 paper 'The spandrels of San Marco and the panglossian paradigm: a critique of the adaptationist program' argued that it is a mistake to seek adaptive explanations for all of phenotypic structure. Some aspects may be accidental side-effects of the evolution of structures that are adaptive, and therefore they cannot be understood as adaptations in their own right, or at least not as the results of the process of adaptation by natural selection. Even an adaptive structure in an organism might have originated as such a non-adaptive structure and only subsequently taken on its present function, a case referred to as 'exaptation'. It is argued from this point of view that the prior belief in the adaptive origin and nature of all aspects of the organism removes the possibility of discovering any features that are actually not adaptive. Logically this is entirely correct, but the problem is that there are no evident criteria for establishing whether a given structure is non-adaptive, short of the impossible task of testing for all possibilities. Failure to discover an adaptive purpose may indeed be because none such exists. But it may also be because the true function has not yet been thought of. Even vestigeal structures in living organisms often turn out to have some residual function, such as a transient role during embryology. Thus, the idea of non-adaptive evolution at the organism level has rather little practical impact on palaeobiology.

Most palaeobiologists, indeed most biologists, have to be content with the woolly view that most aspects of most organisms can be shown to be adaptations for performing particular functions with reasonable, if not necessarily total effectiveness. One cannot at the same time both look for the adaptation of a structure and also test the assumption that it has one! While allowing and bearing in mind the value of Gould and Lewontin's thoughts, in practice it is difficult to see how to avoid a prior assumption that some particular fossil structure did have a function in life, when creating hypotheses about fossil morphology.

In the study of speciation, there is a continuing debate about whether the process of formation of a new species is a result of the differential adaptive evolution between the parent population and the incipient new species, or whether it involves a non-adaptive, chance trigger at least to initiate it. Goldschmidt's theory that speciation is the result of a chromosome mutation causing a radically altered phenotype is an extreme version of the latter view. Less extreme and therefore closer to the neo-Darwinian framework is the concept of the genetic revolution. This important idea was introduced many years ago by Ernst Mayr, and is that the gene pool of a large, well-adapted species is resistant to evolutionary change because of what is termed genetic homeostasis. It is claimed that the interactions between the alleles within an interbreeding gene pool discourage the spread of mutant alleles, for the latter do not fit into harmonious combinations with the former. In order for significant evolutionary change to occur, there has to be a rapid disturbance to the gene pool, called a genetic revolution. A prime candidate for the cause of genetic revolutions is the founder effect, whereby a very small sample of the population becomes isolated. Necessarily, the founder population will be a very biased sample of the original gene pool and the processes in that gene pool maintaining homeostasis will be disrupted. At this point, the theory has it, evolution both by drift and by local selection forces cause the rapid formation of the new species. If genetic revolutions do indeed precede speciation, and if as one must presume founder events are accidents of dispersal of organisms and not in any way adaptive responses to selective pressures, then speciation must be seen as having a non-adaptive trigger. The whole process of speciation cannot therefore be explained solely in neo-Darwinian selection terms.

A new generation of population geneticists is presently modelling the evolutionary process at the level of the species, using much more realistic assumptions than hitherto. One thing that is clear is that a simple adaptive model does not easily explain how change occurs and new species arise, when so many alleles are involved. It may well prove that multiple evolutionary pathways are equally possible, and that the ones taken by real populations depend on the chance of which particular mutations arise, rather than on any predetermined preferred adaptive routes. At any event, there is a great deal of discussion about the existence or otherwise of genetic homeostasis, genetic revolution, and the role of chance in speciation; palaeontological evidence plays a considerable part as will be seen at length later (page 129*ff.*).

The fourth and final area of controversy about whether chance has a large role is macroevolution—the pattern of speciation and extinction of species over geological time spans. This issue first arose when computer-generated 'phylogenies' using simple assumptions about probabilities of the splitting and the extinction of lineages of evolving organisms were created (Fig. 3.6).

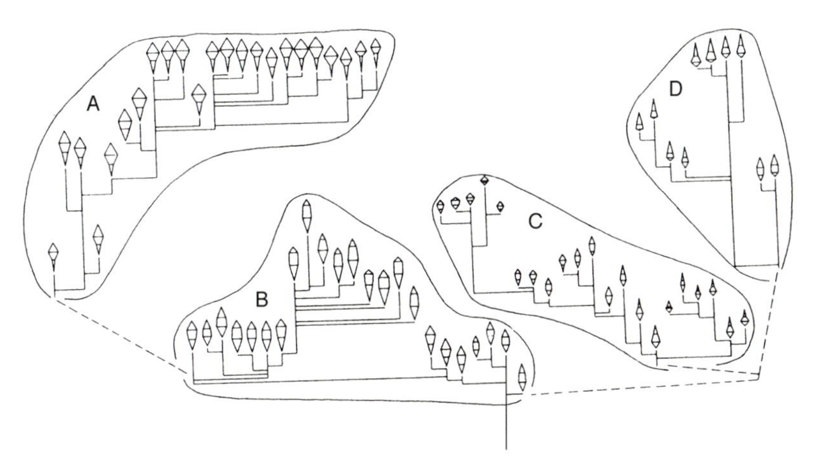


**Fig. 3.6** A computer-generated 'phylogeny' of a hypothetical group of organisms, based on assumed probabilities of lineages splitting or going extinct. (After Raup 1977.)

These frequently appear unnervingly similar to the phylogenies of real organisms as inferred from the fossil record. The standard neo-Darwinian expectation is that the species with the best-adapted members are least likely to become extinct, but the computer simulations suggest by analogy that species behave more like radioactive particles, each having much the same probability of becoming extinct. This matter is of such fundamental importance to palaeobiology that further discussion of it is left until a later chapter (page 156*ff.*).

## The question of units of selection

Selection, whether natural, artificial, or even computer generated, is a simple deterministic process (Fig. 3.7). The inescapable logic of Darwinism is thus:

**If**
(1) there are entities that replicate,
(2) the entities show heritable variation that affects the probability of replication, and
(3) there are limited resources available to the entities;
**then** as time passes there will be a continual increase in the proportion of those entities that have a higher probability of replication, and a continual decrease in the proportion of those entities with a lower probability of replication, up to the point where the former have replaced the latter. This is the basis of selection as a process, and as long as new variation of the right kind arises, the process can keep on going.

The logic of selection has been stated in this sparse fashion to help to make clear that it does not matter what the entities actually are, nor exactly what the mechanisms are by which the premises (1), (2), and (3) are met. As long as they are met, then selection must follow. If they are not met, then selection cannot be occurring.

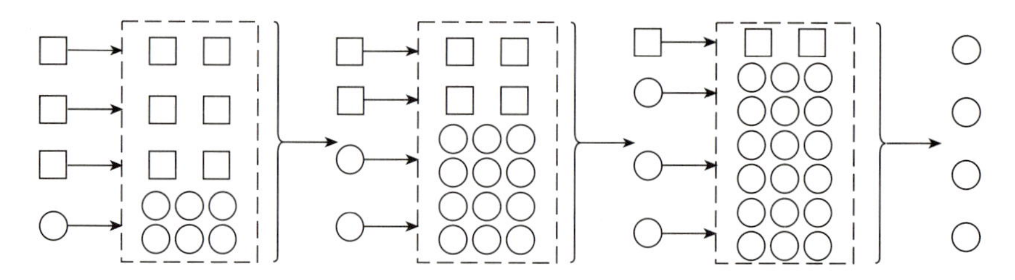

**Fig. 3.7** A model of a unit of selection in action. It is assumed that the shapes replicate producing identical copies of themselves, that the circles replicate three times as fast as the squares, and that the population of shapes is not allowed continually to increase. Given these assumptions, there is an extremely high probability that the squares will in time have been completely replaced by the circles.

As to what actually are the units of selection in real biological evolution, clearly the three premises apply to organisms. Organisms are entities that replicate. They show heritable variation such that some variants are fitter than others and are therefore likely to produce more offspring. And there is not enough ecological space to allow all the offspring to survive, so resources are limited.

But the biological universe is more complicated because organisms represent just one level in a hierarchy of entities. Organisms consist of parts such as cells with nuclei, while organisms also come together as groups and therefore are themselves the parts of populations and species. This hierarchical arrangement has several recognizable levels: DNA molecules; chromosomes; cells; organisms; populations; species; and communities of species. Now, in principle, the entities at any of these levels might be able to behave as units of selection, if the three premises for the selection process are met. Certainly they can all replicate because over time they can divide into more than one copy of themselves: DNA replication, mitosis, cell division, reproduction, dispersal, and speciation, respectively. Moreover, the resources necessary for the process of replication will always be limited by the limited supply of space and raw materials available. So the important question is, which of them show heritable variation that affects how likely they are to replicate?

The idea that genes are units of selection has become popular in recent years. Genes replicate, and one version of a gene may spread at a greater rate through a population than another because it endows the organism in which it occurs with greater fitness. Certainly the premises for selection apply to genes, so they must undergo the process selection. A confusion has arisen about whether the genes are a different unit of selection, at a different level in the biological hierarchy, from the organisms in which they occur. For a gene to spread through a population, its own replication must coincide exactly with the replication of the organism in which it occurs: one copy of each gene in one copy of each organism. Similarly, the different probabilities that respective genes have of replicating is exactly the same as the different probabilities that the respective organisms containing them have of producing offspring. Therefore it is meaningless to regard the genes as different entities from the organisms in this respect. Genes are the organisms' mechanism for replicating; organisms are the genes' mechanism for interacting with the environment. The unit of selection is the genotype–phenotype complex, not one or the other. The expression 'selfish gene' is a persuasive but rather misleading analogy; counting genes is simply a convenient way of analysing some particular case of organism selection.

There is a rather different molecular candidate for the status of unit of selection—silent, apparently functionless DNA molecules. Most organisms carry large amounts of DNA in their nuclei that are neither translated into RNA to become involved in protein synthesis nor used in any known manner for controlling activities within the genome. Often, functional genes themselves occur in multiple copies, up to 20 000 of them, when so far as is known only one is necessary. One explanation for this is the 'selfish DNA' hypothesis. This supposes that, within the nucleus, some DNA molecules have a higher propensity for replication than others. These sequences accumulate without having any effect for good or ill on the functioning of the organism in which they occur. If this is true, then these molecules are behaving as units of selection. They replicate, some sequences replicate faster than others, and, of course, the space within the nucleus is not infinite so resources for DNA production are limited. There is an alternative view, that this kind of DNA does actually have some function for the organism, and therefore accumulates by selection. For example, it may affect cell size, or provide variants of gene sequences in parallel with the functional gene sequence that could prove to be advantageous to the organism. The family of globin molecules in vertebrates appears to have arisen in some such fashion. From an ancestral gene coding for a particular globin protein there has evolved by replication and divergence a series of slightly different genes. Each one codes for a variant globin molecule, such as the range of haemoglobins and the myoglobin, which all function in slightly different ways in the body.

Going up the biological hierarchy from organisms, the possibility of group selection has long interested evolutionary biologists. This is the idea that communities or populations of a species might show differing abilities to survive and flourish as a result of characteristics of the group as a whole. Originally group selection seemed to offer an explanation for the evolution of social behaviour. Groups of organisms that co-operated were fitter than groups that did not co-operate, so the individuals in the co-operating group produced more offspring. This explanation fails because it does not explain why mutant non-co-operators in groups of otherwise co-operating organisms do not increase and spread through the population, even though they ought to be the fittest individual organisms of all: they gain the advantages of living with co-operators but do not pay the costs of co-operating with the others. As these selfish variants increased, the whole co-operative arrangement within the group would decline, and with it the advantage to the group. The short-term interest of the organism would override the long-term interest of the group. It is now appreciated that this kind of group selection could occur only in very restrictive, improbable circumstances.

An even higher possible unit of selection is the species, and this is of particular relevance to palaeobiology. Biologically, a species is an interbreeding population of an organism, whose members are by definition unable successfully to interbreed with members of another species to produce viable offspring. As so defined, a species can certainly replicate, in the sense that it can split into two or more daughter species over time. There are also limited resources for species, in that there is presumably some limit to the number of species that can exist in any area. The controversial question is whether species can show variation that affects the probability with which they speciate, and which can be inherited by daughter species. If the answer is yes, then species must be units of selection. If the answer is no, then they cannot be such, but arise and become extinct only as a consequence of the births and deaths of the individual organisms that constitute them.

First of all, what kinds of variation can apply to species? It is important not to confuse variation at the organism level with that at the species level so, for example, it is meaningless to speak of a *species* of cow that eats grass efficiently differing from a *species* that eats grass inefficiently; species do not eat grass. Eating ability is an attribute of organisms, the individual cows. Therefore, if the second species became extinct, it would be because all the cows that constituted it had died from malnourishment. The way in which two species might differ as species is the size of the population, the population structure, the rate of dispersal, and so on. These are all properties that can be applied only to collections of organisms; individual organisms do not have a population size or structure, or dispersal rate. Only a genuinely species-level character could be the basis of any differences in speciation probability in a true species-selection situation. If the chance of a speciation event occurring is greater in a species because it has a large population size than in one that has a small population size, then the pattern of speciation could result from a species-selection process (Fig. 3.8). For this process to be significant over time, the daughter species would also have to have the same kind of population structure as the parent species.

Species selection is controversial. Increasing numbers of evolutionary biologists accept the theoretical possibility, based on the logic of selection. But they tend to dismiss it as of little actual significance in real evolving organisms. It is argued that it is too slow compared with organism-level selection to make much difference, or that it cannot be responsible for the origin of adaptations of organisms. Others, mainly as it happens palaeobiologists, argue that because of its very slowness, species selection is highly significant in the evolution of aspects of diversity that occur over long periods of time, such as the speciosity of different taxa, and maybe certain simple, long-term morphological trends, such as size increase. This is a matter that will be returned to in depth later (page 146).

A final candidate for unit-of-selection status is the clade, or monophyletic group of species. Clades come and go in the fossil record in a fashion comparable to the formation and extinction of species. Do clades have clade-level characters that affect how likely they are to replicate by splitting into sister clades, or to disappear? It is a tempting thought when looking at higher-level patterns of evolution in the fossil record, but, unfortunately, it

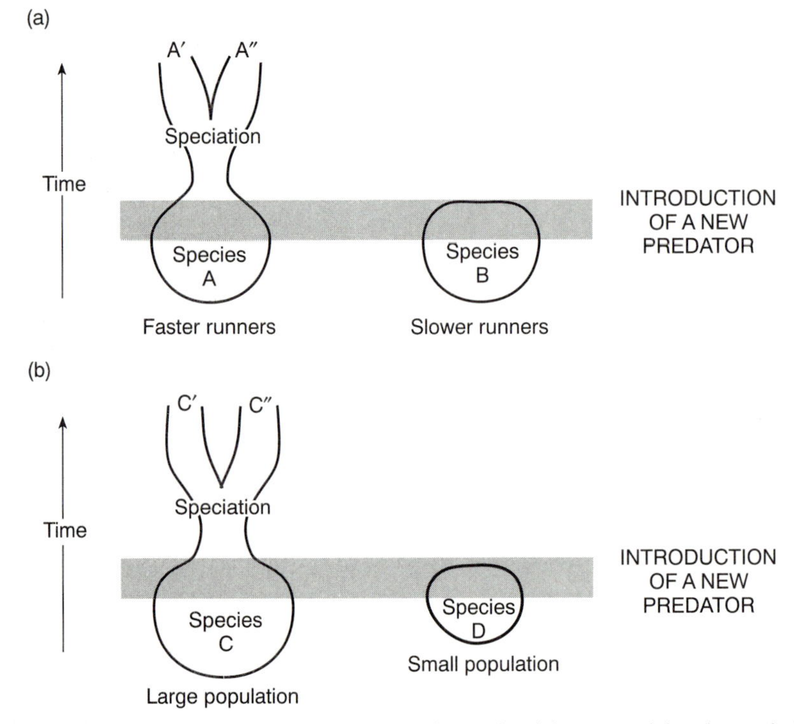


**Fig. 3.8** (a) Organism-level selection leading to species sorting: it is assumed that the survival of species A, and therefore its higher probability of speciating compared with species B, is due to the greater ability of the organisms constituting it to escape predation. (b) Species-level selection leading to species sorting: here it is assumed that the survival of species C, and therefore its higher probability of speciating compared with species D, is due to its larger population size and therefore greater probablity of escaping extinction in the face of predators.

is impossible to think of anything that could be a clade-level character in the sense that it could affect the clade as a whole. Clades are defined and recognized by unique characters, which all the species possess. In the same logical way that eating grass is a character of organisms, not species, so a defining character of a clade is actually a character of the organisms constituting the clade. Only brevity of expression allows the statement 'mammals are hairy'. In reality, a donkey, an elephant, a mouse, etc. are hairy. So a clade seems persistently to be a collection of lower-level entities that share nothing more than a historic origin; and a history alone is not an obvious basis for a mechanism. It will be necessary to return to this issue later, for in some palaeobiological investigations clades such as genera and families have to be used as the units of evolutionary change, if only for practical purposes. It will be seen that there is one possible, but dubious, way in which a taxonomic group might be viewed as a true individual unit in its own right (page 157).

A few years ago the rising interest in alternative units of selection, and therefore of levels in the biological hierarchy at which selection processes act, produced the idea expressed as the hierarchical expansion of evolutionary theory. The two arguments were, first, that indeed molecules below and species above the level of the organism can be selected on the basis of characters that apply or emerge at the respective level. Second, it followed that selection occurring at one level could indirectly cause a change at another level, without true selection happening at the latter. For example, if a certain species is selected over another species by a species-selection process, then there will be a consequential (epiphenomenal) change in the kinds of organisms and genes existing in the biota. Similarly, selection at the organism level can indirectly affect which species become extinct and which do not; selection may be based on grass-eating efficiency of organisms, but if all the inefficient ones die, then the species they constituted will be extinct. Selection of selfish DNA does nevertheless alter the toal amount of DNA to be found at the organism level. In short, observed changes in the entities at any level in the hierarchy might be due to selection or it might be due to what has become termed 'sorting', which is alterations caused indirectly by selection at a different level. Another pair of expressions coined for this idea is upward causation and downward causation, depending, of course, on whether the selection process having the sorting effect is higher or lower in the hierarchy than the level at which the sorting is actually being noted.

Interesting as it is in principle, hierarchical thinking has not so far had much impact on the interpretation of actual evolutionary events, with one major exception. The idea that species selection may be of considerable importance for the interpretation of long-term patterns of species turnover in the fossil record, including possibly certain trends usually attributed to natural selection acting on organisms, will be extensively pursued later.

Other areas of debate, disagreement, and passion in the field of evolutionary biology not mentioned here could be identified, such as the mechanism of speciation, the significance or otherwise of the concept of an adaptive landscape, the role of transposable genetic elements, and the meaning of sex and recombination. In fact, they are all arguments largely based around the three themes that have been discussed: the role of constraints on what is possible; the role of chance; and the role of alternative units of selection.

## Further reading

There is, of course, a vast and ever-growing literature about the theory of evolution, and it is virtually impossible to document the exact dates and authorships of many of the current ideas. Indeed, it is only possible to give a few bibliographic pointers into the various aspects of the subject here.

Dawkins (1986, 1996) describes 'biomorph' evolution, as well as providing an utterly straightforward neo-Darwinian outlook. Gleick (1988) is a readable and non-mathematical introduction to chaos theory, May (1987) offers a specifically biological view, and Depew and Weber (1995) review the current evolutionary perspective, along with the concepts of complexity in biology generally.

For an account of the origin and the nature of the synthetic theory, Mayr (1982b) is utterly absorbing (for more detail, see Mayr and Provine 1980); Depew and Weber (1995) must again be mentioned for their difficult but very important attempt to see how the molecular revolution is beginning to affect evolutionary theory. Marshall and Schopt (1996) have compiled a more accessible account of this development.

D'Arcy Thompson's great 1942 classic *On growth and form* foresaw many of the contemporary arguments about constraints. Schindewolf (1950) and Riedl (1979; see 1983 for a brief version) are two important figures in the heterodox but curiously persistent tradition of extreme structural constraints; Kauffman (1993; see also Depew and Weber 1995) is a modern embodiment. Maynard Smith *et al.* (1985) offer a balanced overview of the constraints problem. The examples used are from Raup (1966), also discussed by Dawkins (1996) and Holder (1983). Work on the molecular basis of development is coming in thick and fast. An older account of homeotic mutations in *Drosophila* is in Raff and Kauffman (1983); recent reviews include Raff (1996).

Sewall Wright's views on the role of genetic drift can most readily be gleaned from Wright (1986), and Fisher's contrast from Leigh (1986, 1987). For a comprehensive overview of studies of natural selection in real, wild populations, see Endler (1986). Gavrilets (1997) has reviewed recent new thinking on randomness and selection at the level of population genetics and speciation.

There is a huge literature on the neutral theory, from Kimura's now classic account of 1983 to Kreitman and Akashi (1995). Ridley (1996) is a particularly clear textbook exposition on this as well as population genetic topics generally. The concept of genetic revolution and the role of founder effects are approved of by Mayr (e.g. 1963) and reviewed with less enthusiasm by Barton (1989).

Description of computer-generated phylogenies are in Raup *et al.* (1973; see also Raup 1977).

Sober (1984) analyses extraordinarily clearly the units of selection issue, and the volume edited by Brandon and Burian (1984) contains papers from several points of view. G. C. Williams (1966) skilfully gave the death blow to group selection as an explanation for altruism and various authors have attempted to do the same for species selection (e.g. Hoffman 1989) but with less success. Maynard Smith (1983) took a more balanced, but still critical view.

Eldredge (1995*b*, 1996) has written extensively on the idea of hierarchy in evolution.

# 4
# Taxonomy: analysing pattern

'Taxonomy', 'classification', and 'systematics' are all terms that refer to the creation of classifications that organize the diversity of living organisms into groups. They have been used interchangeably and inconsistently for so long that any attempt to define them separately would offend many people's long-held usages. As 'taxonomy' does seem to be the most in fashion at present, it will be used here.

At any event, taxonomy is possible and classifications can be constructed and argued over because of a belief that the diversity of organisms, by which is meant the ways in which they variously differ from and resemble one another, is not random. If so, then there must be some sort of organization or pattern to the diversity out there in the real world, waiting to be discovered. There is believed to be what is called a natural classification, consisting of natural groups. Virtually without exception, biologists accept that the cause of organic diversity is evolution, 'descent with modification' as the minimal definition has it. The basis of the natural classification is therefore the real, single phylogenetic tree that connects together all organisms, living and extinct. Natural groups are those groups of organisms that are in some way genealogically related to one another. But it not quite so simple to decide what exactly constitutes an evolution-based natural group because evolution achieves two results—descent (that is, a genealogy) and modification (that is, changes in the organisms with time). These two aspects are not very well correlated with one another, because rates of evolutionary change are not generally constant. Sometimes change is rapid, sometimes slow. Therefore sometimes genealogically closely related organisms are very different from each other because one or both has undergone rapid change. Sometimes distantly related organisms are very similar because neither has actually changed much in the time since they diverged genealogically.

The three ways of dealing with the lack of close correspondence between genealogical relationship and degree of difference between organisms underlie the three 'schools' of taxonomy as described by numerous authors. First, the traditional approach of evolutionary taxonomists is to juggle with both aspects of the evolutionary process. A natural group is conceived of as a collection of organisms that share a common ancestor, but that also have a reasonably high degree of general similarity to one another. This can be illustrated by the phylogeny of the amniote vertebrates, about which there is relatively little dispute (Fig. 4.1). Reptiles, birds, and mammals are inferred to share a single common ancestor and also to constitute all the descendants of that ancestor. There is, however, a great deal of difference between mammals and birds, respectively, compared with the scaly, sprawling-gaited, ectothermic groups such as lizards, crocodiles, and turtles. Traditionally, therefore, a class Reptilia has been recognized for the latter, to the exclusion of the former two, which are placed in the classes Mammalia and Aves, respectively. Where this approach is found wanting is in the absence of objective criteria for deciding how much general similarity should be allowed in the recognition of the groups. It is a somewhat personal opinion to say that birds and mammals are sufficiently different from other amniotes to be removed, leaving behind the Reptilia; yet snakes or turtles are not so different, so they must remain within Reptilia. To the extent that taste, or authority, are the guides to group boundaries, then the groups are effectively human creations rather than real things to be discovered. Different humans will simply prefer to recognize different groups, and therefore to create different classifications.

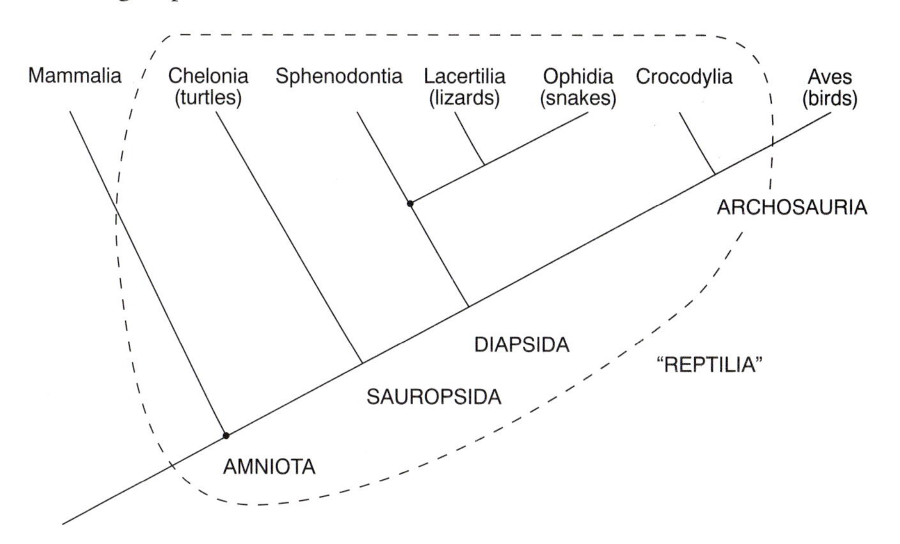


**Fig. 4.1** Phylogeny of the Amniota.

The second 'school' is phenetics, where the concept of a natural group is one consisting of organisms that share more overall similarity with each other than with organisms in other groups. They are particularly associated with the school of numerical taxonomy, which looks for scientific objectivity in taxonomy by developing a measure of overall similarity between organisms that is free of personal opinion. Any assessment of genealogical relationship as such is excluded from the classification. All the characters are used and treated as equally important, and a single number is derived that represents the overall similarity in characters between each respective pair of species. The most similar species are clustered together in the same groups. Then the groups sharing the most similarity of characters are clustered into the same higher groups, and so on. The outcome is a hierarchy of groups called a phenogram from which the formal classification is derived (Fig. 4.2).

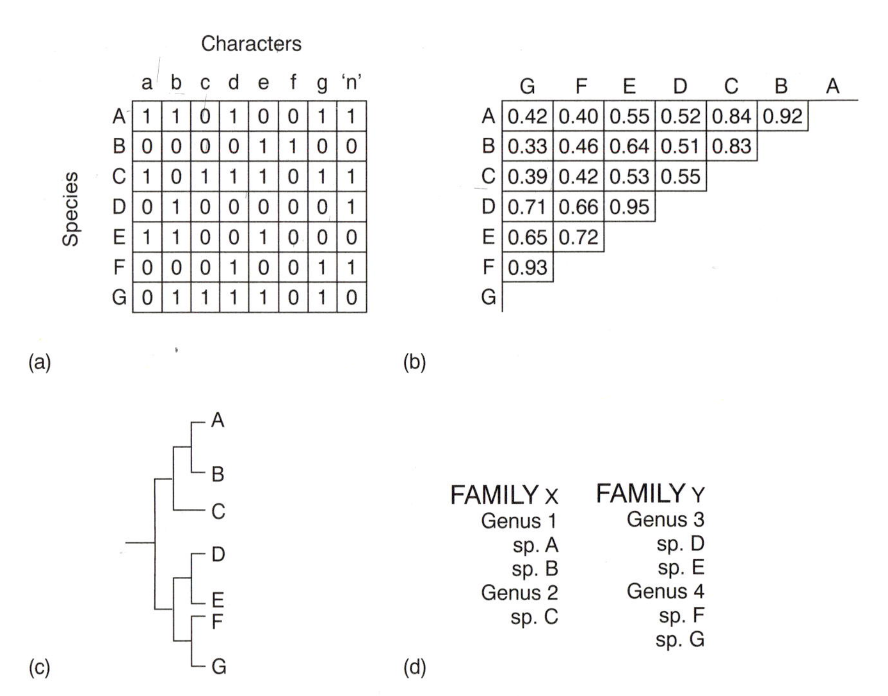


**Fig. 4.2** The steps for creating a phenetic classification. (a) A character matrix, in which the state of all the characters used are expressed for all the species (or other taxa) to be classified. (b) A similarity matrix, in which a coefficient of similarity for every pair of species has been computed. (c) A phenogram, which is a hierarchical diagram in which the most similar species are associated as closely as possible, and the least similar as distantly as possible. (d) A phenetic classification, in which named taxonomic groups are created to express the relative degrees of overall similarity of the species.

The arguments against phenetics are twofold. First, there is no genuinely objective way of measuring and dealing with overall similarity. Characters have to be chosen, and decisions are needed on what is to constitute a unit character. Then decisions have to be made on how exactly the raw data are to be converted to measures of overall similarity, and how this similarity is to be represented as groups. Neither the plexus of similarity nor its carving up into groups correspond to any particular, objectively definable aspects of entities in the real world. The second perceived failing of phenetics lies in the equal use of all characters in measuring overall similarity. Some of the characters contributing to the measure of phenetic similarity of a particular group will also be found in other species that end up outside that group. Other characters will not be found in all the members of the group. Characters in either of these categories do not support the particular groupings that the majority of characters support. They are therefore uninformative, or 'noisy' characters, and yet they still have a role in deriving the measures of similarity, and therefore the creation of the groupings. Unless there are reasons for supposing that these noisy characters are evenly distributed amongst the organisms, they will tend to obscure relationships. Despite the problems in accepting a phenetic approach as the basis of taxonomy, it has a very real importance, not yet adequately formalized, as a means of studying the extent of morphological variation that occurs within and between the members of groups. This property is referred to as disparity, and is discussed shortly (page 63).

The third 'school' of taxonomy is cladistics. Cladists ignore degrees of difference and similarity, because of the lack of objective concordance with real, definable groups; they use only the genealogical relationships as the basis of classification. To them, a natural group consists of all those organisms that share a common ancestor, irrespective of how little or how much any of them have diverged phenetically from the rest. In effect, the classification is based solely upon the branching points that occur in phylogeny. These are accepted as real events, that mark real boundaries and not arbitrary ones. Natural groups conceived in this way may be said to await discovery rather than creation.

## Cladistics and genealogy

Cladistics has now become the standard taxonomic method, and the reason for this is that it lies firmly within the framework of accepted scientific methodology: logically, it consists of proposing testable hypotheses. A postulated taxonomic group is such a hypothesis, and the test is whether all the members of that group possess characters not found in any organisms outside the group. For example, rats, horses, monkeys, pangolins, kangaroos, etc. are the members of the group Mammalia. What this means in cladistic analysis is that the hypothesis is proposed that all those animals placed in the taxon Mammalia shared a common ancestor, and that no other animals shared this same ancestor. It can be predicted that if this is so, then all these animals will share characters inherited from that ancestor, and which no other animal possesses. When it is observed that they all have the characters hair, mammary glands, a single bone in the lower jaw, etc., none of which is found outside this particular set of animals, the hypothesis that the group Mammalia as so constituted is a real, natural, monophyletic group is supported. A hypothesis that a group such as Reptilia is a natural, monophyletic group is not corroborated in this way. As soon as it is accepted that some of the descendants of the common ancestor of the reptiles may be excluded from the group Reptilia (Fig. 4.1), then there need not be any characters that are present in all reptiles, and absent in all other organisms. It follows that if uniquely possessed, or defining, characters need not be present in all members, there is no way of testing the hypothesis that Reptilia is a natural group. For example, the amniote egg of reptiles also occurs in birds and some mammals (monotremes); the sprawling four-legged gait and long tail of reptiles is also present in salamanders.

By insisting that taxonomy should be based on testable hypotheses, cladistics is like other branches of science, and, like them, the procedures go further. Faced with two or more alternative hypotheses of groups, which is to say two conflicting classifications, there is a means of choosing one over the other on the basis of evidence rather than opinion. The hypothesis with the most support from characters is to be preferred over the one with lesser support. Neither of them is taken to be proved true and therefore immutable in any big sense, but one is taken to be a better explanation of the known character distributions. Discovery of new characters, or re-study of the known ones, may well create a situation where the previously rejected hypothesis becomes better supported than the presently accepted one. The latter is consequently abandoned and the former instated.

There are therefore three principles of cladistic analysis that need discussion:

1. *Monophyly*: the only natural groups to be incorporated in a classification are those consisting of a hypothetical ancestor and all its descendants. In practice this generates a hierarchical classification of groups within groups.
2. *Synapomorphy*: monophyletic groups are recognizable by the possession by all the members of shared characters derived from the hypothetical ancestor. In practice, this means characters unique to all the members of the group.
3. *Parsimony*: where there are alternative hypotheses of monophyletic groupings for a particular set of organisms, because of convergence or secondary loss of defining characters, the preferred hypothesis is the one estimated to be the simplest. In practice, the simplest hypothesis is the one requiring the least number of inferred evolutionary changes in characters.

## Monophyly

Strict insistence on monophyly in cladistics leads to the recognition of some unfamiliar groups, and the corresponding abandoning of familiar ones.

Such ancient, revered taxa as Prokaryota (bacteria), Pteridophyta (ferns, etc.), Reptilia, Pongidae (great apes), and Apterygota (wingless insects) are all non-monophyletic because each contains some but not all of the descendants of the respective hypothetical common ancestors. Ancestral or paraphyletic groups like these are rife amongst traditional fossil taxa. However unfortunate it may be that familiar groups are rejected, one can only repeat the logic that if groups are to be scientific hypotheses of genealogical relationships between organisms, then they have to be in the objectively testable form of proposed monophyly, rather than the merely defensible form of paraphyletic groups where defense actually consists of offering an opinion about the general level of similarity of the members that is deemed to be acceptable. What is far more important than worrying about reptiles, apes, and ferns is the effect of allowing monophyletic groups in the classification of unfamiliar taxa. Inspection of an evolutionary classification of, say, a lesser group of fossil bryozoans leaves one completely in the dark about what the evolutionary relationships between the members actually are (Fig. 4.3). Any part of any monophyletic group may have been removed and put into its own group, which has the effect of concealing the relationships of that part to the rest. When only monophyletic groups are allowed, the relationships are transparent. Indeed, the ideal cladistic classification is virtually the diagram of relationships simply turned on its side and the groups given names.

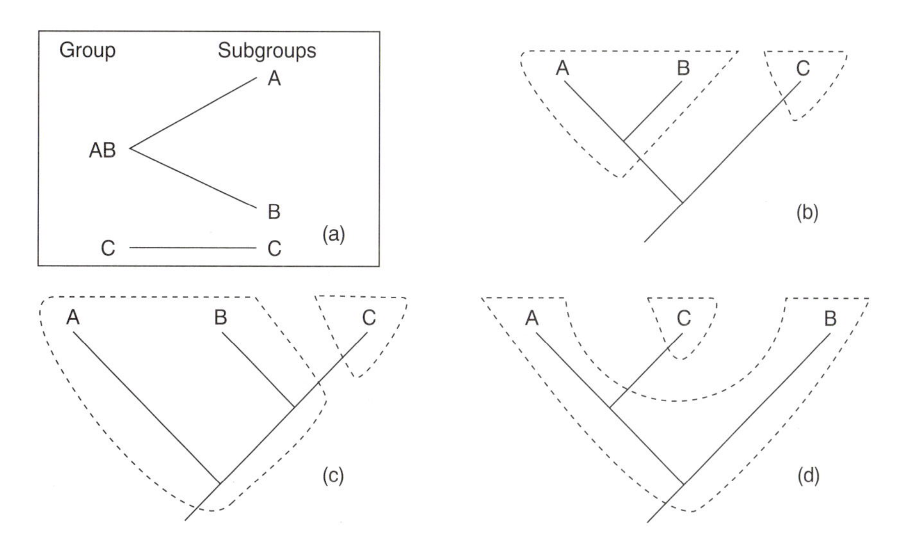


**Fig. 4.3** The absence of phylogenetic information in paraphyletic groups compared with monophyletic groups. (a) A hypothetical non-cladistic classification of the three groups A, B, and C. (b), (c), (d) Possible phylogenies that could be represented by classification (a), as long as the possibility of a paraphyletic group being included exists; if, however, paraphyletic groups are known to be excluded, only phylogeny (b) is consistent with classification (a).

Another awkward aspect of a strictly cladistic classification concerns the matter of rank. Rank refers to the level within a classification of the various taxa, and goes back to the Linnean system of species, genus, family, order, class, and phylum, with subdivisions as appropriate. Within a cladistic classification, two lineages that diverged from a hypothetical ancestor are referred to as sister taxa, or adelphotaxa, and they are regarded as equivalent in rank. Strict observance of this leads to cases of what in traditional terms are exceedingly unbalanced pairs of taxa. At the extreme, a single species might be the sister group of a whole class; strict procedure demands absurdly that the species in question be raised to the rank of class too. For example, the Liassic fossil fish species *Pholidophorus bechei* (Fig. 6.8) is believed to be the sister group of the entire group Teleostei, which contains around 25 000 species. Furthermore, in a detailed cladistic classification of a large group, there may be an extraordinarily large number of levels in the hierarchy, and therefore several new rank levels have to be named. Faced with these two semantically cumbersome difficulties, it is clear that the days of the formal Linnean rankings are coming to an end. Terms such as 'class' and 'subfamily' will no doubt continue to be accepted informally because they are useful in general communication, but a numerical prefix system is likely to replace named rankings in the formal classification (see page 122).

In addition to these questions of convention, there are two related questions of more substance concerning monophyletic taxa. It is customary to seek a cladogram consisting entirely of dichotomies; that is, every monophyletic group at every level has a single sister group. Multiple branching points, polychotomies, are interpreted as lack of resolution of the cladogram due to insufficient information about characters, rather than being a representation of a single hypothetical ancestral species that split into three or more descendant lineages simultaneously. Much was written about this in the early days of cladistics, for it seemed to be assumed that species always split into two daughter species when, clearly, there is no reason why simultaneous multiple splitting should not occur during evolution.

It is now accepted, however, that the assumption about dichotomy is to do with the method of analysis rather than the mode of evolution. If there are three taxa in a group, but no two of them share a character that is absent from the third, then it might be that they did indeed evolve by a trichotomy. But it might also be that two of them are sister groups compared to the third, and that no character supporting this relationship has yet been found. The cladistic argument is that such a character might yet turn up, in which case the apparent trichotomy will be reinterpreted as two successive dichotomies (Fig. 4.4a). Therefore, as new information accumulates, multiple branches may be resolved into series of dichotomies. The reverse cannot occur, for new information (other than correcting mistakes, of course) will not reduce the resolution by creating polychotomies from a

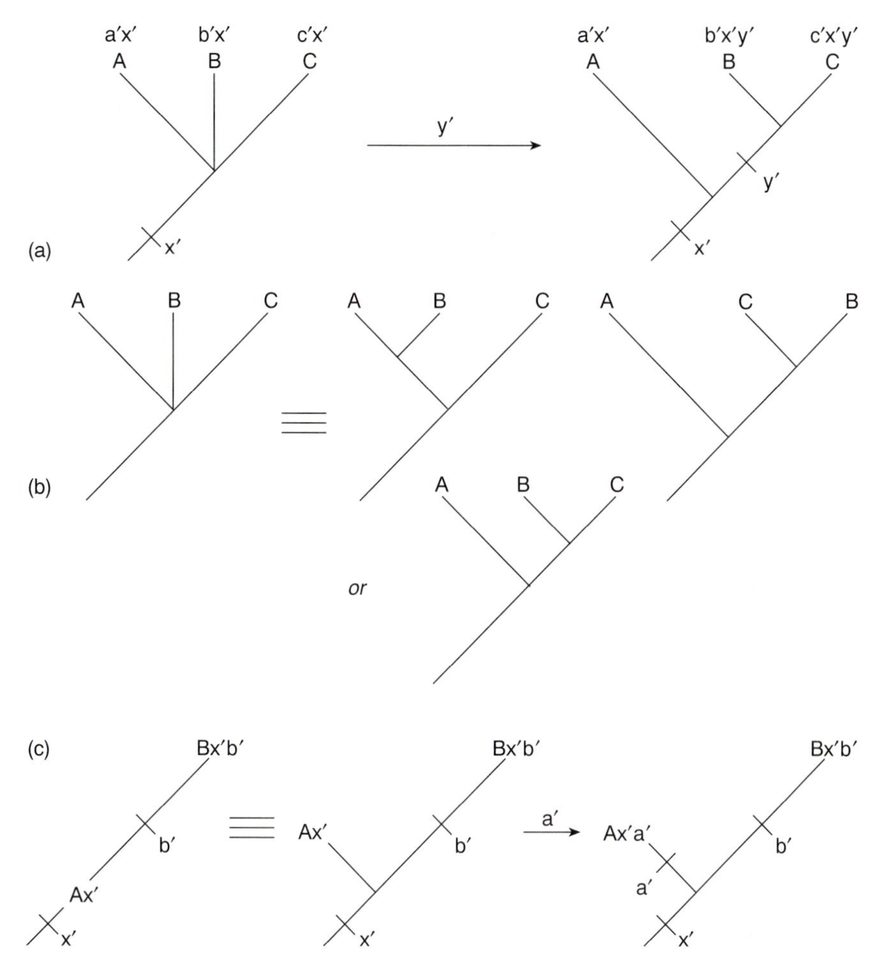


**Fig. 4.4** (a) Resolving a trichotomy into two successive dichotomies by the new discovery of a derived character y′, shared in this case by taxa B and C. (b) Three equally well supported (or, better, equally unsupported) cladograms derived from an unresolved trichotomy. (c) Ancestry as necessarily hypothetical: taxa A and B are related by the synapomorphy x′, while B has an additional derived character b′. Taxon A could be either ancestral to or the sister group of B and it is impossible to know which. Discovery of a new character such as a' will confirm the sister group relationship, but there is no possible discovery that could ever confirm an ancestral relationship of A to B.

series of dichotomies. Because increasing resolution can only be in the direction of increasing the number of demonstrated dichotomies, the ultimate resolution possible for a cladogram is when only dichotomies exist. Even if true multiple evolutionary splits did indeed occur, they cannot actually be demonstrated, unless it is argued that every single character has been studied, which is impossible. A resolving character might turn up after all. The importance of understanding this procedural detail is that it has become common to announce that, in some particular taxonomic exercise, more than one cladogram are equally well supported by the characters. This often turns out to be due to unresolved polychotomies, rather than to convergent acquisition or secondary loss of characters. An unresolved trichotomy can be represented just as well by any of three different successive pairs of dichotomies (Fig. 4.4b).

One of the hardest implications of monophyly to grasp, particularly for palaeontologists, was the question of ancestry. Put simply, an ancestral group, even an ancestral species, by definition can have no unique characters of its own. Every character it possesses must be either the same as in its own immediate ancestor or else the same as in its descendant. It has not diverged so has no divergent characters. The logic of cladism allows the recognition of real groups only by the possession of characters shared exclusively by the members. If there are no unique characters, there can be no monophyly, and therefore no discoverable and testable existence. Ancestral groups are necessarily hypothetical, for logically they cannot be identified with known organisms, living or fossil.

The misconception that this argument meant that cladists did not believe in ancestors, and therefore presumably not in evolution, was not helped by a rather strident accusation by some cladists that palaeontology had been led astray by excessive zeal for ancestor-hunting on the part of its practitioners. In fact, the refusal to recognize real ancestors is another part of the methodology of cladism and has nothing whatever to do with attitude to evolution. To claim that some particular group of known organisms is ancestral to some other is to say that no unique characters have been discovered in the former. One possible explanation is that it was indeed the ancestor. Another is that it is the closest relative, the sister group, and that its unique characters, which would support such a view, have not yet been discovered. If it happens to be a fossil, then it is likely that such unique characters are not even preserved. So, as in the case of polychotomous splits, ancestry cannot be positively tested for, and further resolution by means of new characters can lead only to the identification of sister-group relationships. It is therefore a corollary of cladistic logic that ancestors must be treated as hypothetical, and all known taxa assumed to be divergent from them. In practical terms, characters can have the ancestral state. Any organism that has all its known characters in the ancestral state may be regarded as a perfectly good model for the hypothetical ancestor as far as its characters are known. This is not the same as assuming that all its unknown characters were also in the ancestral state and consequently recognizing it as the ancestor (Fig. 4.4c). No relevant information is lost about evolutionary pathways by adopting this procedure.

## Synapomorphy

If several organisms share some particular character that is known to have evolved only once, and never to have been secondarily lost in any other organism, then it is certain that those organisms, and only those organisms, descended from a common ancestor that itself possessed the character. Synapomorphy proves genealogical relationship and that is why synapomorphy is the test of monophyly. Unfortunately, there is nothing that proves synapomorphy. When it is claimed that a character is a synapomorphy for some particular group of organisms, this is itself a hypothesis whose corroboration depends on the application of certain tests or criteria of synapomorphy.

The initial test is whether the character in question is sufficiently similar in the various organisms that share it to be confident that it really is a version of the same character in all cases. In other words, is it really homologous in the original sense of that word, for only homologous characters are candidates for synapomorphy? The required degree of similarity of the character is normally rather subjective, at least as far as morphological characters are concerned. A comparison of the structure, development, and relative position of the wings of butterflies and dragonflies leaves little doubt that they represent different versions of the same character. Comparing the hind wings of these with the pair of small, sensory halteres behind the single functional wings of dipterans leads to the conclusion that the halteres are a modified version of hindwings, but the substantial difference between the characters in this case means that this hypothesis of homology is rather less well corroborated.

Once an acceptable level of similarity of the character in the different organisms has been established, the next important criterion of synapomorphy is outgroup comparison, which is a slightly misleading term; a better one is the character distribution test. The idea is that organisms that are not part of a monophyletic group will not possess the characters that uniquely define that group. So all that must be done is to look at non-members, referred to as outgroups, and ensure that they do not possess the putative synapomorphy. This procedure appears at first sight to be circular, because it depends on first recognizing an outgroup. Yet selecting an outgroup implies that the in-group and therefore the synapomorphic characters that define the in-group are already known. For a single character occurring in two states, it is impossible to use outgroup comparison to establish which of the states is ancestral and which derived. Aside from this logical difficulty, there is also the possibility that the derived character was evolved independently in an outgroup, or that the derived character was secondarily lost in one or more of the members of the in-group.

In fact, outgroup comparison is not circular in principle at all. An outgroup is usually selected for some particular analysis not on the basis of the characters of immediate interest but on the basis of many other characters, often studied over so many years that they have more or less been forgotten about. An amphibian is an appropriate outgroup for an analysis of amniotes because it has long been established that there are many characters shared by amniotes that are not to be found in amphibians, and vice versa. Therefore a well-corroborated hypothesis about the interrelationships of amphibians, crocodiles, lizards, birds, mammals, etc. has been established (Fig. 4.1). What is being looked at is the distribution of all the characters known, within all the groups of concern including the proposed outgroup. The cladogram that is best supported by these characters is accepted, at which point the outgroup will be apparent. Using this outgroup, new characters of interest can be investigated and the likely synapomorphic state established. If many new characters are discovered whose distribution is incongruent with the existing cladogram, then a point may come when a re-analysis of all the characters together, old and new, leads to a better supported cladogram in which a different group emerges as the outgroup. In practice, of course, a taxonomist will usually feel sufficiently confident of the existing cladogram not to make explicit this logic. So what is termed outgroup comparison actually consists of reference to a cladogram of all the relevant groups based on the pattern of distribution of all the characters ever studied. With the advent of computerized cladistics, where very large numbers of characters can be utilized, the limited concept of the outgroup is becoming outdated. Synapomorphies are tested in effect by whether they emerge as the characters that define the groups that are generated in the best-supported cladogram.

Some other criteria of synapomorphy are sometimes used, none of which has the logical rigour of the character-distribution criterion. Ontogeny has been much heralded as a method of recognizing character polarity that is independent of the character-distribution method. It is based on von Baer's rule that the more widely distributed character states appear earlier in development than the less widely distributed states. In other words, the more general characters—those defining a larger group—tend to appear during embryonic development before the less-general characters that define the subgroups contained within it. In the case of mammals, for example, the vertebrae can be distinguished in the embryo at an earlier stage than can the limbs, and these in turn precede the ear ossicles. Therefore, the vertebral column is a character of the larger group, Vertebrata; the limb is a character of a subgroup of vertebrates, Tetrapoda; ear ossicles is a character of a subgroup of tetrapods, Mammalia. Using this principle as a test of whether a given character is synapomorphic for a particular group involves showing that it appears later in development than the alternative state of the character. Unfortunately, the scope for adaptive evolutionary modification of embryos is such that in practice character polarity based on embryology inevitably requires testing against the character-distribution criterion, and where the two kinds of evidence disagree, it is most unlikely that the embryology would be regarded as the more reliable.

The stratigraphic criterion is more interesting, both because it concerns fossil evidence and because it is at least based on an unarguable theoretical principle. The ancestral character state must have occurred earlier than the corresponding derived character state. Therefore with a complete fossil record, the ancestral state must appear lower in the stratigraphic column. Unfortunately, the fossil record is far from complete so the possibility always exists that the derived character state has been found in earlier fossils than has the ancestral state because a sequence of even earlier organisms bearing the ancestral state happens to be unpreserved. How incomplete the fossil record is in this regard is discussed later (page 109). In any case, the stratigraphic criterion is likely to be tested against the character-distribution criterion and be found to have relatively little effect on its own.

A final criterion of no more than historic interest is biogeographic. On the basis of one particular, untestable model of evolution, it has been claimed that ancestral characters are more likely to occur near the centre of origin of a group than further away. Apart from the obvious unlikeliness that this is even a general rule because of extinction and migration, there is the problem that centres of origins of groups are more or less impossible to discover with any confidence in the first place.

Up to this point, characters as hypotheses of synapomorphy have been seen to be tested mainly by the similarity test and by the character-distribution test. Some putative synapomorphies will be better supported than others, which is to say that they are more strongly corroborated hypotheses. The third test is called the congruence test. It consists of noting how many well-supported synapomorphies support a hypothesized group. The larger the number that do so, the better supported is that group and therefore the better supported as a true synapomorphy is each of the characters that defines it. Congruent characters lend weight to each other, which leads to the third principle of cladistics, parsimony.

## Parsimony

There are two reasons why a certain character may not agree with a cladogam and therefore support a different grouping. The character may have evolved separately in two different lineages, in which case it seems to define a group that is different from the group defined by uniquely evolved characters. Or the character may have been secondarily lost in some members of a group, in which case that character will appear to define as a group what is only part of the group defined by characters that have not undergone such loss in any members. Characters believed to belong to either of these categories are termed homoplasies. If convergent evolution of the same character in different taxa is possible, which it is, and also secondary loss of a character in any descendant taxa is also possible, which it is, then in principle absolutely any hypothesized cladogram *could* be a true representation of the actual phylogeny. Dogs *could* be the sister group of elephants, and cats only distantly related, if enough evolutionary convergence and loss of characters had occurred. Why then is the hypothesis that dogs and cats are sister groups, and elephants their more distant relative much preferred?

All branches of science need a principle to allow a choice between alternative explanations of a given set of observations, and it is invariably based on some version of simplicity or Occam's razor. William of Occam (1300–1349?) wrote: *Frustra fit per plura quod potest fieri per pauciora* (It is unnecessary to do by many things that which can be done by fewer).

As applied to modern science, this effectively means that if there are two hypotheses that both explain some set of observations, but one of them is demonstrably simpler than the other, that one should be accepted and the second one rejected as the best explanation. Simplicity, or parsimony of explanation as it may be called, is measured by how much additional, untested assumptions have to be made in order for the hypothesis to be true, that is, how many extra *ad hoc* assumptions are needed to defend it (see page 17).

As far as taxonomy in particular is concerned, the usual way in which the principle of simplicity is applied is to state that any implied evolutionary change in a character is an *ad hoc* assumption because that event cannot be tested by direct observation, but only inferred from the distribution of character states in organisms. Therefore the simplest set of groupings—namely the simplest cladogram—is the one requiring the least number of implied evolutionary changes. This is minimum-evolution parsimony, and will generally result in minimizing the number of inferred homoplasic characters, because each one implies an extra evolutionary change, either an independent acquisition or a secondary reversal of the character state in question.

In fact, the relationship between minimum-evolution parsimony and the congruence test for synapomorphy of a character mentioned earlier is rather confused, because the two are sometimes supposed to be identical but actually are not. Maximum congruence as a test consists of preferring the grouping supported by the largest number of congruently distributed characters, and completely rejecting the remaining non-congruent characters. But, as Fig 4.5 illustrates, minimum-evolution parsimony can sometimes give preference to a different cladogram from the one supported by maximum-congruence parsimony. Which version of simplicity should be used?

The pattern cladist view (page 14) is that minimum-evolution parsimony should not be used because it entails assumptions about evolutionary processes; namely that characters actually do transform from one state to

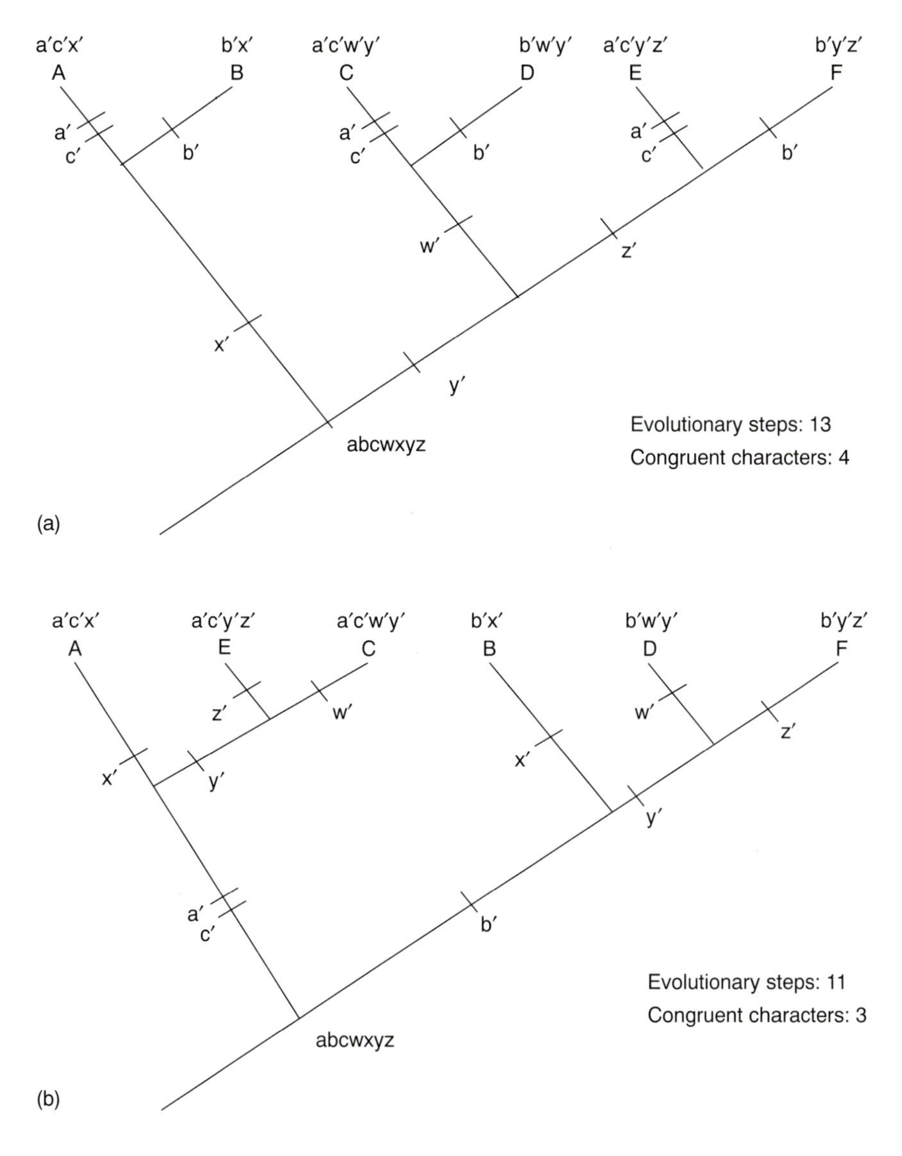


**Fig. 4.5** A character distribution in which one cladogram (a) is preferred using the principle of maximum congruence, but a different one (b) is preferred using the principle of minimum evolution parsimony. (a) A cladogram of taxa A to F with derived character states shown both for the taxa and in the best positions that they could have evolved if the cladogram is read as a phylogeny. There are 13 evolutionary steps, and four congruent characters (w′, x′, y′, z′). (b) Another cladogram of the same taxa with the same character states. In this case there are 11 evolutionary steps, but only three congruent characters (a′, b′, c′).

another. Possession of unique characters alone is regarded as the definition of, and therefore as the test of proposed natural groups. From this point of view, the larger the number of defining characters, the better supported are the groups constituting the cladogram, and therefore maximum congruence must be used. But in the spirit of the approach developed here, a classification is a theory about the cause as well as the effect of character distributions and therefore minimum-evolution parsimony is appropriate. The cause of a cladogram is taken to be character transformation by evolution, and therefore the most-acceptable cladogram is the one that minimizes such transformations. Strictly speaking, from this perspective the number of congruent characters defining a group is not a test of the synapomorphic state of individual characters at all, despite the assertion made earlier (page 57). Rather, the appropriate test is whether a character hypothesized as a synapomorphy is found to be a defining character of a group that is contained in the minimum-evolution cladogram. If it is, then its status is corroborated; if not, it is rejected as non-synapomorphic. In practice, minimum-evolution and maximum parsimony often lead to the same groups, but by no means always. Congruence is thereforereasonable as an approximate test, but should not be used in a rigorous exercise, especially where many characters are available and high levels of homoplasy present.

This has been a very simple account of the general idea of parsimony, because it has been assumed that in principle all the characters can contribute equally to the determination of relationship, which in turn implies that all evolutionary changes from one character state to another are equally probable. Clearly this is an exceedingly unlikely state of affairs. Some character changes must need more underlying genetic modification than others and are therefore may be less probable. The more that is known about exactly how evolution goes about its business, the more realistically could be determined the relative values of different kinds of characters for indicating relationships. For example, those highly prone to evolutionary change would be given less weight because they are more liable to convergence or reversal. In the days of evolutionary taxonomy, such a priori character weighting was uncontroversial, and sets of rules were published, prescribing, for example, that complex characters should be given high weight, and that adaptive characters, simple characters, and character losses should be given low weight, and so on.

A priori character weighting is a veritable minefield, however, because at present the only way to test whether some rule about weighting is correct is to look at the results of evolution. This is a pattern-versus-process situation. If, say, colour was believed to be an excellent indication of relationships, all the red organisms would be classified together and all the blue ones placed in another group. Inspecting the results of evolution as indicated in the resulting phylogenetic classification would show that colour is indeed a good guide, but, of course, this is an entirely circular argument.

So the impasse is reached where there is a powerful feeling of obviousness that some characters and character changes are better guides to relationship than others on the one hand, but, on the other, the apparent impossibility of independently discovering which they are. This is why the simple version of parsimony that assumes equal probability of change tends to be preferred, particularly when the characters are morphological. Unless and until a great deal more is known about the genetic changes underlying character changes, there is no justification for these kinds of authoritarian prescriptions. But the explicit non-weighting of characters is itself open to unconscious (or even conscious) abuse for the simple reason that there is no guide to what should be a unit character. The pattern of bones on the head of a fish could be regarded as one character, or each of a dozen or more bones taken separately, effectively converting the bone pattern into a dozen characters. Satisfactorily dividing morphology into independent unit characters is one of the most pressing problems in morphological cladistics.

The whole issue of character weighting and parsimony is undergoing a revolution in the hands of molecular taxonomists, and this new thinking is filtering into palaeobiology. When the characters in question are nucleotide bases in DNA or RNA molecules, then a unit character can be defined objectively as a single nucleotide at a specified locus on the molecule. There are certain difficulties in handling deletions and insertions of bases into sequences, but nevertheless compared with morphological characters the situation is vastly less ambiguous. Furthermore, there is also the potential to discover the probabilities of the various possible base substitutions, using rules derived from nucleic acid chemistry and not preconceived phylogenetic beliefs. For example, due to stereochemical reasons transitions, which occur between the pyrimidine bases or between the purine bases respectively, have a higher probability than transversions, which occur between a pyrimidine and a purine. Therefore a more realistic model of evolution can be used in which the less probable transversions are given more weight in the phylogenetic estimation than are the more probable transitions. This is indeed a priori weighting, but is not a circular argument because here the weighting is based on information entirely independent of the phylogenetic relationships of the organisms concerned. It is not yet clear whether morphological, and therefore fossilizable characters could ever be the subject of rules of weighting that are independent of phylogeny in this way. One distinct hope is that increasing understanding of the genetical basis of the development of morphology may provide at least some idea of which character modifications are most and which least subject to convergence, based on the relative probability of the necessary genetic mutations actually ocurring in the organism.

## Limits of cladistics

Cladistics is a method for discovering the branching pattern that links taxa together in the real, unique phylogeny generated by evolution. There are, however, circumstances where cladistics is inadequate for answering taxonomic questions. The first of these is the 'bush' problem (Fig. 4.6a). There are numerous cases where cladistics has been unable to resolve the relationships amongst a set of groups because of a lack of convincing synapomorphies distinguishing sister groupings. The orders of placental mammals, fossil and living, is one such example. At the very best, three or four weakly supported relationships have been proposed. This leads to a basal polychotomy of some nine or more lineages that cannot be resolved further. Notwithstanding the comments about resolving polychotomies into dichotomies made earlier, to the effect that the taxonomist is never in a position to give up and assume that evolution did in fact generate a polychotomy, this particular group has been so extensively researched, including much molecular work, that it does not seem likely that its resolution requires only the discovery of further characters. What seems a great deal more likely is that the divergence from one another of what are now recognized as the mammal orders occurred at a very low taxonomic level, based on the small character differences that typically distinguish between genera or, at most, families of modern mammals. These original distinguishing synapomorphies have since been further modified out of recognition by subsequent divergence within the separate lineages, and are no longer discoverable. In other words, the pattern of evolution was indistinguishable in practice from a bush-like pattern. If that is indeed what did happen, then no amount of refined cladistic analysis of the end products is going to produce the well-resolved cladogram that is sought by the method.

At the other extreme of the taxonomic range, there can be considerable problems in applying cladistics to the species level (Fig. 4.6b). Suppose a population of organisms is known that has all the characteristics expected of the ancestor of another known species. All its characters appear to be in the ancestral state compared with the descendant, it has an appropriate biogeographical distribution and, if a fossil, an appropriate stratigraphic range. Strict cladistic principles do not allow it to be interpreted as an ancestral species, and indeed do not even allow its recognition as a taxonomic group because its members lack synapomorphies. But given all the information that is known about it, the hypothesis that it is an ancestral species may actually be simpler than the hypothesis that it is not. Cladistics is clearly not an appropriate means of analysing such a situation and should, therefore, not be applied. Cases like this are certainly known, and neontologists have coined the expression 'paraspecies' to cover them. With appropriate caution, a palaeobiologist could in principle do likewise for a fossil population.

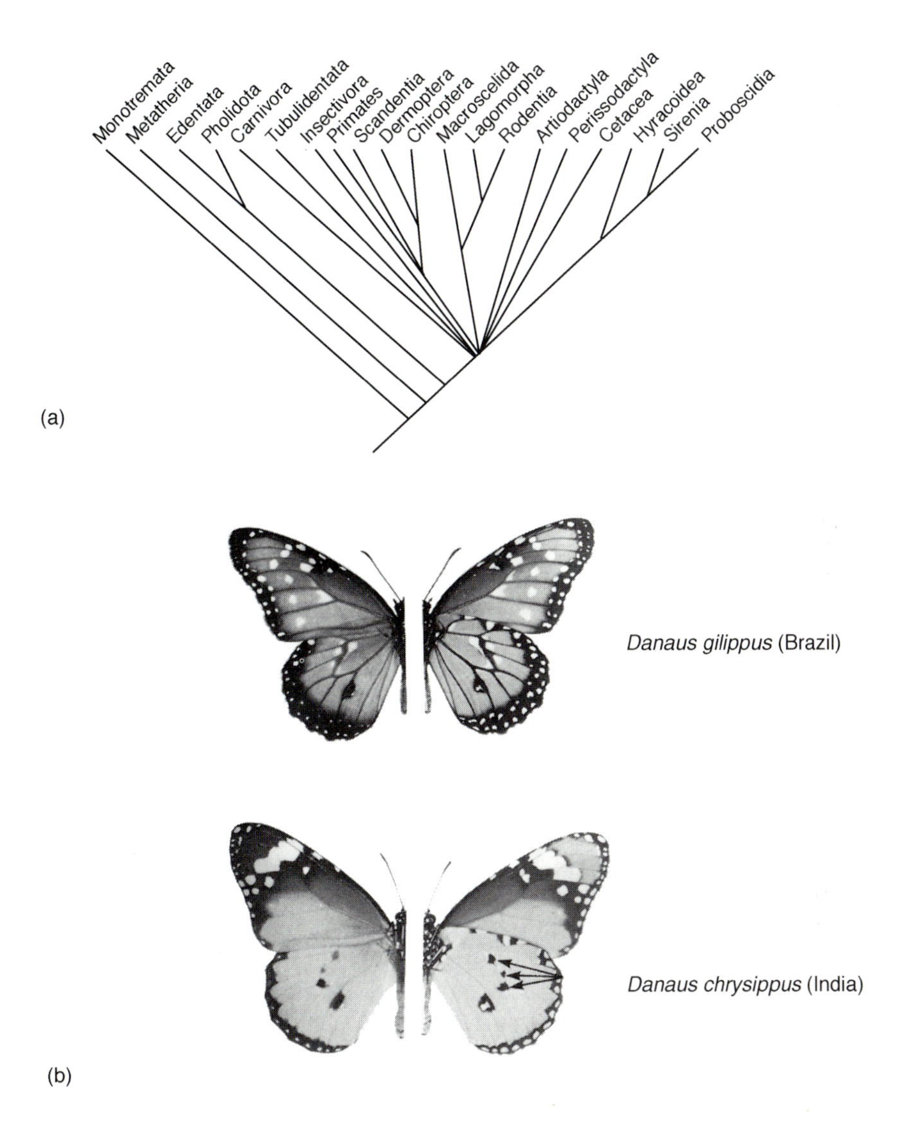


**Fig. 4.6** (a) Modern mammalian orders as a phylogenetic bush that is largely unresolvable by cladistics; above the Metatheria (marsupial mammals), even the groupings proposed are not strongly supported. (b) A paraspecies situation unresolvable by cladistics. The two butterfly species *Danaus chrysippus* and *D. gilippus* share two apomorphies so form a cladistic group, while *D. chrysippus* has three further derived characters. But *D. gilippus* has no known derived characters and cannot formally be recognized by cladistics; it is referred to as a 'paraspecies' with the implication that it may include the ancestor of *D. chrysippus*. ((a) After Novacek *et al.* 1988; (b) after Ackery and Vane-Wright 1984.)

A final word of caution about the application of cladistics concerns confidence levels. While it is entirely proper in principle that the most parsimonious cladogram should be preferred, it is also in keeping with good scientific practice that the level of confidence in it should also be expressed. All too often a study results in the publication of a best cladogram, but scrutiny shows that it is not significantly better than one or more alternatives.

## Phenetics and disparity

Cladistics has become the dominant methodology for taxonomy because of its claim to scientific objectivity and its reliance on real, discrete events (phylogenetic branching points) as the basis of its natural groups. But by abandoning the methods of phenetics, where measures of degree of similarity between organisms and groups of organisms are calculated, there are some questions about the pattern of evolution of morphology that cannot be addressed. Evolution generates changes within lineages as well as splitting of lineages. Obviously there is some broad meaning to the idea that different amounts of change may occur in different lineages, producing taxa with greater or lesser differences in morphology between the members.

If all the morphological variables of an organism were to be measured and plotted on a multidimensional graph along as many axes as there are characters, then that organism could be described as occupying some particular point in multidimensional morphospace that is different from the point occupied by any other organism. Furthermore, a group of organisms could be described as occupying that volume of the morphospace that contains all its members. From this point of view, there are some very interesting questions about the taxonomic pattern that could be investigated. Is the volume of morphospace occupied by a group simply proportional to the number of species in the group, or do some groups have species more widely dispersed in morphospace than other groups? Do different groups, such as two respective sister groups, sometimes evolve to occupy different-sized volumes to one another? Does the volume occupied by some particular group change as it evolves over time? How rapidly is morphospace invaded or evacuated over geological time, and what factors might determine these rates?

The property of a group that needs to be measured in order to tackle such questions as these has recently become termed 'disparity', which is defined as the extent of the variation in overall morphology amongst the members of a taxonomic group of organisms. It is distinct from diversity, which is a term that is now restricted to the simple number of subtaxa contained within the group, particularly the number of species or genera although it could apply to any rank. Diversity is a property that is measured by strictly cladistic techniques, whereas disparity needs phenetic methods

Considerable interest has been generated recently by the concept of disparity and possible means of measuring it. The spark was a debate about the significance of the Burgess Shale fossil arthropods (see page 240). This is an extremely well-preserved fauna from the Middle Cambrian, and it includes a considerable number of higher arthropod taxa not found in modern faunas. The question posed is whether the disparity of the Cambrian arthropod fauna was greater than the disparity of modern arthropods, and what this might mean for evolutionary processes. Another area in which disparity measurements can be applied concerns the different contributions that the respective subgroups make to the overall disparity of a higher taxon, and how this changes with time (Fig. 4.7). No doubt many more questions will be addressed, some old ones such as the differences in rates and extents of morphological evolution often found between pairs of sister groups, and some new ones such as comparisons between rates of morphological and of molecular evolution within a single group.

Desirable as these investigations might be, the difficulties in establishing a satisfactorily objective measurement of overall morphological range within taxa or of differences between them remain profound. Evolution-generated difference is a multidimensional continuum, not a series of discrete, objectively discoverable phenomena. Therefore there is an arbitrariness about any boundaries that are drawn between differing morphologies, and consequently about any quantification of those differences. Nonetheless, there is one important difference between present problems and the comparable problems that faced the pheneticists of 30 years ago, during their efforts to establish a phenetic method for classification (page 50). This is that the taxonomic units within and between which estimates of disparity are being made can at least be defended as natural, so long as they are cladistically determined monophyletic groups. It does mean something potentially answerable by hypothesis and test to ask which of two monophyletic sister groups is the more disparate, or whether a given monophyletic group has increased or decreased in disparity over time. For example, it would be vague to the point of meaningless to ask if, say, birds were more disparate than reptiles because of the act of pure subjectivity that removed birds, and mammals, from the monophyletic group Amniota in order to create the 'Reptilia' (Fig. 4.1). On the other hand, it is perfectly proper to ask if the Mammalia (including the stem-group mammals) is more or less disparate than the Sauropsida, which is its sister group and includes birds, living reptiles, and various fossil groups. First, both groups are objectively determined realities, and second, by virtue of being sister groups they share a hypothetical common ancestor and so must be of exactly the same age. If a measure of the difference in disparity *can* be found, then at least it is a measure of a property of something real.

Methods of actually achieving useful measurements of disparity are very much in their infancy. The basic idea is to measure, or code for a large

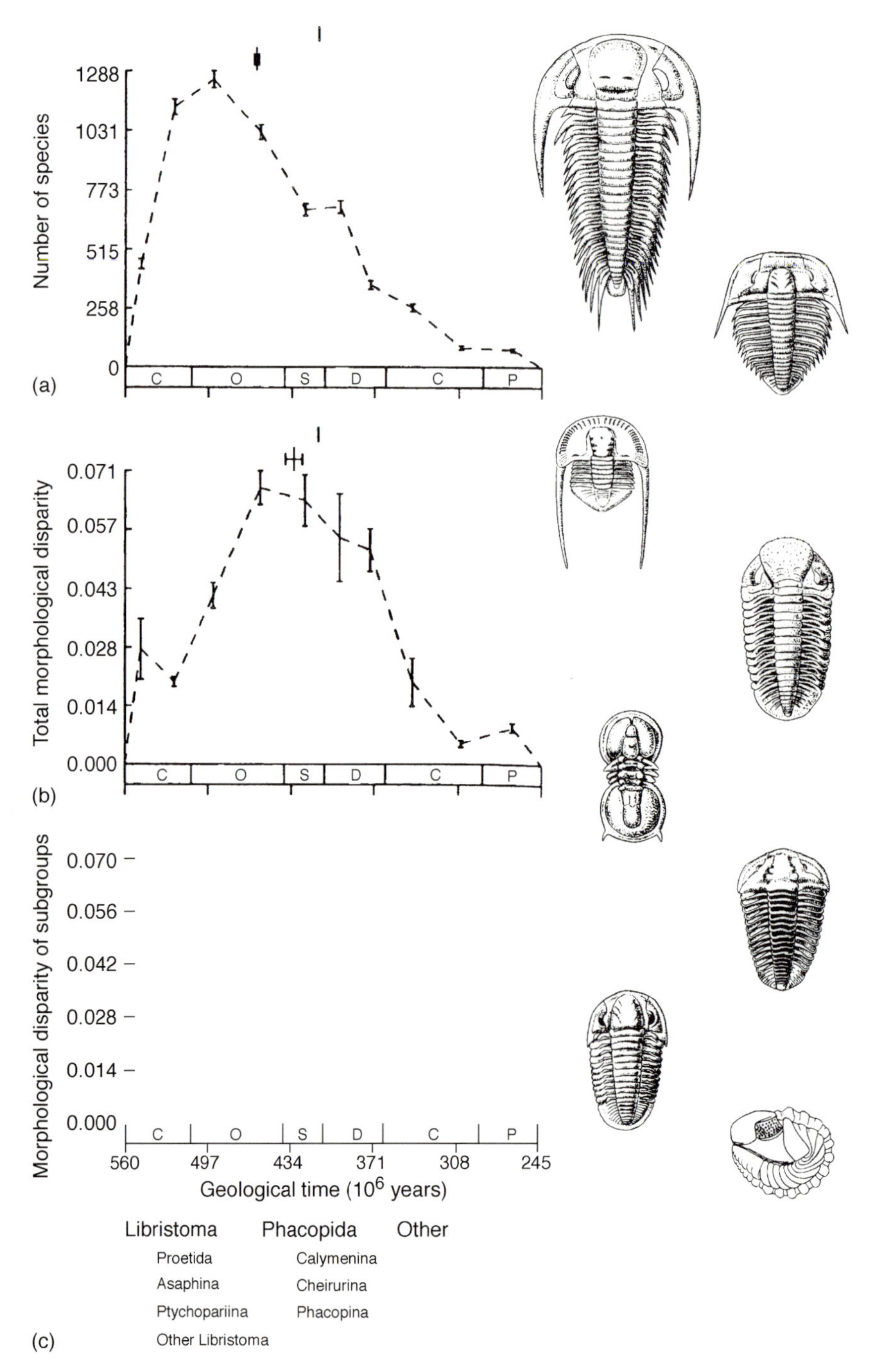


**Fig. 4.7** Disparity in trilobites. (a) Plot of the diversity of trilobites expressed as the number of species against geological time. (b) Plot of a measurement of disparity of the same organisms over their history showing that the two plots do not coincide. (c) Plot of the changing contribution of the main subtaxa of trilobites to the overall disparity of the group over time. (Plots after Foote 1993*a,b*; illustrations of an assortment of trilobites after Kuhn-Schnyder and Reiber 1986.)

number of morphological characters, treat each character as an axis in morphospace, and plot the positions occupied by the organisms in question within this space. Many refinements and alternative procedures, such as principle component and principle coordinate methods are already available from earlier phenetic theory, while statistical tests of hypotheses particular to this field are emerging. The objection that measuring disparity will always have a subjective component need not be a deterrent, any more than the way in which choice of unit morphological characters for cladistic analysis need be a deterrent to aiming for scientific rigour in cladistics. The questions that disparity measurements seek to address about monophyletic groups concern real, and often extremely interesting evolutionary phenomena.

## Further reading

As with Chapter 3, on evolutionary theory, so here, too, the literature on the history, philosophy, theory, and practice of taxonomy is vast and overlapping. It would be pointless, not to say difficult, to unearth the original author of many of the important ideas that have been developed in the past 30 years or so that has seen the phenetic revolution come and go and the cladistic revolution come and grow, not to mention the rapid development of molecular techniques. Even now, there is an explosion of new interest in taxonomy, connected on the one hand with the arrival of molecular techniques and the need to handle increasingly vast amounts of character data, and on the other to the concern about conservation and recording of the diversity of the world's biota before it is too late.

For a short, general overview of the subject see Kemp (1985*a*) or the simpler Ridley (1986). Hull (1988) gives a detailed account of the recent history and Panchen (1992) is an excellent in-depth investigation of the philosophical structure of taxonomy. Sober (1988) explains parsimony clearly and in detail.

Much the clearest account of the phenetic approach is still the classic *Numerical taxonomy* by Sneath and Sokal (1973), which also discusses many issues in taxonomy as relevant today as then. For the principles of cladistics, Hennig's (1966) classic *Phylogenetic systematics* is worth looking at, although much of it appears very old-fashioned to the modern reader. One of the best and clearest papers describing the principles is Farris (1983), while a standard textbook account is that of Eldredge and Cracraft (1980). Smith (1994) is a very good review of the whole subject from the palaeontological perspective. Forey *et al.* (1992) describe more directly how to do cladistics. Accounts of the state of play in molecular taxonomy are usefully presented by Avise (1994) and Hillis and Moritz (1996).

The example of mammal evolution can be followed up in the review by Novacek *et al.* (1988).

The developing study of disparity can be followed though a correspondence between Gould (1993) and Ridley (1993), with commentaries by McShea (1993), and Wills *et al.* (1994), who include a discussion about the measurement of disparity. Foote (1993*a,b*, 1997) discusses the general question.

# 5
# Incompleteness and what to do about it

All scientific information is incomplete at some level; no one actually counts the number of leaves on a tree or pebbles on a beach. What matters is whether the information is adequate for generating theories about some particular question to hand, not whether it can answer all conceivable questions. So it is important to have some idea of the extent and nature of the incompleteness of the information available, because these dictate what potentially that information can be used for.

The observations that are made about fossils are deemed to be an incomplete picture of what really existed or happened over the geological time period in question when they do not correspond to the expectations held about how biological objects should be, or should behave. A fossil locality that yielded nothing but one species of animal would be judged to be an incomplete representation of the palaeocommunity because of the belief that while alive there must have been something for those individuals to eat, and probably some predators as well. Even recognizing a fossil shell, say, as merely a part of a mollusc depends on the obvious idea that only a whole animal with lots of other parts could have had such a shell-like object. Clearly the principle in dealing with incompleteness is to interpret the actual observations in the light of rules governing the nature, ecology, and evolution of living organisms generally. *Ad hoc* assumptions must be made to create compatibility between the two. In other words, dealing with incompleteness is another aspect of spanning the epistemological gap (page 21).

The study of incompleteness of data involves the use of methods to estimate the extent of the incompleteness—for example, how much of some particular stratigraphic record is actually missing or how many species are unpreserved at some locality. It also encompasses reconstructing the missing data—for example, by assessing the functional aspects of the preserved anatomy—or the probable original geographical range of some taxon. Within such a general framework, four categories of incompleteness can be recognized.

1. *Organismic incompleteness* is the failure of most of the attributes of most organisms to fossilize.
2. *Stratigraphic incompleteness* refers to gaps representing unpreserved

segments of time in a sequence of sediments, and therefore gaps in the potential fossil record, due either to failure of deposition in the first place or to subsequent erosion of formerly existing rocks.

3. *Ecological incompleteness* is failure of a preserved fossil community to match accurately the formerly living community that it represents, due to such factors as the different preservation probabilities and transport rates of different species.
4. *Biogeographical incompleteness* is the failure of organisms to be preserved as fossils throughout their former full geographic range, primarily due to local stratigraphic incompleteness but also involving such factors as differing preservation in different environments and the subduction of former landmasses.

## Organismic incompleteness

Most tissues of organisms decay by bacterial activity or are consumed by predators and scavengers, leaving at best the hard, mineralized parts to be fossilized. A typical fossil consists of a very small percentage of the total anatomical features of the living organism, and, of course, all the behavioural and physiological attributes are also unpreserved. On the face of it, this so limits the usefulness of the fossil record for studying evolutionary and palaeoecological questions that enormous efforts are made to reconstruct as far as is possible the missing information about the unpreserved characters.

There are several cases where soft parts of fossils have been preserved, often in the most exquisite detail. Sedimentary strata yielding such forms are called *Lagerstätten*. Although the details differ greatly from example to example, *Lagerstätten* preservation typically involved the sudden, catastrophic burial of organisms, a process termed obrution. Perhaps the most renowned case is that of the Burgess Shale, from the Middle Cambrian of Canada, where a massive submarine mud-slip carried the bottom-living organisms into anoxic conditions too quickly for aerobic bacterial decay to break them up. After the initial event, anaerobic bacterial activity seems to have been important in preserving soft tissues in many *Lagerstätten*. For example, anaerobic sulphur bacteria can produce hydrogen sulphide, which in turn reacts with iron salts to produce the mineral pyrite. The pyrite can replace the soft tissues such as muscle or plant cellulose, leading to the preservation of extremely fine details of the anatomy. In other cases, a chemical process of phosphatization of tissues such as arthropod cuticle can lead to preservation of a perfect replica of the original form.

The great majority of fossils, however, have experienced no such fortuitous circumstances, and consist solely of hard skeletal parts. Soft tissues like muscles, blood vessels, glands, and brains are only represented to the extent that their former presence is indicated by markings such as grooves, spaces, and processes showing on the skeletal elements themselves (Fig. 5.1a). As well as missing soft parts, even the skeleton is not often completely preserved in some groups—notably vertebrates in which isolated skulls and disarticulated limb bones are common. Furthermore, fossils have frequently been subjected to the effects of compression of the sediments in which they lie. They may be distorted in various ways, and are often reduced to a completely two-dimensional object (Fig 5.1b). The result is that even a published illustration of the skeleton of a fossil represents an initial hypothesis about what it looked like, namely a reconstruction.

Such reconstructions of the skeletal anatomy depend initially upon the use of several specimens, each of which is likely to have different parts well preserved. After that, direct comparisons with living relatives must be made, so far as the latter exist at all. The addition of soft tissues to the reconstruction—such as muscles, nerve courses, and brains—also requires a comparison with living forms. The latter have provided a number of rules about general relationships of skeletal to non-skeletal structures, such as the anatomy of muscle attachment sites, or of surfaces covered with chitin or keratin.

## Functional anatomy

The relationship between the anatomy of an organism and its functioning has always interested biologists, and the fact that different organisms are generally well designed to perform different respective functions is the single most important notion in biology. In terms of the whole history of biology, the process of adaptation by natural selection is really just the latest, but by far the best explanation available to account for this correspondence between form and function.

The phenomenon is particularly focused in palaeobiology, firstly because it is generally supposed that it is changes in function that have the actual significance during evolutionary changes in anatomy. The second reason is the challenge that the concept makes to palaeobiologists. In the case of living organisms, the form–function relationship is studied by describing and measuring the morphology on the one hand, and observing and experimentally manipulating the function on the other. The two can be compared directly and shown to complement one another as the case may be. For example, the thickness of a mollusc shell can be measured and its geometry described. The magnitude of forces to which the living mollusc is exposed in its environment can also be measured. It becomes simple then to show that the shell's form is explicable as a suitable design to resist those kinds of forces. For fossils, however, only form can be studied directly; function has to be inferred indirectly from the form because the organisms cannot

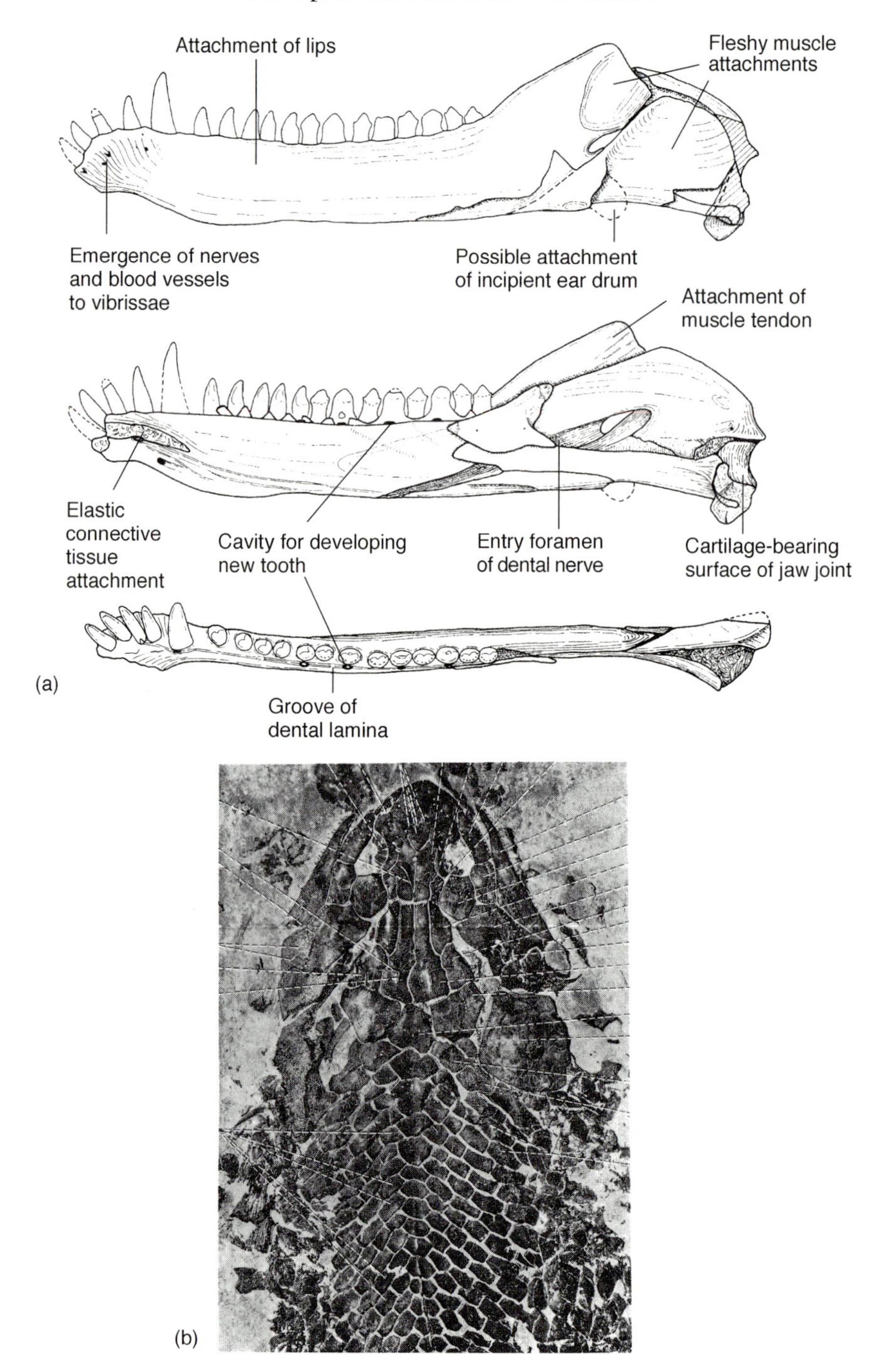


**Fig. 5.1** Fossil preservation. (a) Signs of various unpreserved soft tissues impressed on the lower jaw bones of the mammal-like reptile *Procynosuchus*. (b) The front part of a lobe-finned fish *Osteolepis* to show dorso-ventral compression causing completely two-dimensional preservation. (a) After Kemp 1979; (b) after Save-Sonderbergh 1933.

be observed actually performing in their environments. Therefore methods based on rules about the relationship between form and function must be applied, rules derived from living organisms and with no way of ensuring for certain that they are appropriate to the particular fossil organism being studied.

Three general methods are available for inferring function from form—the design test, the comparative test, and the phylogenetic test.

### Design test

If a part of the anatomy of an organism is designed to perform a particular function efficiently, then the recognizable design features of that structure should indicate what it is designed to do. Design features include such things as the shape of cross-sections of skeletal elements, the orientation of the joint surfaces between adjacent bones or shell valves, the size and disposition of anatomical indications of muscle attachments, the nature of dentitions, and the shapes of leaves. From such observations of a fossil, inferences can be drawn about the mode of locomotion, the kind of food consumed, the habitat in which the organisms lived, and so on.

For example, there is a group of Lower Triassic mammal-like reptiles called the Therocephalia with quite distinctive features of their hind legs (Fig. 5.2a). The femur has a slightly S-shaped curvature and also a slight propeller-like twisting along its length. The lower leg consists of the tibia and the fibula that attach to separate ankle bones, the calcaneum and astragalus, respectively. There is a well-formed joint between the latter two bones, giving a quite unexpected mobility within the ankle itself. Manipulation of the separated bones of a well-preserved specimen suggests that the limb was designed to operate in more than one mode. It could behave like a typical reptile, with the femur sticking out sideways, or it could operate in a mammalian mode, with the knee turned forwards and the femur close to the body. Further corroboration of this optional dual-gait hypothesis comes from the bone shapes and the processes for muscle attachments, which indicate that the single set of hindlimb muscles present would have been positioned in just such a way as to permit either of the two gaits to be adopted. Although there is no way of directly seeing if this is how therocephalians used their hindlimbs in real life, the hypothesis does explain all the anatomical features as a good design for achieving the dual-gait system.

Interest in the design test has led to more sophisticated versions. The paradigm method was proposed in the 1960s by Martin Rudwick. The initial step in the formal version of this approach is to postulate possible functions for some particular structure of the fossil. Then, for each such function, a model or paradigm is created. This consists of an engineering design, in principle created with no further reference to the fossil, that would most effectively carry out the proposed function. Finally, the actual fossil structure is compared with each of the paradigm structures in turn and the one

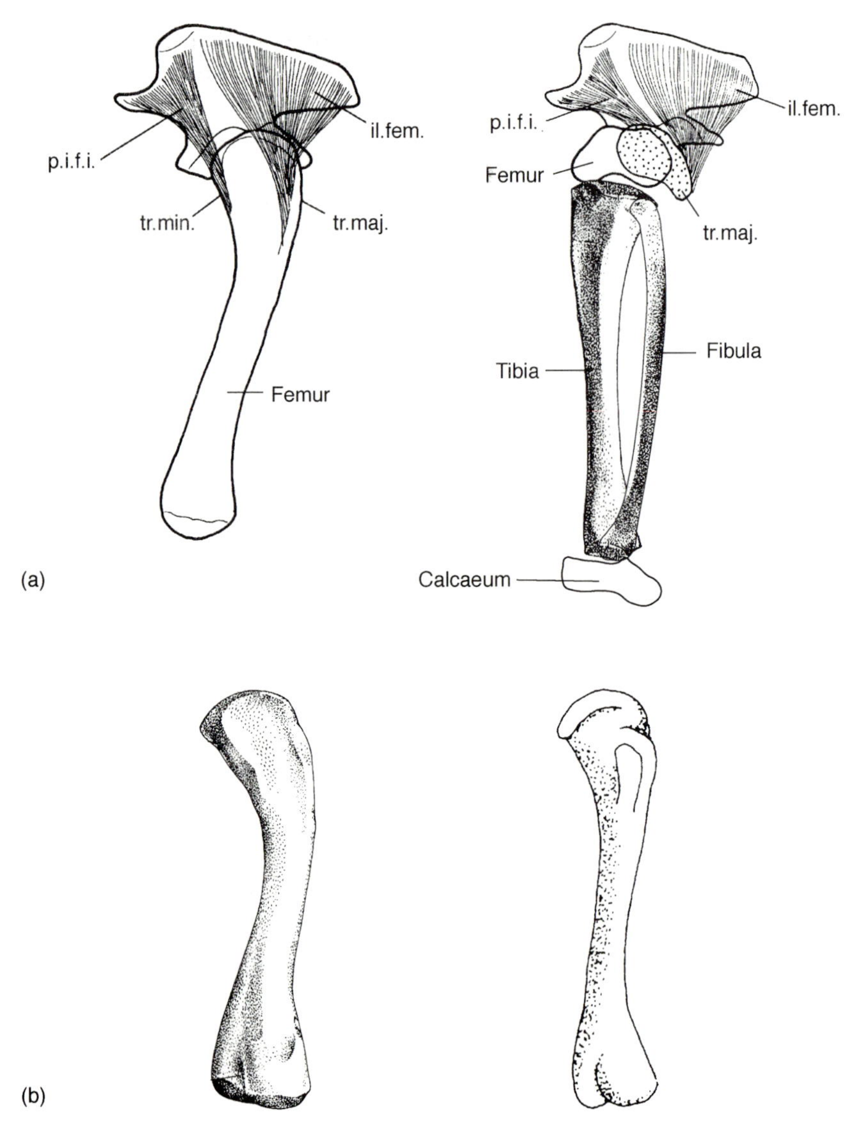


**Fig. 5.2** (a) Therocephalian hindlimb structure and function: (left) the femur in the 'mammalian' position, extending downwards from the hip joint; (right) the femur in the 'reptilian' position, extending laterally from the hip joint, to the tibia and fibula. Note that the muscles (p.i.f.i. = pubo-ischio-femoralis internus; il.fem. = ilio-femoralis) and their attachments (tr.min. = trochanter minor; tr.maj. = trochanter major) are positioned such that they can operate in either gait. (b) Comparison of therocephalian (left) and crocodilian (right) femora.

that it most closely matches is taken to be the best hypotheses for the function. It is claimed that this method is properly scientific, in that the hypothesis proposed is potentially testable and refutable. In principle, one could always think up a new paradigm, and see if it matches the fossil even more closely than the best one so far discovered. Actually, for all its apparently greater rigour, the paradigm test is just formalizing what any intelligent application of the design test achieves.

Rather more different, and of greater importance, is the method known as constructional morphology, a concept introduced by Adolf Seilacher. Here it is recognized that function is not the only determinant of form in organisms, but that form is also constrained by other factors (see page 28). One of these is the previous phylogenetic history of the organism. The form of any organism can only be a modification of the form of its ancestor, so there are historical constraints on what natural selection can actually achieve that may result in the evolution of less than optimal structure for carrying out the function in question. Furthermore, there are constraints due to the limited range of biological materials and processes available for creating the structure. A drill tipped with tungsten carbide might well be the best design for a boring mollusc to use, but no metabolic pathway is available to organisms for producing such a material, and so a softer material, far more subject to abrasive wear, has to be utilized. The constructional morphological approach allows for these other factors when interpreting function from anatomy in a fossil. What it suffers from, though, is a lack of methodological rigour when disentangling the combined causes of a particular form from each other. There can be several different reasons for apparent sub-optimality of design of a structure, apart from the possibility that the hypothesized function itself is not the true one.

In fact, all versions of the design test suffer from a major difficulty, which is that form and function do not have such a simple relationship that the one can be read unambiguously from the other. Plenty of cases are known amongst living organisms where, in design terms, the structure is a bit of a botch up but works well enough for the organism. Such sub-optimal structures can be recognized in a living organism but would not be so apparent in a fossil. Other cases are known where a virtually identical structure performs different functions in different organisms, as, for example, the different uses to which a long, slender beak is put in different species of bird. Some use them for collecting insects from crevices, some for molluscs on sandy shores, and yet others for nectar feeding; all that these functions have in common is that they are forms of feeding, which is true of beaks of all shapes. Conversely, often quite different designs perform the same function in different organisms, as witnessed by the variety of ways of filter-feeding found in near-shore animals. Perhaps the most confusing cases of all concern structures that have more than one function. The anatomy will be a compromise between the design requirements of the two or more respective functions, and only in the light of knowing for sure what the functions are can there be any hope of recognizing the complex, subtle nature of the compromise. In frogs, for example, the mouth cavity functions both as a respiratory pumping system to fill and empty the lungs with air, and for capturing and manipulating food. On top of these, many frog species also use it as part of the sound-producing system. What *is* the optimal design for such a multipurpose organ? The usefulness or otherwise of the design test is returned to later.

### Comparative test

Although used in an unformulated way by functional anatomists for years, it is in the context of studying the adaptation of animal behaviour that the comparative method has achieved prominence. The principle is that if independent cases of some particular attribute of organisms are found to be associated with some particular environmental feature, then this is evidence for the hypothesis that that particular attribute is an adaptation for solving the problem posed by that particular environmental feature. As applied to the unknown function of fossil structures, the test would be that if a similar structure is found in an unrelated living organism where the function is known, then it can be hypothesized that the same function was performed by the fossil version of the structure. For example, to continue with the story of the therocephalians, there is a remarkably close similarity between their hindlimbs and those of modern crocodiles (Fig. 5.2b), including a hinge between the calcaneum and astragalus. As is well known, crocodiles can adopt alternative gaits, either a sprawling 'low walk' when moving slowly around river banks or an upright, almost mammal-like 'high walk' when moving fast or far. Yet crocodiles and therocephalian mammal-like reptiles are not at all closely related and there is no doubt at all that they evolved their hindlimb modifications independently from a very much more primitive state. The comparative test therefore supports the hypothesis, based initially on the design test, that therocephalian hindlimbs could operate facultatively by two different gaits.

Strictly, the comparative method requires that the cases compared should be independent of one another, in the sense that the attribute has evolved separately rather than having been inherited from a common ancestor. It is also true that the more cases found of a coincidental occurrence of the attribute and the function, then the greater the probability that there is a structural–functional relationship between the two. The limitations of the comparative test include those already referred to in the context of the design test—namely, that a similar structure may have different functions, and different structures the same function. It is also true, as, for example, in the limbs of therocephalians and crocodiles, that the structure will not usually be completely identical in the independent instances, and the differences could well reflect significant functional differences. Naturally the test can hardly apply in those cases of fossil structure where there simply are no modern analogues to compare them with, such as the various enigmatic Cambrian groups of vaguely arthropod-like construction (Fig. 1.2b) or the remarkable neck of *Tanystropheus* (Fig. 1.2d).

## *Phylogenetic test*

The rise and spread of cladistic methods has extended into the field of functional anatomy because functional attributes of organisms can be considered as character states. It is after all the business of cladistics to analyse the presence or absence of particular character states in particular organisms. The application of the method to functional analysis consists in principle of first constructing a cladogram of the fossil and its immediate extant relatives, based on as many characters as possible (Fig. 5.3a). The distribution of the functional character states is noted in the extant taxa, after which the most likely state of the functional character in the fossil can be decided on the basis of parsimony (page 59).

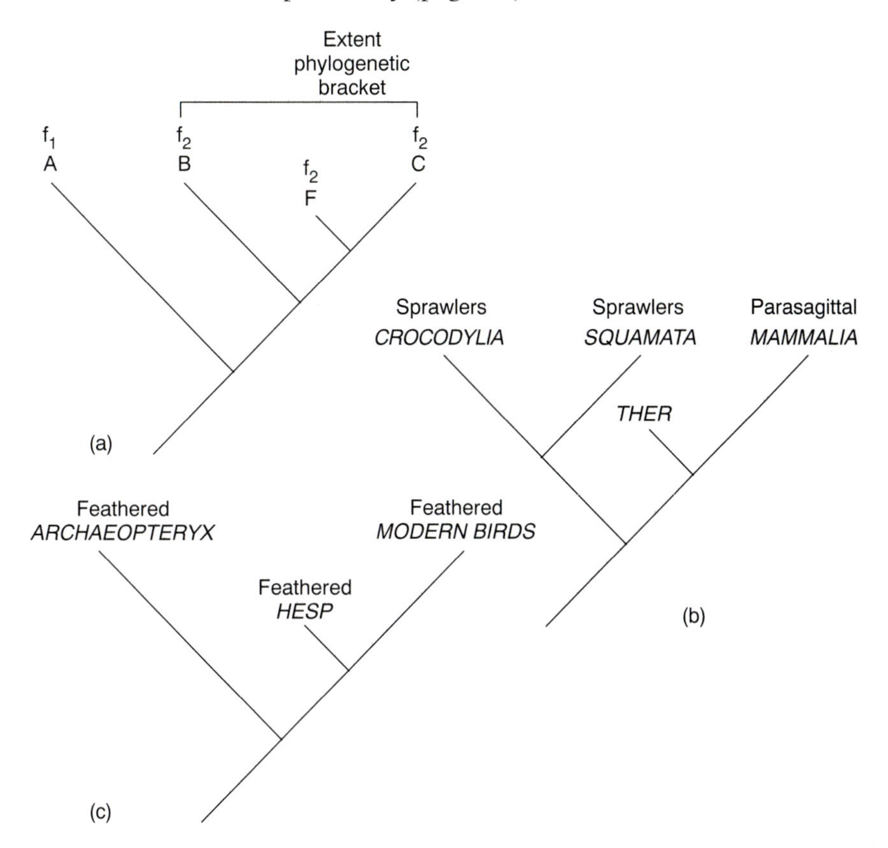


**Fig. 5.3** (a) The principle of the extant phylogenetic bracket: the fossil taxon F lies cladistically between the modern taxa B and C, which form its extant bracket. Because both B and C have the functional character state $f_2$, it is inferred that F also has $f_2$. (b) The extant phylogenetic bracket of the therocephalian mammal-like reptiles (THER): in this case it is impossible to infer whether THER had the locomotory function of 'sprawler' or of 'parasagittal' because its extant bracket consists of taxa with different respective functions. (c) The phylogenetic bracket of the Cretaceous toothed bird *Hesperornis* (*HESP*) as evidence that it was feathered.

The main virtue of this method is that, as with all cladistic procedures, it has explicit logic and a testable form. In practice, however, it has severe limitations. The most obvious one is that the fossil may have a function not represented in any of the living relatives—in cladistic terms, an autapomorphic character state. There is no way in which this could be determined positively. There is also the frequent possibility of an ambiguous result arising when the two nearest extant groups, the 'phylogenetic bracket' as it has been termed, differ in functional character state, while no other extant taxa help to resolve the matter. To return yet again to the therocephalian hindlimb, the extant sister group of Therocephalia is the Mammalia, and the living sister group of these two together is the Sauropsida, which consists of all living reptiles and birds (Fig. 5.3b). These two, therefore, comprise the phylogenetic bracket of the Therocephalia. As far as locomotor function of the respective hindlimbs is concerned, the reptiles plus birds group has, primitively, a sprawling gait; mammals have a gait with the femur turned in close to the body. It is impossible to tell from parsimony analysis of the states of the hindlimb function which of these, if either, the Therocephalia possessed. Indeed, as the earlier tests imply, they probably had neither (or, more precisely, both). Indeed, it is hard to find an example where the phylogenetic test is any use at all. In reviewing and promoting the method, Lawrence Witmer (1995) managed only a single concrete, but trivial, example—namely the demonstration that the Cretaceous flightless bird *Hesperornis* probably had feathers because it is bracketed by *Archaeopteryx* and extant birds, both of which are feathered (Fig. 5.3c). Design and comparative criteria already support this hypothesis rather firmly anyway.

## *Conclusion*

There is no simple, unambiguous relationship between form and function in organisms. Factors in addition to function are involved in determining the anatomy of an organism, and every organism is a complex entity with many functions that affect each other. As for fossil anatomy, there is the additional handicap that functional hypotheses cannot be tested by observation and experiment anyway. One possible conclusion from this is the scepticism of some modern palaeontologists (for example, George Lauder, 1995) that at best no more than broad general inferences about function in fossils can ever be defended, at about the level of whether an organism is a herbivore or carnivore.

Most palaeontologists do nonetheless accept that functional hypotheses about fossil structure are acceptable provided they are created and defended as testable hypotheses and not just subjective assertions. Of course, they then proceed to differ from one another about what actually constitutes a 'proper' test. Proponents of the modern phylogenetic test, such as Lawrence Witmer, and Harold Bryant and Anthony Russell, argue that the objective nature of cladistics allows a properly formulated hypothesis to be erected on the basis of character-state distribution. The hypothesis may then be tested by application of the non-historical tests of design and comparison. Such authors clearly feel that the virtues of the phylogenetic test mean that it should always have priority. At the other end of the spectrum, several palaeontologists (for example, Carol Hickman) have considerable faith in the design test for generating hypotheses of function from structure. She, not surprisingly perhaps, is interested in enigmatic organisms with no modern relatives or close analogues.

The most thoughtful position is that all three kinds of tests are logically valid because they all derive from expectations about the relationship between structure and function derived from the study of living organisms. David Weishampel, for example, has recently taken this view. Non-historically, anatomy is demonstrably designed, to at least a reasonable extent, for function; similar structures in modern organisms do usually perform reasonably similar functions; function does occur as ancestral and derived character states, the distribution of which tends to follow phylogenetic relationships. There is no obvious overarching principle that points to any one of the tests having universal priority over the others. A case where all three tests, or even two of them, point to the same functional conclusion would lead to a good, well-corroborated hypothesis. The example of the therocephalian hindlimb (Fig. 5.2) seems to be such a case. Conversely, a case where the tests produced conflicting conclusions about function would generate one or more weakly corroborated hypotheses. More work is, however, always possible: a more extensive consideration of design; a wider-ranging literature search for modern analogies; a better cladogram. Any of these could increase the support for one of the hypotheses over the others.

## Recognizing fossil species

A further suite of characteristics of organisms that are unpreservable are those concerned with the behaviour and genetics of interbreeding between them, which is to say those to do with recognizing a population of organisms as members of the same species. In fact, because the breeding behaviour and genetics of the vast majority of modern species are also unknown, palaeobiologists in practice are often little worse off than their neontological colleagues when trying to recognize conspecificity. The difference is that when questions of evolutionary processes are being considered, the neontologist can go ahead and find out about the genetic structure of a population, and so on, an option unavailable to the palaeontologist.

The argument about what is the proper species concept persistently remains unresolved, despite enormous efforts. At least there is widespread agreement that the species is important in evolution and taxonomy, and this is not surprising because at its basic level the term 'species' means 'kind' of organism, and there is no doubt that the process of evolution produces recognizably different kinds of organisms. There are, however, two fundamentally incompatible ways of thinking about how to tell one species of organism from another. The first is to use the observable features of the organisms, namely their phenotypic characters such as morphology. A species is a collection of organisms with the same phenotypic characters. The importance of this type of concept is that it gives a method for recognizing species that is not prejudiced by any ideas about how species do or do not arise; it is based on patterns of characters. The second approach is to use the mechanisms by which a species arises and is maintained separate from other species. In this case, a species is conceptualized as a population of organisms that interbreed with each other, exchanging genes and thereby sharing a common gene pool. The importance of this second type of concept is that it is based on the idea of the minimum irreversible evolutionary step, because once new species have diverged from each other, they cannot exchange genes and therefore cannot recombine or revert to the ancestral species; the concept is based upon evolutionary processes.

The superficial reason why there is a debate about which of the two is the better concept is because they do not completely coincide. There are plenty of examples amongst modern populations of organisms of polytypic species in which interbreeding occurs between organisms that are phenotypically recognizably different, but whose offspring are perfectly viable. Conversely, sibling species are sets of organisms that are phenotypically indistinguishable but which behave as separate species with no interbreeding between them. In addition, all manner of patterns of geographical variation of form, and differing degrees of hybridization between distinct populations, cause confusion between the two approaches to species recognition. There is also the matter of asexual species, which are quite common and which appear to consist of morphologically distinguishable kinds of organisms of no less reality than sexually reproducing species. All of these problems are not difficult to resolve in principle, because they are not mysterious but are, on the contrary, perfectly explicable by evolutionary theory.

There is, however, a more profound reason why the two types of concept can never be compatible, and is the real reason why the argument between the respective proponents of the two is ultimately unresolvable. A concept that is based on observable characters of organisms means that an organism is placed in some particular species if it possesses the characters that have been agreed to define that species. If it does not, then it is placed in another species, or a new species is erected to contain it. A species is like a box into which those organisms that have the necessary defining characters are put. Now the point about defining characters is that they cannot logically be changed. If an organism does not have the defining characters, then it cannot go into that species box. If defining characters were allowed to change—in other words, if an organism that had lost them nevertheless was put into the box—then the characters in question would not actually be defining at all, and consistency of species recognition would be lost. The great advantage that this morphological or phenotypic species concept has is that any organism can in principle be classified as a member of one species or another. But the great disadvantage is that it cannot easily accommodate the idea of evolutionary change.

Evolution implies that characteristics of species can and do change, and therefore there can be no absolute defining characters. The alternative to relying on characters is the breeding, genetic, or biological species concept, which is based on gene flow by interbreeding, and which is completely compatible with the idea of evolutionary change of characters within a continuously evolving species. If a population of interbreeding organisms shows gradual evolutionary change, then any one of its members could alter one of its phenotypic characters while continuing to be a member of the interbreeding population. In time the modified character may well spread to the whole population without at any time the species as a whole ceasing to exist. The interbreeding type of definition suffers, however, from the converse difficulty, which is that, without defining characters, it is impossible in principle to decide to which species any particular organism looked at in isolation actually belongs. So the central impasse is reached. A phenotypic type of concept allows the classification of organisms, but is incompatible with the idea of evolutionary change. A genetical type of concept allows for evolutionary change, but is incompatible with a formal classification of organisms into species as taxa.

This is, of course, an extremely refined view of the distinction, but it does indicate that differences in philosophical viewpoint as much as practical difficulties fuel the species-concept debate. Several attempts to reconcile the two alternative types of concept have been made. One way is to create a sort of umbrella concept that allows either one to be acceptable for the definition of a species. For example, a version called the Cohesion Concept says, in effect, that a species is a collection of organisms that remains cohesive (distinct) but that the mechanism maintaining this cohesion can be natural selection, which implies a phenotypic kind of definition, or exchange of genes, which implies a genetic type of definition. Such arguments evade rather than resolve the central dilemma. Another device is to create a concept that is genetic in principle but phenotypic in practice. The way to achieve this is to decide which particular phenotypic characters are most likely to reflect the underlying interbreeding behaviour of a species, and use them for species recognition. The Recognition Concept of Hugh Paterson is one such version. Here it is those phenotypic characters that the members of an interbreeding species use to recognize one another that are used in species definitions. Each species must have a specific mate-recognition system (SMRS) unique to itself, be it morphological, behavioural, chemical, or whatever. If the SMRS for a species can be identified, then any organism with that characteristic must be a member of that species.

Turning to fossils and the problem of recognizing fossil species, clearly there is no option but to use those phenotypic characters that are preserved. A palaeospecies can be defined as those fossils that are sufficiently alike to deserve the same species name. This completely pattern-based concept is not controversial as a basis for species as taxa in a classification. There are plenty of arguments about just how similar members of a single species should be, of course, and major taxonomic revision of fossil groups consist very largely of these sorts of decisions. Such discussion becomes very muddied, however, when gradual change in morphology over geological time is perceived. George Gaylord Simpson, for example, developed the chronospecies concept, by which a range of specimens through time that showed sufficiently small change could be recognized as a single species undergoing evolutionary change. No criterion of degree of change allowed is given, and indeed it is left unclear what criteria should be applied to show that the case in question did indeed consist of a single, interbreeding population through time.

Reaction to the subjectivity necessarily involved in delimiting palaeospecies and chronospecies has led many palaeobiologists to adopt a rigorously cladistic approach of one sort or another. If a species is regarded as a taxonomic group, then by the principles of cladistic analysis it must be demonstrably monophyletic and defined by uniquely derived characters (page 52). Furthermore, if it is to be the lowest category of taxon, a species must be the smallest group that can be so defined. This pays no attention to evolutionary change through time; if a group can be diagnosed in this minimal way, then it is a separate species and that is that. It is rather confusing if such a taxonomic unit is termed a 'species', and Andrew Smith has proposed that they be termed a 'phenon' instead, restricting the term 'species' to the biologically interbreeding groups of neontology.

Phena as so conceived cannot be used for palaeobiological considerations of evolutionary change, because of their essentially unchangeable, defined nature. If the fossil record is to be used to investigate evolution, then fossils themselves must be considered as members of potentially evolvable units—that is, as genetic or biological species. It may be argued that there are no possible means for converting pattern-based phena that have been defined by characters into process-based species that can be described as evolvable, and therefore that the fossil record is no use for studying change within populations. The sort of questions regarded as unanswerable by such critics are these. How can polytypic species be distinguished from multiple monotypic ones, and sibling species from single species? How can successive fossil populations through geological time be shown to form continuous interbreeding populations rather than combinations of local extinctions and geographical migrations? Questions of this nature require more information than simply the characters seen in fossils. They require knowledge about the processes of change that were affecting the living organisms. Once again the pattern-versus-process syndrome has appeared, and, once again, the response is that patterns and processes can be combined into hypotheses (page 15). In this case, a hypothesis about which fossils belong to the same species can be derived from predictions about what are the most likely patterns of characters amongst the fossilized remains of members of a one-time single species.

Analogy with modern organisms indicates that members of the same fossil species will generally show:

(1) close morphological similarity;
(2) a statistically normal distribution of variation in character states such as size;
(3) equal numbers of male and female members where sexes are dimorphic; and
(4) continuous growth stages.

The species population will also be expected to show both spatial and temporal continuity, to the extent that these can be determined. Thus a collection of fossils from a particular locality and stratigraphic age that are all pretty much alike may be hypothesized to be members of a single species (Fig. 5.4). If it really represents two sibling species, or only one part of a polytypic species, then the hypothesis is wrong. But the hypothesis is also testable to the extent that new characters and new specimens may show that the expectations about the morphological structure of a single species cease to be well met, and an alternative grouping into more than one species becomes preferable.

## Stratigraphic incompleteness

Most sedimentary rocks have formed from particles of weathered rock, and mineralized remains of organisms, that have sunk to the bottom of a body of water and settled there. The particles are typically carried some distance first, by flowing water such as rivers, ocean currents, and floods, although wind-borne sediments also exist. Convenient as it would be for the palaeontologist, the world's rivers, lakes, and oceans are not subject to a continual, steady rain of particles. Instead, deposition rate varies from prodigiously high, such as when a tsunami or a flash flood instantaneously unloads material many metres thick, to zero as in the case of a fast-flowing river or strong submarine current. Worse still, sediments and already formed sedimentary rocks are subject to erosion, which removes material, creating in effect a negative rate of deposition. Causes of erosion are many. They include

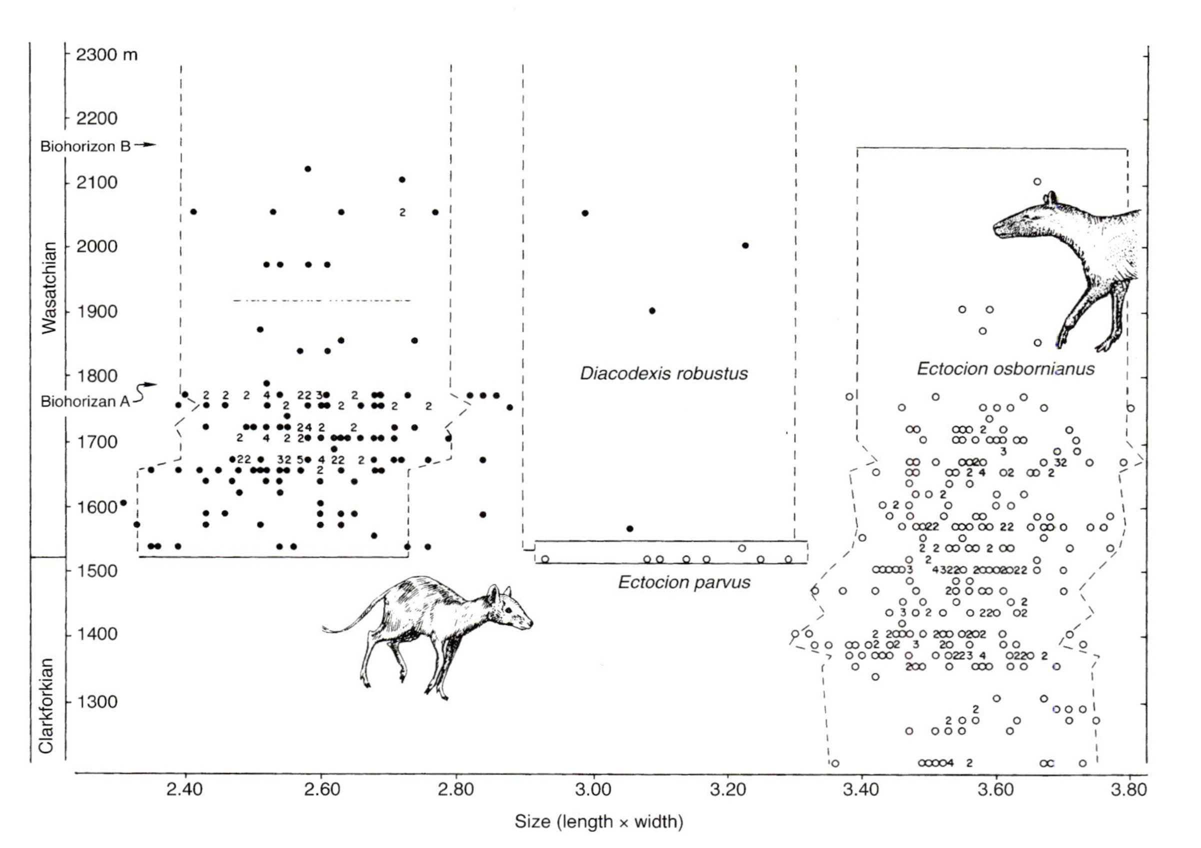


**Fig. 5.4** An example of well-corroborated fossil species: fossil mammal species of two genera of early Eocene mammals from Clark's Fork Basin, Wyoming. The size of the lower first molar tooth is plotted against time, and the species *Diacodexis metsiacus* (artiodactyl) and *Ectocion osbornianus* (condylarth), respectively, show a normal distribution of the character in a population that is continuous in time and space. Numbers on the plot indicate multiple specimens occupying a particular point. (After Gingerich 1985.)

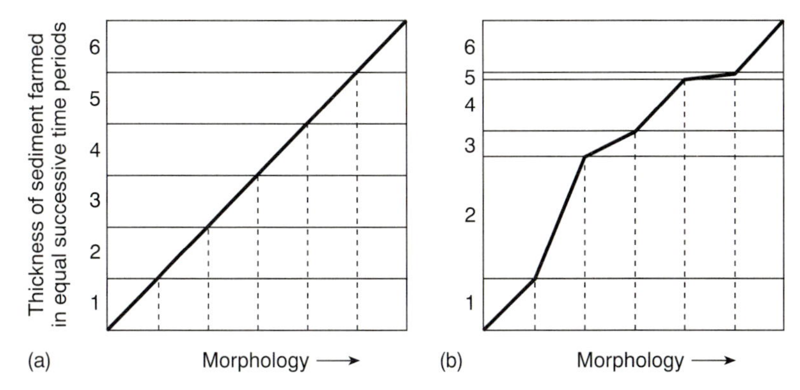


**Fig. 5.5** The effect of varying rates of deposition on apparent rate of change in the fossil record. The numbers refer to equal lengths of time. In (a) the rate of deposition is constant, and the assumed constant rate of change in morphology appears constant. In (b) different thicknesses of sediment are laid down in the equal time intervals with a large amount in interval 2, a small amount in interval 3, and a hiatus with no deposition at all in interval 5. Although the rate of actual morphological change is still assumed to be constant, the apparent rate varies with deposition rate, from lowest in interval 2, to high in interval 3, to a quantum shift in interval 5.

the scouring effect of altered patterns of water circulation in the sea acting upon submerged sediments and rocks, and the weathering effect of wind, rain, and ice acting on rocks exposed at the surface. The net result of all these processes is that the sedimentary rocks that constitute any geological section will be an incomplete and distorted record of the geological processes that actually occurred during its formation. Any fossil record that the rocks possess will be a correspondingly distorted and incomplete picture of the history of the biota at that particular place and period in time.

An example of the kind of distortion that would follow from different rates of deposition at different times would be the apparent rate of evolutionary change that occurred through some particular section (Fig. 5.6). A high rate of deposition would produce a thicker layer of rock and changes occurring within the time period over which the layer was formed would seem slow. Conversely, if the rate of deposition was low, a thinner rock layer would represent the same length of time, and therefore events recorded within it would appear to be faster. A period of zero deposition would have the effect of making change appear instantaneous, and a period of erosion when record was actually removed could cause all manner of artefactual effects. It is therefore very important to attempt to assess the incompleteness of the stratigraphic record before reading the enclosed palaeontological record too literally.

Incompleteness cannot be defined simply as the percentage of the total rock laid down in a section that happens to be still preserved. During periods of zero deposition there would have been no rock laid down in the first place, so the section could be complete in this sense but still be an incomplete record of the events of the period. Completeness clearly has to

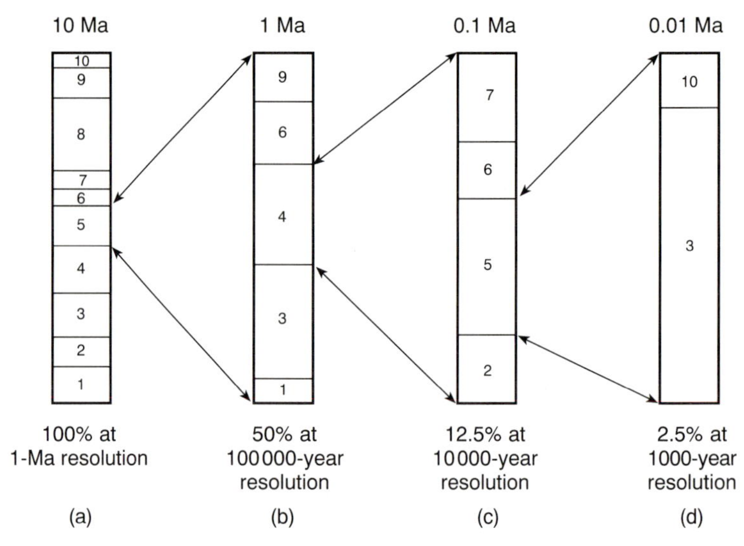


**Fig. 5.6** Relationship between the selected time interval and the percentage completion of a stratigraphic section. In (a) the section represents 10 million years (Ma) of deposition, and every successive 1-Ma interval in that time is represented by some sediment: the section is 100% complete at the 1-Ma resolution. In (b) the record for the fifth 1-Ma interval of (a) is shown enlarged and within it only half of the successive 100 000-year intervals are represented by sediment: the section is estimated to be 50% complete at the 100 000-year resolution. In (c) one of the 100 000-year sections of (b) is shown enlarged, and only half of its successive 10 000-year intervals is represented by sediment: the section is an estimated 50% 50% = 25% complete. In (d), with only two of the successive 1000-year intervals represented by sediment, the section is an estimated 25% 20% = 5% complete at the 1000-year resolution.

refer to this period of time over which the section formed, and therefore could be defined as the percentage of that time represented by preserved rock. This is better, but is still not very useful. Is it necessary for every minute, every day, every month, every year, to be represented by rock for the section to be regarded as complete? The concept of completeness also has to refer to some specified time interval within the total period of formation of the section. So completeness is defined as the percentage of the successive time intervals of a specified length that are represented by sediment in the section (Fig 5.6). Thus a section could be 100 per cent complete at the million-year interval, which means that every successive one million-year interval is represented by some sediment. But the same section might be only 50 per cent complete at the 100 000-year interval, because only half of the successive 100 000-year intervals are represented by some sediment. Finally, this section could be a mere 12.5 per cent complete at the 10 000-year interval, and maybe only 2.5 per cent complete at the 100-year interval.

Having decided that completeness is a relative and not an absolute measure of the quality of a stratigraphic section, the next idea is that of resolution. This is a measure of the smallest time interval that can be distinguished as temporally separate from an adjacent interval of the same size. In practice, it refers to how far apart in time two fossils, or recorded events in the section, must be in order for them to be distinguished as successive and not contemporaneous. Resolution is therefore the level at which the rock sequence is 100 per cent complete. The resolution will differ from place to place throughout a typical section, as some parts of the record will be better preserved than other parts.

For palaeobiology, the important quality of a given fossil record was termed 'adequacy' by C. R. C. Paul. Adequacy is the relationship between the temporal resolution of the fossil record and the time-course of the process being studied. If the process occurs in steps of, say, 10 000 years at a time, then a fossil record with a resolution of 10 000 years will be able to throw light on that process, because each step will be preserved distinct from the successive step. A record with a temporal resolution no better than, say, 100 000 years will be inadequate for studying the process, because in this case 10 successive steps will have the appearance of a single large one. In practice, there can be no absolute limits to adequacy, or even much of an objective measurement of it. Generally speaking, a fossil record with a reasonably high level of completeness at a time interval sufficiently less than the time-course of some particular process of interest will have a resolution that is adequate for developing hypotheses about that process.

The actual measurement of completeness of stratigraphic sections is difficult and is always an estimate rather than an actual measurement. The most obvious and direct way would be to measure the rate at which sediment accumulates in a modern environment and then to compare this figure with the thickness of sediment accumulated over a known length of time in a comparable ecological setting. Unfortunately the actual measured rates of sedimentation in modern environments are so enormously variable both within a single geographical area, and between areas, as to make any average measure virtually meaningless. David Schindel quotes figures between 400 mm and 2 740 000 mm per thousand years for deltaic systems, and between 35 mm and 1 460 000 mm for coastal reefs and carbonate shoals. On top of this, different patterns of bioturbination by organisms living within the sediments, secondary erosion by water currents, and compaction as the sediments turn to rock all affect the final thickness of rock actually preserved over a given length of geological time.

A more direct approach is to measure the net accumulation rate that has actually occurred between absolutely dated points in a stratigraphic section, in order to find an average appropriate for some particular kind of deposit. The most important example of this is in the case of the Deep Sea Drilling Project cores taken from the sea bed in many areas of the world. It is found, rather surprisingly at first sight, that the net accumulation rate depends among other things on the total time between the bottom and the top of the section being measured (Fig. 5.7a). A section that spans

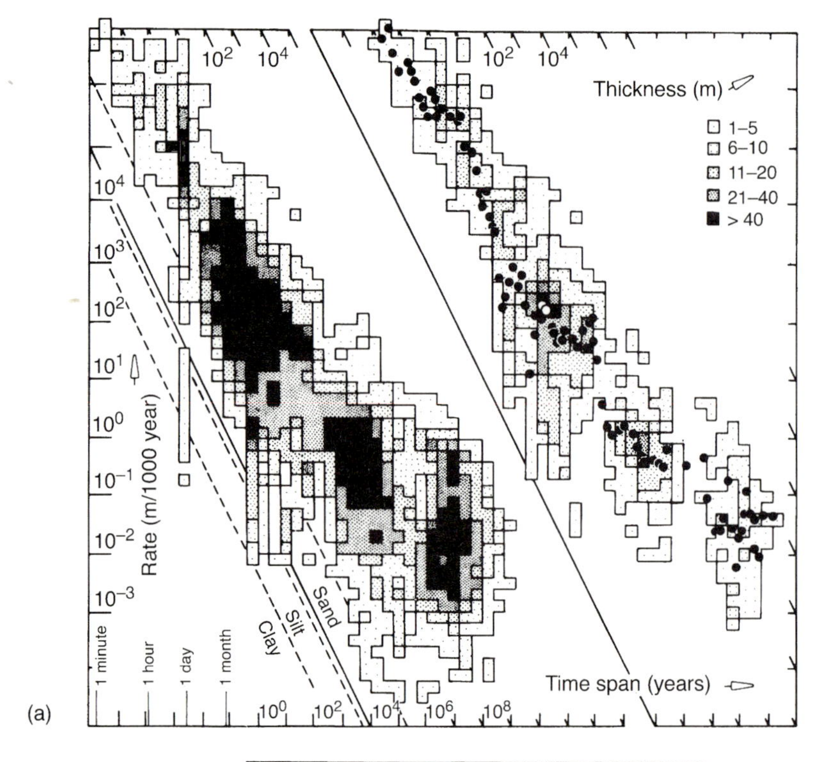
Thickness (m)
1–5
6–10
11–20
21–40
> 40
Rate (m/1000 year)
Time span (years)
Sand
Silt
Clay
1 minute
1 hour
1 day
1 month
(a)

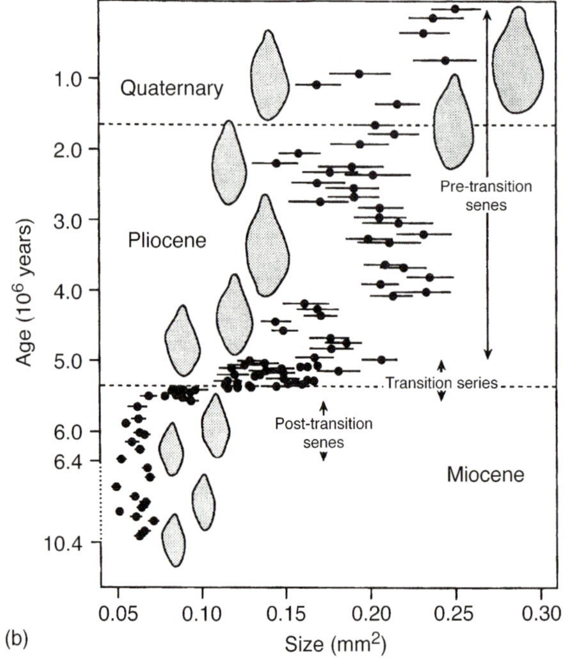
Quaternary
Pliocene
Miocene
Pre-transition senes
Transition series
Post-transition senes
Age ($10^6$ years)
Size (mm$^2$)
(b)

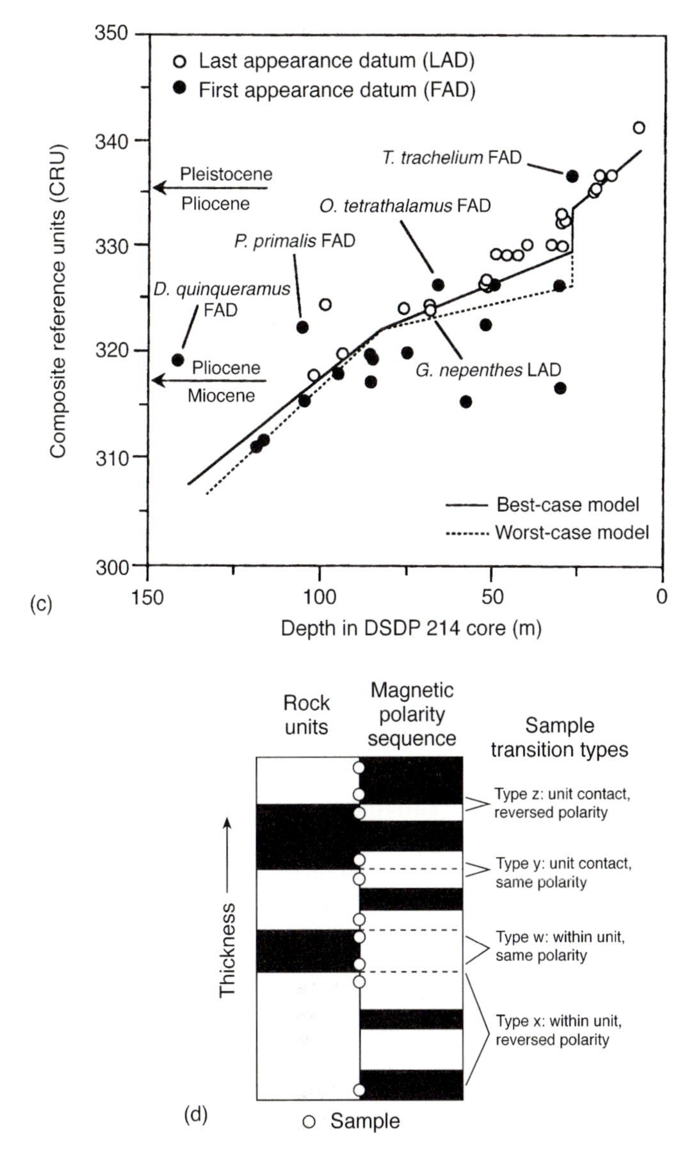


**Fig. 5.7** (a) The apparent rate of deposition against the time span of the section studied: on the left are all depositional environments; on the right are fluvial environments. The density scale refers to the number of samples at each point on the graph. (b) Evolution of the foraminiferan *Globorotalia tumida* in the Deep Sea Drilling Project core DSDP 214 as interpreted by Malmgren *et al.* (1983), indicating the apparently significant transition in the early Pliocene. (c) First appearance (FADs) and last appearance (LADs) datums of DSDP 214 plotted against the same datums in the composite reference section, indicating several changes in the rate of accumulation of sediment in DSDP 214. In particular, there was a marked increase in rate in mid-Pliocene, followed by a complete cessation in the late Pliocene. (d) Palaeomagnetic method for estimating completeness. Given a suitable model of the average rate of magnetic polarity switches, the completeness of the stratigraphic section can be estimated from the ratio of type X transitions to type Z transitions. ((a) After Sadler 1981; (b), (c) after MacLeod 1991; (d) after Algeo 1993.)

a large time period has a lower net rate of accumulation than an otherwise similar-looking section spanning a short time. The reason for this is that as the section was being formed, there were periods when there was no sedimentation, so that the accumulation rate was zero. At other times, erosion caused a negative accumulation rate. These zero and negative rates are invisible in the actual record, and so are not counted towards the average accumulation rate. The longer the time period over which the whole section formed, the more numerous will have been these periods, and therefore the higher the percentage of 'missing' rock, reducing the total thickness of the section. From the plot of rate of accumulation against time, the expected net accumulation rate for any selected time interval can be found. Comparing this with an actual record of that time span gives an estimate of its completeness. This is a very simple model for assessing the completeness of a section being studied.

More sophisticated models can be created based on the idea that completeness depends on three general aspects of the depositional processes that occurred. Two are those already mentioned—namely the total time span of the section and the time interval for which the completeness estimate is sought. The third aspect is the nature of the variability of the rate of accumulation, its unsteadiness. To allow for this, various assumptions need to be made and incorporated into the model, including whether the variation is random or periodic. Using such correction factors, improved estimates of completion are possible although, like any comparable modelling technique, they tend to be biased towards the parameters of the model rather than towards direct empirical observations of the sequence in question.

Other, quite different ways of assessing stratigraphic completeness have been developed that do use direct empirical observations, two of which may be mentioned as potentially of particular practical value. The graphic correlation method consists of taking two or more stratigraphic sections from the same time period and noting particular stratigraphic indicators in each. These may be the first or the last appearances of contained fossil taxa, magnetic reversals, radiometric markers, signs of volcanic activity, and so on. By cross-referencing shared markers, a single composite reference section can be built up. Any single section can then be compared with the composite and assessment of its completeness determined. As an example, Norman MacLeod used this method to restudy a well-known case of rapid anagenetic (non-splitting) evolution in the foraminiferan *Globorotalia tumida* (Fig. 5.7b). This event is preserved in the Deep Sea Drilling Project (DSDP) core 214 from the Indian Ocean. By comparing the core with a composite reference section (CRS) for this period of time, it is found that the rate of accumulation in DSDP 214 varied from time to time (Fig. 5.7c), and that there is a complete hiatus with zero accumulation of sediment during the late Pliocene. If the varying rate and the hiatus are allowed for, the apparent anagenetic shift cannot be shown to differ significantly in degree from the random variation in evolutionary rate of the lineage as a whole.

Another, rather ingenious method using palaeomagnetic reversal data alone has been proposed by T. J. Algeo (Fig. 5.7d). It is based on the fact that magnetic reversals occur independently of any stratigraphic events, typically at an average rate of around one every 5000 years. By close inspection, a stratigraphic section can be divided into recognizable units, separated by hiatuses indicated by sudden changes in the lithology of the rock, and which represent the times of non-accumulation of sediment. The more complete a section is, the greater the probability that a magnetic reversal will occur during a period of accumulation of sediment and therefore will be recorded within a sedimentary unit. Conversely, the less complete the section is, the greater the probability that a magnetic reversal will have occurred during a period of non-accumulation or erosion of sediment, in which case it will occur at a hiatus between units. From the ratio of reversals within units to those at hiatuses between units, the completeness can be calculated.

The present state of analysis of stratigraphic completeness is still rather rudimentary. Theoretical ideas and models are being developed, but achieving convincing estimates for actual sections is proving highly elusive because of the growing awareness of the size of the confounding features, mainly the enormous range of variation in sedimentation rates, the complications of bioturbation and compaction, and the relative inaccuracy over shorter periods of time of the absolute radiometric dates needed for calculating total time spans of sections. For example, in 1984, Lowell. Dingus claimed that completenesses at the 100-year time resolution of a series of sections that spanned the boundary between the Cretaceous and the Tertiary varied from 0.17 to 10 per cent. By attempting to correct for various factors, however, Mark Anders and colleagues calculated in 1987 that there was 100 per cent completeness at the 100-year level, and even at the 10-year level. But they themselves cautioned that for various reasons their figures are also unreliable. They do believe that the temporal resolution of pelagic sediments is much greater than had been thought, suggesting that a typical 1-million-year-old section may be at least 65 per cent complete at the 100-year resolution.

W. D. Allmon's 1989 study of Tertiary molluscs illustrates a seriously worked-out case of completeness measurements, for a fairly typical example of what can be regarded as a good fossil record. Using the stratigraphic modelling technique, he estimated that the Palaeocene–Eocene fossil record of the US Gulf and Atlantic coastal plain is 30–50 per cent complete at a resolution of 100 000–1 000 000 years. In terms of adequacy, he suggests that this level of completeness is not adequate for studying the process of speciation but it is adequate for contributing useful relative dates of species in phylogenetic analysis of the fossilized groups concerned.

## Ecological incompleteness

All of the biological processes of interest to the palaeobiologist, such as speciation, extinction, invasion of new adaptive zones, mass extinctions, adaptation, etc., occurred within particular ecological settings and must have involved the interactions between species and their biotic and abiotic environments. The more that is known about the contemporary physical environmental setting and the community within it, the more detailed and well supported any palaeobiological hypothesis proposed might be. Unfortunately the preserved parameters of the environment are likely to be a very incomplete and distorted representation of the actual environment in which the fossilized organisms once lived,

It is easier to list reasons for the ecological incompleteness of a fossil-bearing locality than to remedy them in a palaeoecological reconstruction. Different kinds of organisms have very different probabilities of preservation because of the differing nature of their hard parts. They also undergo differential transport rates. In flowing water, the smaller and less-dense organisms will tend to be carried further, while the larger, denser corpses will tend to be deposited on the bed sooner. Other large-scale distortions arise from what is termed time-averaging, which is where fossils of organisms that lived at different times—tens, hundreds, and even thousands of years apart—end up in the same fossil sample and therefore appear to have formed part of the same living community. Similarly, space-averaging occurs where organisms from a wide geographical area and range of habitats are transported to the same place after death. Taken together, these factors ensure that the death assemblage is not likely to be closely representative of any particular once-living community. As for the abiotic aspect of the former habitat in which the community lived, not much at all leaves any indication in the rocks. Thus palaeoecological reconstruction beyond the very gross level is the major stumbling block to the finer levels of interpretation of biological processes undergone by the erstwhile living organisms.

Nevertheless, there is an enormous literature devoted to palaeoecology and the reconstruction of palaeoenvironments. It is very much an interdisciplinary field, using evidence from the anatomy of the fossils themselves, experimental work on the decay and transport of dead organisms, the lithological characteristics of the sediments, sophisticated geochemistry, and palaeometeorology. Connecting it all together is the assumption that ancient ecosystems operated in much the same fashion as modern ones—an ecological version of uniformitarianism. Patterns of energy flow, trophic levels, and competitive and predator–prey relationships must all be supposed to have existed in a fashion sufficiently similar to the arrangements in equivalent modern ecosystems to be able to take much of the reconstruction of the palaeoecology for granted. This is fine, and justified for as long as it is borne in mind that a palaeoecological reconstruction underpinned by the uniformitarian assumption cannot then be used to demonstrate that the ecology of that time and place actually was 'normal'. Here once again is the pattern-and-process question.

The essential thrust of the main areas of investigation bearing upon palaeoecological reconstruction can be very briefly discussed, to give a flavour of this active area of palaeobiological research.

## Structural sedimentology

The general nature of the environment, or facies, in which the particular deposition occurred is indicated by many features of the sedimentary rocks themselves. The rate of flow of water, referred to as the energy level, is shown by the size of particles of which the rock is composed; fine ones as in mudstones and silt stones represent a slow flowing environment, while larger ones as in sandstones and gravels are associated with faster flow. Such broad physical features as the position of the shoreline, depth of water, and direction of flow all leave characteristic signs in the gross structure of the rock. All this information bears, of course, only on the place where the fossils occur, which certainly may not be the place where the organsms lived. Judgements on whether and how far the dead organisms may have been transported to their final site of deposition depend amongst other evidence on the conceptual reconstruction of the site itself.

## Geochemistry

A few broad indications of environmental features of the site of deposition are given by simple chemistry, such as the presence of oxidized iron in redbeds, which is indicative of the strongly oxidizing conditions of seasonal aridity in a terrestrial setting. Conversely, the high carbon content of black shales is a good indication of anoxic conditions in the environment where the sediments were accumulating.

The most exciting current applications of geochemistry involve measurements of the ratios of stable isotopes, particularly those of oxygen, carbon, and sulphur (Fig 9.3). The ratio of $^{18}O$ to $^{16}O$ in modern sea water is roughly 500:l. Water molecules containing $^{18}O$ evaporate more slowly than those containing $^{16}O$ because of the differences in their respective molecular weights. What is more, the difference in the two rates increases as the temperature falls. Therefore the cooler the water is, the relatively more $H_2{}^{18}O$ it contains. The ratio of these isotopes (usually expressed as a change in the ratio, $\delta^{18}O‰$) contained in the calcium carbonate of fossilized shells is the same as in the sea water in which the shells were originally formed. Therefore, the $\delta^{18}O$ of shells of foraminiferans or molluscs, as measured by a mass spectroscope, gives a measure of the temperature of the water in which they lived.

Another consequence of the different evaporation rates is that fresh water, which is derived entirely from evaporated water falling as rain, has a higher $H_2{}^{16}O$ fraction. Knowing the value of the ratio for sea water and fully freshwater samples of the time, the salinity level of the habitat of a population of fossils can be estimated.

The stable isotopes of carbon can be used to estimate the amount of photosynthesis that was occurring in the immediate vicinity of preserved fossils: $^{12}C$ is preferentially utilized over $^{13}C$ in the process of photosynthesis, so water in which higher levels of photosynthetic activity are occurring becomes relatively enriched in the $^{13}C$ left behind. Meanwhile the calcium carbonate of the shells of organisms living in the water involves the use of $CO_2$ dissolved in the water, but here there is no differential utilization of one isotope over the other. Therefore the $\delta^{13}C$ of the shell calcium carbonate is the same as in the surrounding water. The technique can be used to estimate the depth at which a shelled organism was living in the oceans because at lower depths primary productivity is absent, and so $\delta^{13}C$ is lower than in the surface waters. It is also possible to estimate geographical differences or changes over time in productivity in the surface layers, by analysing shells of planktonic organisms.

The isotope of sulphur of palaeobiological interest is $^{34}S$, because during the reduction of sulphate to sulphide by sulphur-reducing bacteria, $^{32}S$ is preferentially used, leaving behind an enrichment of $^{34}S$. This process occurs in reducing muds and therefore the $\delta^{34}S$ found in contemporary deposits can indicate anoxic conditions.

## Trace fossils

Ichnology is the study of those fossils that consist only of the traces of the activities of organisms, rather than any part of their actual anatomy. Worm casts and burrows, arthropod trails, dinosaur trackways, and bite marks on shells all come into this category and often give clues to the palaeoecology of the facies in which they occur. Trace fossils are potentially important because they are often the only fossils in the rocks, the bodies having been lost by post-mortem transport or post-fossilization dissolution. They also normally occur at the place in which the organisms that created them actually lived. Their footprints are not transportable, even if the skeletons of the dinosaurs that made them are.

One use of trace fossils has been to estimate oxygen levels in different parts of the preserved sea-bed. From comparison with modern habitats, it is known that different kinds and population densities of soft-bodied organisms are active in different conditions, between fully oxygenated and anoxic. Another use is to give an indication of how deep below the surface of the sea-bed a particular sediment had been, because again a characteristic soft-bodied fauna can be identified with different depths (Fig. 5.8a). Burrows below the substrate surface can give an indication of the rate of sedimentation. Where it has been high, organisms have not had time to form burrows, and so a sudden disappearance of signs of burrowing activity could indicate an episode of catastrophic deposition, or alternatively a period of post-fossilization erosion. In freshwater conditions, the characteristic aestivation chambers of lungfish indicate strongly seasonal tropical conditions, while burrows of the terrestrial dicynodont mammal-like reptile *Diictodon* suggest a climate in its Late Permian South African habitat that required hibernation.

A whole area of ichnology is devoted to the study of footprints and trails left in mud and sand (Fig. 5.8b). These can tell a great deal about the environment, and the habits of the species that made them.

## Taphonomy

The study of the processes that lead from a freshly dead organism to a preserved fossil is called taphonomy. Much of the knowledge accumulated by taphonomists has been from experiments using modern material. Rates and patterns of decay in different conditions, and the effect of different strengths of water current on transport of complete and partial corpses of a variety of organisms have led to a much clearer picture of how fossil assemblages form, which in turn helps appreciation of the extent and manner of distortion that has occurred in the transition from the life asemblage, the biocoenosis, to the fossilized death assemblage, the thanatocoenosis.

A first consideration is the loss of organisms and therefore of species and higher taxa from the fossilized community. Estimates of the fraction of species constituting the biocoenosis that are potentially fossilizable lie at around 25–30 per cent, based on modern communities and assuming that organisms lacking mineral skeletons, such as small crustaceans and worms, are unlikely to be represented as fossils. They will be destroyed by predation, bacterial decay, and physical disturbance within the forming sediments by infaunal organisms. A second major cause of loss is dissolution of skeletons after fossilization, an example of a diagenetic process. This occurs particularly in porous rocks such as sandstones, simply due to the action over long periods of permeating water containing $CO_2$. In well-consolidated rocks a cavity may remain in the form of a natural cast of the organism. At the other extreme, soft tissues can be preserved, usually by being replaced by inorganic molecules in a mineralization process. This characterizes the *Lagerstätten* mentioned earlier (page 72), and these uncommon but very important deposits can reveal a much fuller picture of the original species composition.

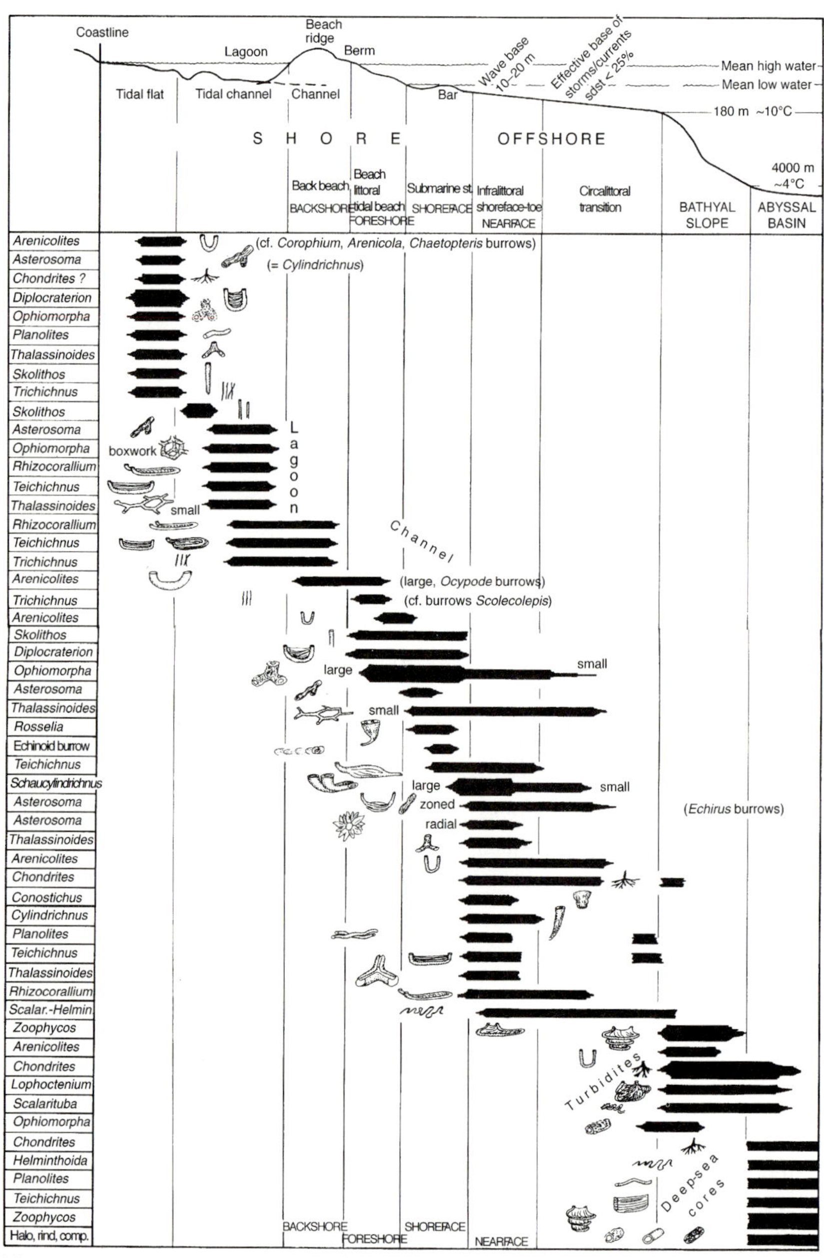
Coastline
Beach ridge
Lagoon
Berm
Wave base 10–20 m
Effective base of storms/currents sdst < 25%
Mean high water
Mean low water
Tidal flat
Tidal channel
Channel
Bar
180 m ~10°C
S H O R E
OFFSHORE
4000 m ~4°C
Back beach
Beach littoral
Submarine st.
Infralittoral
Circalittoral
BACKSHORE
tidal beach
FORESHORE
SHOREFACE
shoreface-toe
NEARFACE
transition
BATHYAL SLOPE
ABYSSAL BASIN
Arenicolites
Asterosoma
Chondrites ?
Diplocraterion
Ophiomorpha
Planolites
Thalassinoides
Skolithos
Trichichnus
Skolithos
Asterosoma
Ophiomorpha
Rhizocorallium
Teichichnus
Thalassinoides
Rhizocorallium
Teichichnus
Trichichnus
Arenicolites
Trichichnus
Arenicolites
Skolithos
Diplocraterion
Ophiomorpha
Asterosoma
Thalassinoides
Rosselia
Echinoid burrow
Teichichnus
Schaucylindrichnus
Asterosoma
Asterosoma
Thalassinoides
Arenicolites
Chondrites
Conostichus
Cylindrichnus
Planolites
Teichichnus
Thalassinoides
Rhizocorallium
Scalar.-Helmin.
Zoophycos
Arenicolites
Chondrites
Lophoctenium
Scalarituba
Ophiomorpha
Chondrites
Helminthoida
Planolites
Teichichnus
Zoophycos
Halo, rind, comp.
(cf. Corophium, Arenicola, Chaetopteris burrows)
(= Cylindrichnus)
boxwork
small
Lagoon
Channel
(large, Ocypode burrows)
(cf. burrows Scolecolepis)
large
small
small
large
small
zoned
radial
(Echirus burrows)
Turbidites
Deep-sea cores
BACKSHORE
FORESHORE
SHOREFACE
NEARFACE

(a)

(b)

**Fig. 5.8** Trace fossils. (a) Environmental distribution of trace fossils commonly occurring in marine cores. (b) Dinosaur footprints of several species, from the Lower Jurassic of Lesotho. ((a) After Dodd and Stanton 1990; (b) after Thulborn 1990.)

A second consideration is the way in which the surviving remains of organisms accumulate in what is to become the fossil deposit. This may involve very little transport, but rather a gentle deposition onto the substrate. Signs of this are that the skeletons of organisms such as crinoid echinoderms, vertebrates (Fig. 5.9a), and plants may remain articulated, while small as well as large specimens occur together with no apparent sorting (Fig. 5.9a). On the other hand, signs of extensive transport by water include disarticulation of skeletons into their separate and unassociated parts, a small range of size and density of specimens, and abrasive damage to the fossils. The faster the rate of flow of the transporting water, the more sorting into size categories is to be expected.

Interesting as it is, and ingenious as many of its techniques are, taphonomy is a rather depressing science, for it mainly indicates how extensive a distortion the fossil record is of the original community. The extent to which it can help in positively reconstructing the original is very limited.

### Palaeoautecology

All modern species have their particular ecological limits, and may be used to infer aspects of the habitat from which they came. In principle, a fossil species can be used in the same way, either if it is sufficiently closely related to a living species of known environmental predilection, or if it shares structures with living species that imply that it had similar ecological requirements. A great deal of palaeoecological reconstruction uses such indicator species. As a simple example, corals form reefs that cannot extend below the photic zone in depth. Reptiles and amphibians—at least those related to modern groups—are likely to be much more abundant in tropical than in temperate terrestrial zones.

Fossilized pollen and spores (Fig. 5.10) constitute a particularly important category of palaeoecological indicators for terrestrial palaeoenvironments. Pollen grains and spores of ferns are widely transported by wind and often end up preserved in nearby aquatic deposits, being very resistant to decomposition. Although they can seldom be identified to species level, they are readily recognizable at the genus or family level.

## Biogeographic incompleteness

The biogeographic distribution of species and higher monophyletic taxa is potentially important information for understanding many evolutionary processes. At the very lowest level of simply recognizing species in the fossil record, spatial continuity is one theoretical criterion (see page 85). In the past, there has been a strong traditional reluctance amongst palaeontologists to regard similar but widely separated fossil specimens as members of the same species, for little reason other than the spatial gap between them. Yet undoubtedly gaps can be the result of biogeographic incompleteness, so the erstwhile intervening specimens are unpreserved. As far as evolution is concerned, there is the possibility of an entirely false appearance of gradual morphological change through time because of missing biogeographic information (Fig. 5.11). Virtually all widely distributed Holocene species show adaptive clines. These are where local adaptation to different regions in the species' range takes the form of gradually changing morphology with distance. If the environmental conditions as a

**Fig. 5.9** *(Opposite)* (a) Virtually complete, articulated skeleton of the theropod dinosaur *Compsognathus*, a small, delicately built theropod dinosaur from the Upper Jurassic Lithographic Limestone of Solnhofen, Germany. As well as remaining articulated, there is a skeleton of a small lizard within the rib cage presumably from the intestine, and apparent casts of a dozen or so eggs, presumably spilled from the abdomen after death. (b) Brachiopod shells showing how transport and rolling causes loss of the thinner regions, and general abrasion. ((a) After Ostrom 1978 and Griffiths 1993; (b) after Brett 1990.)

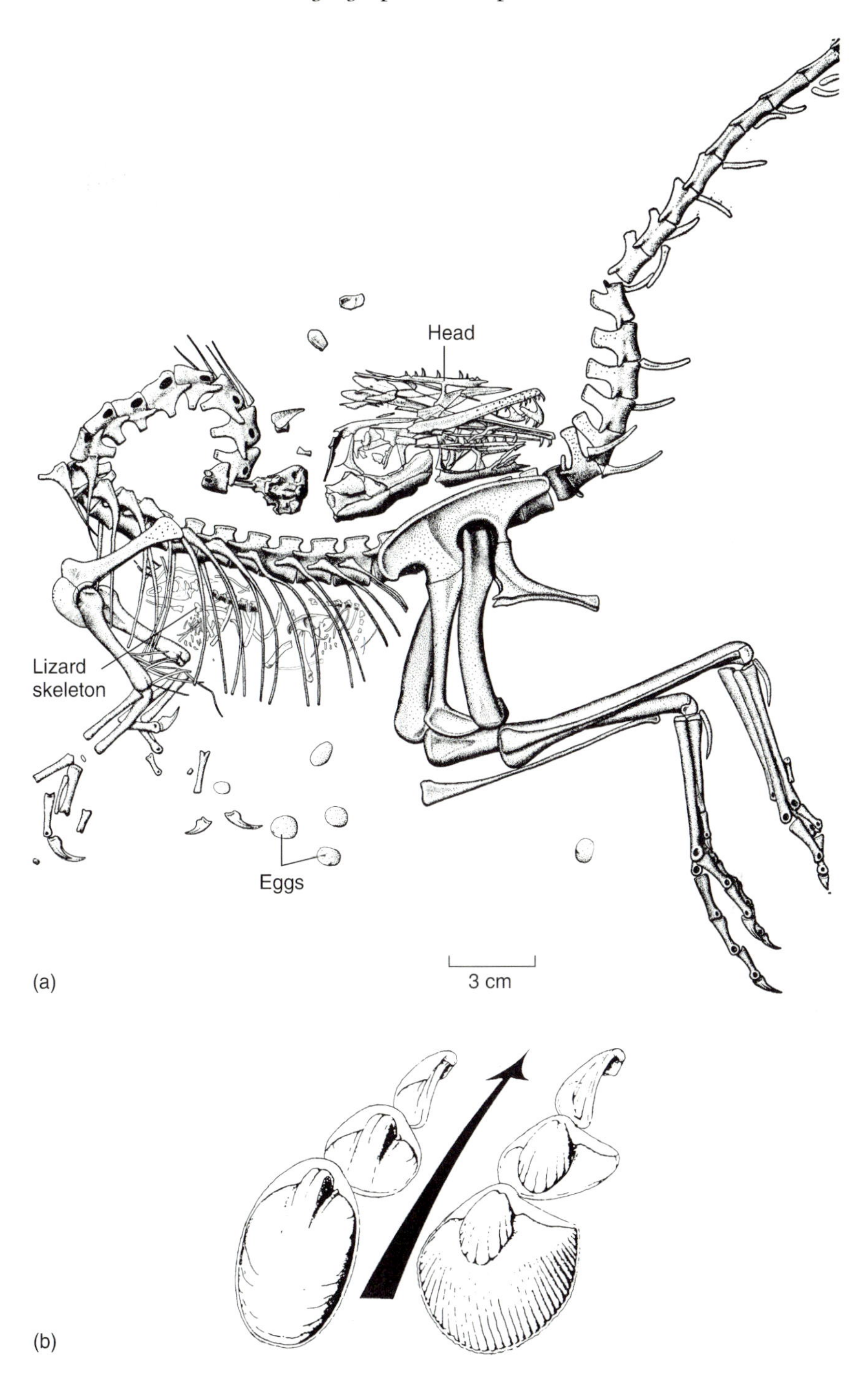
Head
Lizard skeleton
Eggs
3 cm
(a)
(b)

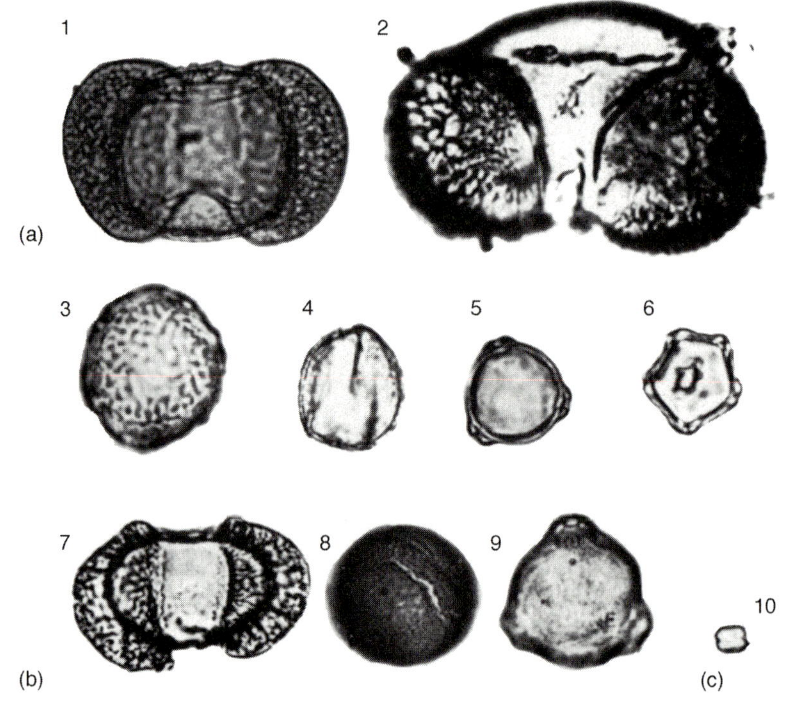


**Fig. 5.10** Some plant pollens are associated with particular kinds of environments: (a) northern mid-latitude forests (1 = *Pinus*; 2 = *Picea*; 3 = *Ulmus*; 4 = *Quercus*; 5 = *Betula*; 6 = *Alnus*); (b) low-latitude forest (7 = Araucariaceae; 8 = *Nothofagus*; 9 = Casuarinaceae; (c) mangrove swamp (10 = Rhizophoraceae). (After Williams *et al.* 1993.)

whole change gradually, as for example decreasing temperature during the approach of an ice age, then the cline as a whole may shift geographically with time. Fossils collected from a stratigraphic section at one spot may then appear to show gradual evolutionary change, when in fact no evolution of the species as a whole had occurred at all.

The debate about whether the process of speciation occurs in large, widely distributed populations or in small isolated ones would be much aided by reliable distribution data, as would the further question of whether the population structure of a species affects its probability of speciation and extinction. At higher levels of process, there are intriguing possible relationships between biogeographic distribution and susceptibility to mass extinctions (page 200).

Unfortunately, biogeographic incompleteness is too extensive in nearly all cases for these desirable pieces of information to be knowable. Reasons for a disjunct distribution of a fossil species can include lack of rocks of the appropriate age or lack of suitable conditions for preservation, even if there are sediments of the right date exposed in the intervening areas. Separation of once-contiguous areas by continental drift, loss of areas by

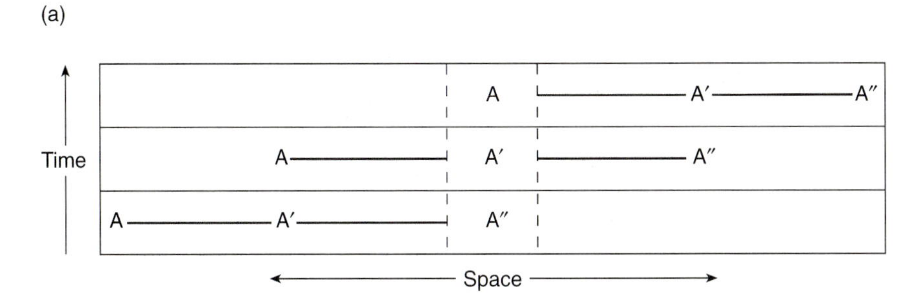


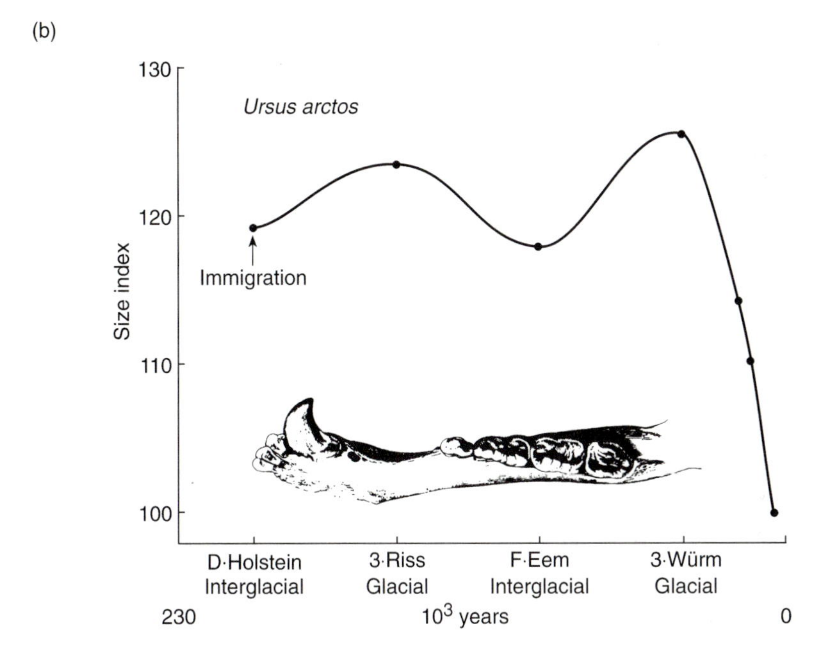


**Fig. 5.11** (a) An ecological cline (A, A′, and A″ as morphological variation) shifting through time and creating the appearance of gradual evolutionary change (A″ to A′ to A) in a stratigraphic sequence at one place. (b) Change in size of molar teeth, and by correlation the bady size of the Pleistocene Brown Bear during successive glacial and interglacial periods. While not certain, it is very likely that this represents a cline in body size that shifts with climatic change, rather than evolutionary change. (After Kurten 1968).

subduction, and destruction by volcanism may all cause secondary discontinuity of the fossil distribution of a once non-disjunct species. Although all these geological effects may be known about and well understood, that does not solve the dilemma that a disjunct distribution of a particular fossil taxon may nevertheless have been because the distribution in life was disjunct.

## Further reading

Thomason (1995) contains some excellent papers on functional morphology, including general overviews by Lauder (1995), which is rather sceptical, and, much more optimistically, Weishampel (1995). Rudwick (1964) describes the paradigm method, and an example of the application of constructional morphology may be found in Seilacher (1990). For the application of the design test to difficult organisms, see Hickman (1988), while Harvey and Pagel (1991) review the comparative method with respect to modern organisms. Wake (1992) also considers the comparative approach in the course of a useful review of the relationship between form and function. The development of the phylogenetic test can be followed in Coddington (1988), Bryant and Russell (1992), and Witmer (1995).

Accounts of the theory and application of the species concept include several of the papers collected by Ereshefsky (1992), and which cover several distinct points of view. Smith (1994) deals well with the problem of the species in the fossil record, and Eldredge (1995*a*) is interesting on this topic.

Berggren *et al.* (1979) give a general account of the fossilization process and taphonomy, and many aspects of this area and palaeoecology are covered in the contributions to Briggs and Crowther (1990). Doyle *et al.* (1994) is a readable explanation of stratigraphy; for more detail see the essays edited by Doyle and Bennet (1998*a*). Ager (1993) takes a delightful, somewhat iconoclastic look at completeness and uniformitarianism. Paul (1990) discusses the idea of adequacy, and other discussions of the nature and limits of information in the fossil record include those by Jablonski *et al.* (1986), and McShea and Raup (1986). More details of principles and good examples of completeness analysis are in Donovan and Paul's (1998) edited collection of papers. Estimation of completeness by stratigraphic modelling can be explored in Dingus and Sadler (1982), and Sadler and Strauss (1990). MacLeod discusses and uses the graphic correlation method, and the reference to use of palaeomagnetic data for completeness studies is Algeo (1993).

Dingus (1984) McShea and Raup (1986), and Anders *et al.* (1987) consider the actual completeness of real examples of stratigraphic sequences; Allmon (1989) is the source of a worked example of the completeness of a fossil mollusc record, showing well how the method is actually applied.

For detailed accounts of palaeoecology see Dodd and Stanton (1990). Doyle and Bennett (1998*b*) discuss fossils as environmental indicators. For trace fossils see Bromley's (1996) textbook and the essays in Donovan (1994). Various aspects of taphonomy are discussed in the papers in Allison and Briggs (1991). Kidwell and Ilessa (1995) have reviewed the reconstruction of life assemblages, with very optimistic conclusions.

# Part II

# PRACTICES

# 6
# Fossils and phylogeny: if only we had more fossils

There was a time when statements could be read to the effect that the evolutionary relationships of some such group of living organisms could not be discovered because it had no fossil record. The underlying, rather naive assumption was that the discovery of appropriate fossils would lead to nice lineages of evolving organisms going backwards in the stratigraphic record from living organisms to their more primitive ancestors, and ultimately to the common ancestors that members of different groups shared. This left groups such as annelid worms and mosses, which have notoriously poor fossil records because of their general shortage of potentially fossilizable parts, with little hope of ever being correctly classified on the basis of their phylogenies.

Another problem of fossils and phylogeny that was also much discussed at one time was that those modern groups that did have a good fossil record lost their well-defined group boundaries. The discovery of fossils intermediate between fish and tetrapods, or between reptiles and mammals, for example, which were coming to light during the middle years of the twentieth century confused the definitions of these classic living groups. Eventually, no one knew what a mammal was, at least in the formal taxonomic sense. Ironically, therefore, too little was known of fossils for phylogenetic classification of some groups, whereas too much was known for phylogenetic classification of others. It is not surprising that many of the early enthusiasts for the then new cladistic methodology were palaeontologists, anxious to make more explicit, precise, and objective the role that fossils have in taxonomy. Both difficulties can be resolved by replacing the idea of seeking ancestors in the fossil record by that of looking for closest relatives based on shared derived characters. Nevertheless there are still important disagreements about exactly what information the fossil record usefully conveys for classifying modern groups.

## Fossils and phylogenetic analysis

Although no modern palaeobiologist seriously expects fossils to be found that connect modern groups to their extinct common ancestors in much detail, some have continued to argue that fossils are uniquely important indicators of phylogenetic relationships because of their temporal relationships. Ancestral groups can be identified from amongst fossils. For example, Harper (1976) proposed that an important part of the evidence for a proposed phylogeny should be the relative dates of the fossils. Gingerich (1979) developed a method that he called stratophenetics for phylogeny reconstruction, which involves a combination of phenetic clustering of the fossil taxa, and ordering them in stratigraphic sequence, to discern the actually evolutionary lineages. He applied the method to species-level cases such as the hominids as well as to higher taxa such as the interrelationships of the tetrapods.

At the same time, other palaeobiologists such as Schaeffer *et al.* (1972) were applying the principles of cladistics to the fossil record and reaching a very different conclusion. As discussed in detail earlier (page 52), the main point of cladistics is that natural or monophyletic groups of organisms can be proposed and tested only by the presence of uniquely derived characters shared by the members. It means that ancestral groups cannot logically be recognized because by definition the members of an ancestral group do not share any uniquely derived characters (page 57). All their features are either the same as in their own ancestors or the same as in their descendants. Therefore, fossils cannot be identified as ancestors of other taxa. They can be identified only as relatives of each other and of any modern organisms. If a particular fossil or group of fossils is claimed to be ancestral, then the principles of cladistics have been breached and all the potential for imprecise, subjective assertions re-appear.

The conclusion from this line of reasoning is, therefore, that fossil groups have to be treated as terminal branches on a cladogram exactly as if they were extant groups. Furthermore, the only test of a cladogram is the distribution of characters amongst the groups, and therefore the relative dates of the fossils are irrelevant as direct evidence of which fossil is related to which. The only possible role of stratigraphy in a strictly cladistic exercise is as one of the criteria for assessing character polarity; that is, which character state is ancestral and which derived. Evolution dictates that the ancestral state must have occurred prior to the derived state into which it changed. But incompleteness of the fossil record in turn dictates that this criterion has to be treated with great caution because of the possibility of the ancestral state not being preserved in any fossil earlier than the first appearance in the record of the derived state. Certainly a well-supported cladogram seems unlikely to be overturned by a few fossils, consisting of limited numbers of characters, that had seemingly anomalous dates, as concluded by Schaeffer *et al.* (1972).

In terms of what information fossils add to a phylogeny, it does not actually matter if they cannot be designated ancestral groups. Classically 'ancestral' fossils, such as *Archaeopteryx* (see Fig. 10.8) or the phylloceratid ammonites, possess that status because all the characters that they are *known* to have are believed to be either ancestral or else derived characters that are also found in their supposed descendants. They have no unique characters, or specializations, of their own. Therefore they could indeed be ancestral. But they could also be sister groups whose own unique characters happen not to be preserved. In practice, given the tiny number of characters actually preserved in a fossil the chances are very high indeed that there were specializations present in the erstwhile living organism, so the sister-group relationship is probably correct anyway. But even if it really was the ancestor, what actually happens is that a hypothetical ancestor is reconstructed on the basis of the characters of the two related groups. If one of these two forms is ancestral in all its known characters, then the hypothetical ancestor will be reconstructed just like it. As far as conceiving the ancestor of modern birds is concerned, it does not matter whether *Archaeopteryx* is the ancestor or simply looks like that ancestor in all the characters that happen to be preserved, but really had some peculiar specializations of, say, its lungs. So to maintain the cladistic principles, it is necessary to accept the formality that all fossils had their own unique characters, even if these have not yet been discovered. It is instructive in this context that when the braincase of *Archaeopteryx* was prepared at the Natural History Museum in London, it proved indeed to have some unexpected and distinctly non-bird-like features (Whetstone 1983). These newly discovered characters show that on the basis of parsimony, it cannot have been the ancestor of modern birds anyway.

In 1981, Colin Patterson wrote a paper in which he described asking his taxonomic colleagues for particular examples where the classification of a group of modern organisms had been critically influenced by fossil evidence (Patterson 1981). No one could provide him with a single such case, and he concluded that fossils were irrelevant for that purpose, primarily because they have too few preserved characters compared with modern forms. Smith (1984) considered in detail the classification of echinoderms, a group with one of the best of all fossil records (Fig. 10.11), and concluded that the fossils were not only of virtually no help in establishing relationships amongst the modern forms but could be positively misleading at times. Schultze (1994) found that fossils did not help significantly in the analysis of the interrelationships of coelacanths, lungfish, and tetrapods, despite the extensive use that has been made of palaeontological evidence in attempting to find which of the fish groups is the closest relative of the tetrapods (Fig. 6.2).

It is easy to move from these conclusions to a hardline view that fossils not only *do* not contribute to the phylogenetic analysis of modern organisms but indeed *should* not be used for the purpose, even in principle. First, because they lack too many characters to be safely used as evidence of the most parsimonious distribution of characters. Second, because their stratigraphic relationships are too unreliable. Ax (1987) argued strongly and at length that a cladogram must be constructed from the wealth of characters known in modern organisms. Only then can the fossils be placed on the cladogram, at the positions determined by their relatively tiny numbers of synapomorphies—rather like hanging ornaments on the branches of a Christmas tree. Others, such as Gardiner (1982), took an even greater exception to the use of fossils in taxonomy, on the grounds that in many cases fossil groups do not have any known characters unique to themselves. In the absence of any such defining characters, they cannot even be designated as monophyletic groups in their own right. Fossils of this kind are all those species and groups classically described as ancestral—Phylloceratida and *Archaeopteryx* again, the synapsid reptiles and the theropod dinosaurs, the lingulid brachiopods, and so on. If, continues this confused argument, these fossils cannot even be designated monophyletic groups, then they cannot possibly be incorporated into cladograms in such a way as to affect the pattern of relationships of all the good monophyletic groups expressed by that cladogram. Indeed, palaeontology has been guilty of the fruitless and logically indefensible task of ancestor-seeking to the exclusion of all else.

Others, however, have moved to a more optimistic line, claiming that these ancestral (paraphyletic) groups are often the result of poor taxonomic practice in creating taxa. They regard the use of fossils as certainly justifiable in principle, and in some cases demonstrably significant in practice. Kemp (1988*a*) argued that even if a fossil taxon did not have any known derived characters of its own, it could still be placed within a larger group, the characters of which it did possess. It may not be possible to detect exactly whereabouts within that group it fits, but nevertheless it can contribute to the overall diagnosis of the larger group. It also happens that most fossils do actually have their own autapomorphies anyway, as, for example, *Archaeopteryx* as now known, and all the mammal-like reptile taxa that are at least reasonably well preserved (Kemp 1982*a*). Once it is accepted that fossils can legitimately be placed on cladograms, the argument of Ax (1987) and others that fossils cannot in principle affect a cladogram of modern organisms also fails.

A cladogram is the most parsimonious pattern of distribution of characters amongst the taxa involved. Suppose that there are two possible cladograms, each implying exactly the same amount of homoplasy, and that a fossil is discovered with some particular combination of character states (Fig. 6.1a). Imagine that placing the fossil in one particular place on one of the competing cladograms causes no increase in the amount of homoplasy, but that the effect of placing it in even the best position that can be found for it in the second cladogram is to cause an additional instance of homoplasy. The first cladogram has now become the preferred one of the two on the basis of its greater level of parsimony. The fossil has therefore affected the phylogenetic analysis in a rigorously logical and therefore perfectly acceptable way, showing that in principle fossils certainly can play a part

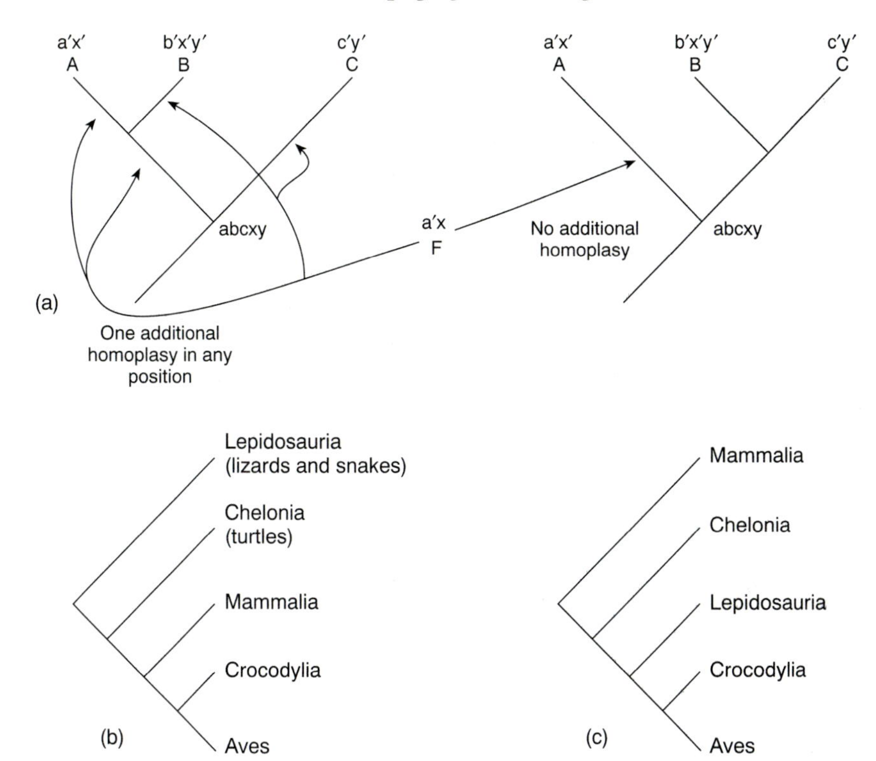


**Fig. 6.1** (a) Illustration of the principle by which a fossil can lead to a preference for one of two otherwise equally well-supported cladograms of modern taxa: the fossil taxon F combines the plesiomorphic character x with the derived character a′. If it is placed at any position in the left-hand cladogram an additional homoplasy is created, but it fits in the position shown on the right-hand cladogram without additional homoplasy. (b) Most-parsimonious cladogram of amniotes without using fossils. (c) Most-parsimonious cladogram when characters of the fossils are included. ((b) After Gauthier *et al.* 1988.)

in the analysis of modern organisms. In fact, in so far as a fossil has a particular combination of characters not already found in any of the existing groups in the cladogram, it acts exactly as if it were a newly discovered living group, with the same potential for further resolving or for rejecting the cladogram.

A few studies have now demonstrated that this is not just a fine point of principle, but that in certain cases at least fossils do have a crucial role in deciding amongst alternative cladograms. Gauthier *et al.* (1988) attacked the problem of amniote phylogeny in an important and widely quoted study. When they performed a computer analysis on numerous characters of a series of recent amniotes, they found that the most parsimonious cladogram linked the mammals as the sister group of the crocodiles plus birds (Fig. 6.1b). But when they extended the analysis to include many fossil amniotes, the most parsimonious cladogram that emerged had the mammals as the sister group of all the other amniotes together—that is, the chelonians, lepidosaurs, crocodiles, and birds (Fig. 6.1c).

Results such as these, however, require a note of caution at present, as is illustrated by continuing with Schultze's (1994) consideration of the interrelationships of the tetrapods and the main groups of sarcopterygian fishes, living and fossil. He compared no less than seven published cladograms, amongst which were examples of all three possible arrangements of the three modern groups, Tetrapoda, Dipnoi (lungfish), and Actinistia (coelacanths) (Fig. 6.2b). There is evidence to a varying extent supporting all of them. What he found was that the differences between different authors's cladograms could be accounted for to a large extent by two things. One is the selection of characters used, suggesting an unconscious bias towards selecting characters favouring one particular outcome. The second is the different hypotheses of homology that are used. For example (Fig. 6.2a), Rosen *et al.* (1981) considered that the internal opening in the lungfish snout is the homologue of the choana (internal nostril) of tetrapods, and, furthermore, that osteolepiform fish such as *Eusthenopteron* lack this structure. This is evidence, therefore, for the tetrapod–lungfish relationship. Others, however, notably Schultze (1987) himself, and Panchen and Smithson (1987) concluded that it was *Eusthenopteron* that possessed the tetrapod-like choana, and that lungfish lacked it. Part of the grounds for thinking this is that the fossil lobed-fin fish *Diabolepis* has the primitive fish condition of two external and no internal nasal openings alongside some lungfish characters, thus suggesting a grouping of tetrapods with *Eusthenopteron*. Forey *et al.* (1991) countered with the proposal that the choana is unique to tetrapods, and therefore this character cannot be used at all to link tetrapods with either of the lobed-fin fish groups.

Obviously the fossils of *Eusthenopteron* and *Diabololepis* are potentially important, but without agreement on the pattern of homology of characters such as the nostrils, they will in practice continue to give ambiguous results. This difficulty is not unique to fossil material, of course, but it is generally more severe because of the absence of soft structures. The test of a hypothesis of homology (page 57) includes the structural, developmental, and positional similarity of the putatively homologous structures, which are likely to be far less easy to assess in fossil than in living organisms.

In principle, then, and with care, fossils that consist of combinations of characters that are not found amongst modern groups can contribute towards phylogenetic analysis. In particular, fossils that lie near the base of the cladogram are likely to be critical, because these will tend to have retained more ancestral characters that have been modified in the later fossils and the modern descendants (Donaghue *et al.* 1989; Wilson 1992). By effectively breaking up long, unbranched lineages into shorter segments marked by branching points, a higher resolution of character distributions becomes possible. Novacek (1992), for example, is concerned with the

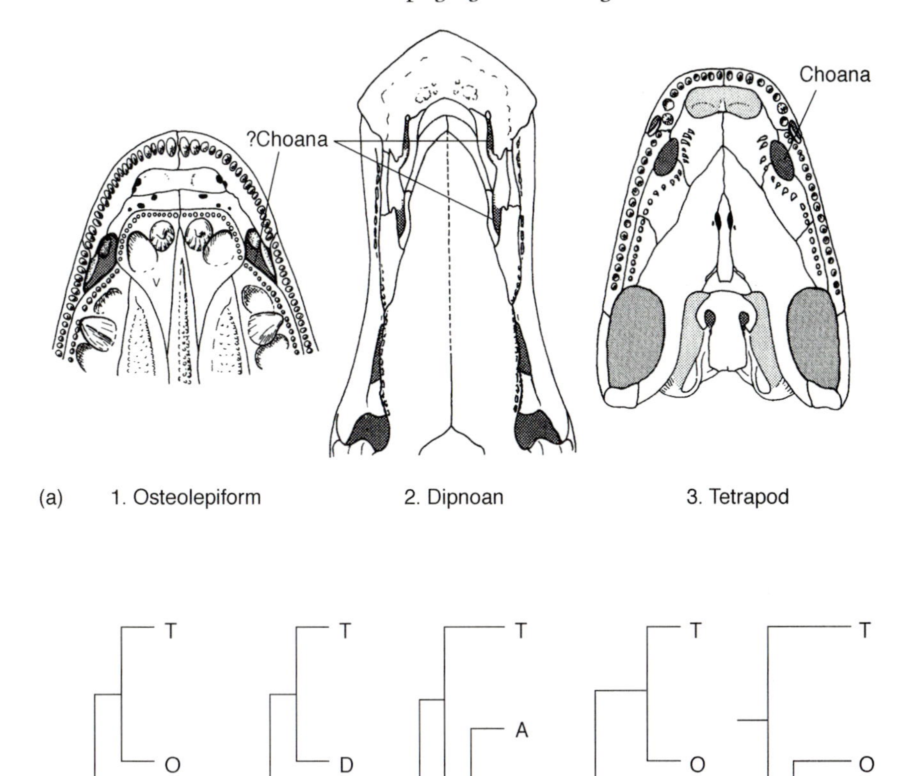


**Fig. 6.2** (a) The disputed homologue of the choana (internal nostril) in (1) the osteolepiform fish *Eusthenopteron*, (2) the fossil dipnoan *Griphognathus*, and (3) the Devonian tetrapod *Ichthyostega*. (b) Alternative cladograms of the Tetrapoda, Osteolepiformes, Actinistia (coelacanths), and Dipnoi (lungfish). ((a) After Panchen and Smithson 1987; (b) after Schultze 1994.)

very difficult matter of the interrelationships of the mammalian orders (page 64). Most fossil mammals are members of derived orders, and the difficulty is that these have extremely few, if any, characters shared with any other orders; there are virtually no characters that reliably indicate supra-ordinal groups. There are, however, some primitive fossils with a combination of characters that support the idea that hyracoids (hyraxes) are related to sirenians (dugongs and manatees) and elephants, rather than to perissodactyls, as had been suggested on the basis of the modern animals (Fig. 6.3).

There is far more scepticism about whether the relative dates of fossils can provide useful information in phylogenetic reconstruction. Kemp (1988*a*) thought that in principle they can, on the rather arcane grounds that if there were two cladograms equally well supported by the characters, but that one implied a greater number of missing fossil forms than the second, then the first cladogram is preferable. This is an extension of the general scientific principle of parsimony of explanation (page 59). If every lineage

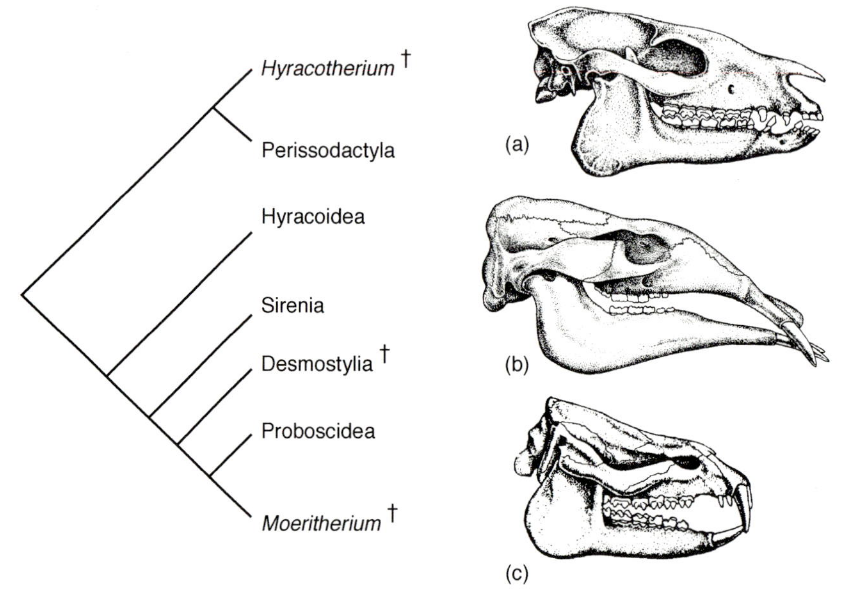


**Fig. 6.3** Fossils supporting a paenungulate clade, placed on a cladogram of elephants, hyracoids, and sirenians: (a) *Hyracotherium*, (b) *Desmostylia*, and (c) *Moeritherium*. (a–c After Romer 1966.)

of missing fossils implied by a cladogram is regarded as an *ad hoc* assumption, then the cladogram that implies the least number of missing fossils has the least such *ad hoc* content, and is therefore to be accepted. Applied to the controversy about the relationships of mammals, birds, and crocodiles (Fig. 6.4a), the hypothesis that mammals and birds are sister groups, as supported by Gardiner (1982), requires a considerably larger chunk of missing fossil record than does the conventional view that birds and crocodiles are sister groups. Fisher (1994; Clyde and Fisher 1997) has developed a similar argument. This is that stratigraphic evidence is particularly important because it is independent of the evolutionary process; it therefore offers an independent test of a cladogram. He termed his approach 'stratocladistics', and in effect it includes stratigraphic inconsistences as well as

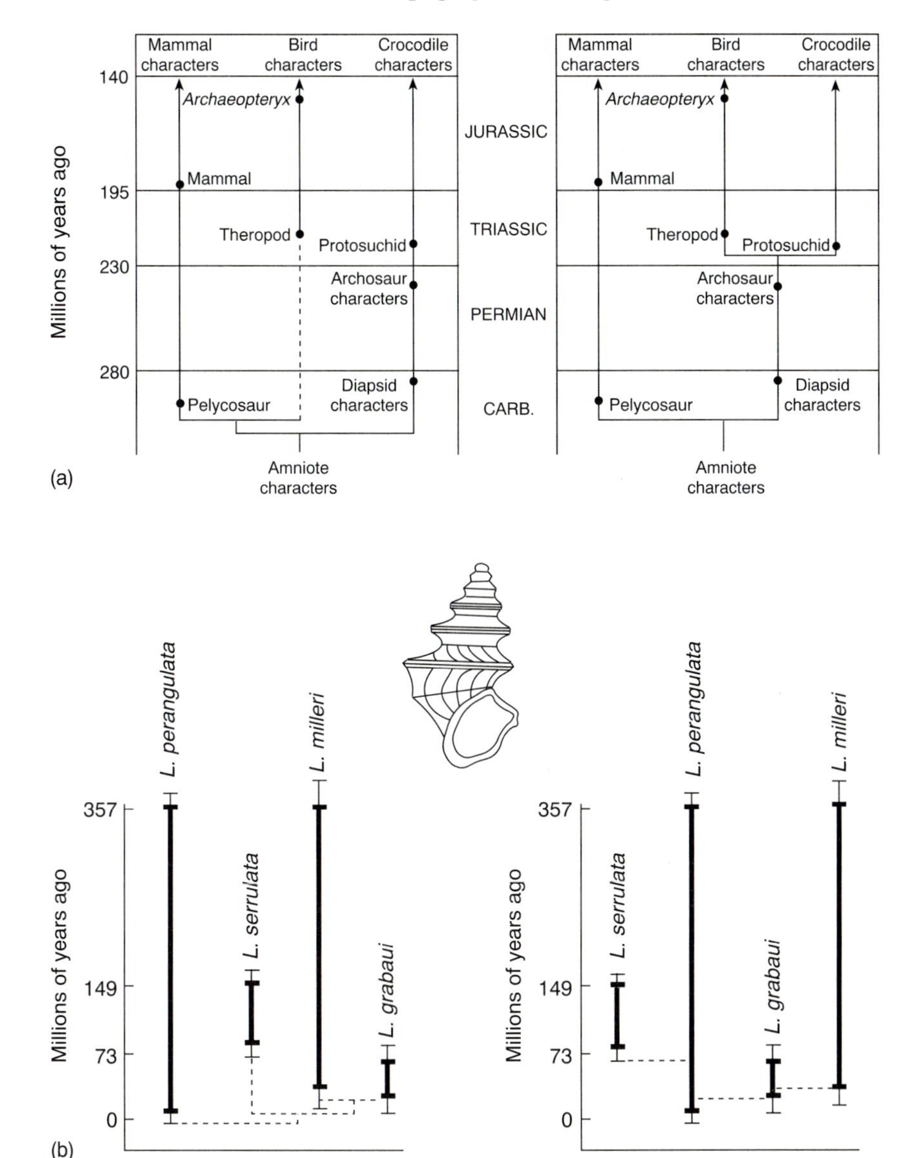


**Fig. 6.4** (a) Missing fossils (ghost lineages) implied by Gardiner's cladograms of mammals, birds, and crocodiles indicated by the dashed line on the left diagram. The standard cladogram on the right implies no such ghost lineages. (b) Wagner's (1995) phylogeny of Ordovician lophospirid molluscs, using stratigraphic as well as morphological data. The cladogram on the left is based exclusively on morphological characters; that on the right is consistent with stratigraphic relationships, allowing for the species *Lophospira perangulata* to be recognized as ancestral to *L. serrulata*. The thick lines are the known stratigraphic ranges, and the thin extension lines are confidence intervals on the ranges. ((a) After Kemp 1988*a*; (b) after Wagner 1995.)

character incongruences as part of the overall comparison of alternative cladistic hypotheses.

Wagner (1995) was even bolder in his belief that stratigraphy should be a major test of phylogeny at the level of species. He pointed out that a discrepancy between a cladogram and the temporal ranges of the species involved could be due to three things. The cladogram may be wrong, there may be missing fossils that are appropriately referred to as 'ghost' lineages and taxa, or a species may be the ancestor of a second species, in which case it will be expected earlier in the fossil record rather than simultaneously with its closest relative, the descendant species. If it is allowed that the most parsimonious cladogram is not necessarily the true one, and that recognition of ancestor–descendant relationships is permitted, then the shortest tree implying no significant gaps in the fossil record can be found and proposed as the best phylogenetic hypothesis. Wagner applied the method to a group of molluscs, the Lophospiridae ('Archaeogastropoda'), which occurs in about 350 Ordovician horizons world-wide (Fig. 6.4b). The result is that the shortest tree consistent with the stratigraphic ranges of the species is only seven character-change steps longer than the most parsimonious tree derived from characters alone, and it includes seven ancestor–descendant relationships. These are certainly perfectly realistic looking features as far as evolutionary theory is concerned, although Norell (1996) criticized the mixing of different kinds of evidence, and particularly the attempt to recognize what to him are still undefinable entities, the ancestral species.

The problem is that there is no clear principle underpinning a method of combining information from character distribution with information from relative stratigraphic position to generate an overall most parsimonious phylogeny. With what scale can a judgement be made about how many character homoplasies is equivalent to some length of ghost lineage? To get a qualitative estimate of the match between them, Norell and Novacek (1992) explicitly determined how closely stratigraphy matches particular cladograms (Fig. 6.5). They took a cladogram that included plenty of fossil groups. For each group they noted its clade-rank, which is the order in which it branches off the cladogram, and its age-rank, which is the order that it appears in the fossil record. For a complete fossil record and a correct cladogram, these two must be completely correlated, so the plot of clade-rank against age-rank would have a 45° slope. Any deviation from that would be an indication of incongruence between the taxonomy and the stratigraphy. For mammal groups they found that in general there was a positive correlation, but the goodness of fit was highly variable. Horses show an excellent fit, but primates and artiodactyls, for example, have a non-significant correlation. It appears, therefore, that sometimes stratigraphic position of fossils does contain useful information about relationships, but sometimes it does not.

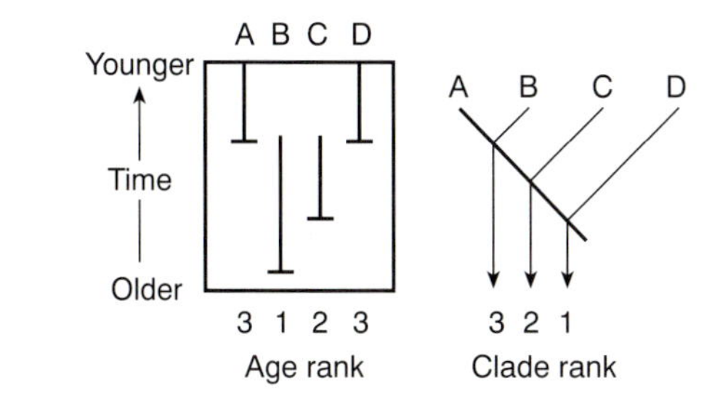


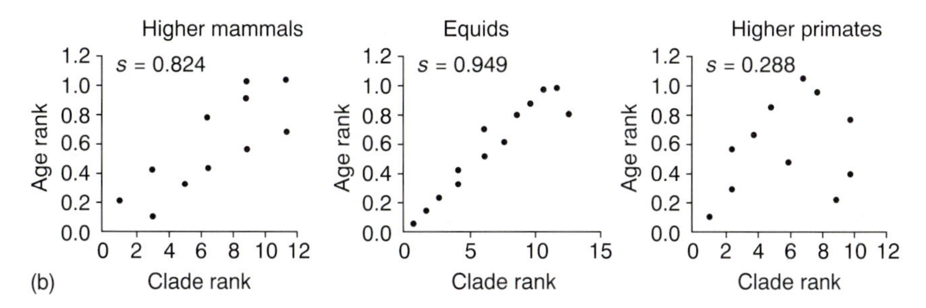


**Fig. 6.5** Plots of clade-rank against age-rank. (a) Calculation of age-rank as order of appearance in time, and clade-rank as order of branching from the cladogram. (b) Plots for higher mammals and for equid horses that show significant correlation, and for higher primates that shows no significant correlation. (After Norrell and Novacek 1992.)

Methods are now beginning to be developed to improve the reliability of extraction of significant data from the stratigraphic sequence. Clyde and Fisher (1997) propose to use a 'retention index' method, which is effectively a measure of the extent to which a particular hypothesis falls short of the best possible hypothesis, given as a figure between one and zero. In the case of character analysis, the units used are evolutionary character transitions, and the best hypothesis is the most parsimonious. For stratigraphy, the units are instances of non-preservation of a species in a particular stratigraphic unit, and the best hypothesis here is the one with the least number of such ghost taxa. There is no implication of exactly equal information value in the two respective units, but at least it is a quantifiable way of comparing them. The aim is to seek the stratophyletic hypothesis with the maximum overall retention index.

In a completely different approach, Huelsenbeck and Rannala (1997) investigated the use of the statistical method of maximum likelihood, based on an explicit model of the fossil-preservation process. The model they use at this stage assumes a process for the preservation of fossil species that produces a Poisson distribution at different stratigraphic intervals. Although the model is unrealistically simple, they do show that if enough is known about the actual probability of preservation of fossils of particular kinds of species under different circumstances, then a more sophisticated and realistic model of the process could be created. In turn, this would permit more reliable use to be made of the relative stratigraphic positions of fossils in estimation of phylogeny.

## Fossils and molecules

The controversy about the use of fossils has become part of a larger issue in taxonomy, which concerns the use of different kinds of data for phylogenetic analysis. If one kind of data leads to one cladogram and another kind to another, should one try to decide which data set is giving the best answer or try to combine the two, leading to yet a third cladogram? In the latter case, the question must then be addressed of how different kinds of information can be combined. For example, in the case of morphological versus stratigraphic data just discussed, the difficulty lies in developing ways of combining into one analytical method incommensurate information from two entirely different fields. No one doubts that in principle at least, the cladogram and the order of appearance of its contained taxa in the fossil record should be congruent, but how should measurements of character support and measurements of stratigraphic position be made such that both kinds of information can be combined into one hypothesis?

The focus about the use of fossils in taxonomy must now shift to the question of the relationship between molecular data and morphological data. The advent of molecular character information and in particular of nucleotide sequences has turned from a tentative trickle 20 years ago into a veritable flood today (Avise 1994; Harvey *et al.* 1996). Parallel with the rise in this information has been a fall in confidence that molecules are about to solve all the taxonomic problems, leaving morphology irrelevant to phylogenetic analysis. First of all there are frequently incongruities between cladograms based on molecular evidence and those based on traditional morphological evidence. It might be supposed that this is because the morphological evidence is misleading and should be abandoned, but unfortunately there is often just as much if not more incongruence between different molecular data sets themselves. The cladogram that is generated depends on which particular DNA molecule is sequenced, which particular species in a taxon are sampled, and which particular methods of statistical analysis are applied to build the cladogram.

Patterson *et al.* (1993) reviewed many of the main groups of living organisms that have been subjected to molecular-based phylogenetic reconstruction and found both the practical and the theoretical problems associated with sequence data unresolved, to the extent that they regard molecules as ‘successful’ only in cases where there is little serious challenge from morphological data, or, in the case of the higher primates, where an enormous amount of effort over a diverse range of molecules and techniques has cracked the problem by sheer weight of data. Therefore they are pessimistic that the great outstanding problems of relationships amongst other modern organisms will be solved until they, too, have been subjected to far more extensive and detailed analysis.

Suppose there are two or more data sets—for example, the sequences of two different DNA molecules plus the morphology of the living and fossil organisms—each of which sets leads, respectively, to a most parsimonious cladogram. As all these characteristics, molecular and morphological, share a common phylogenetic history, ideally they would each point to the same cladistic relationships of the organisms in question (Swofford 1991). If they indeed do, then obviously that particular hypothesis of relationships is the best one attainable on the evidence. But if they do not, then at least one of the data sets must be leading to an incorrect cladogram, and the question arises of whether all the data sets should be combined in some way to produce a compromise cladogram, or whether one or more of the data sets should be ignored in the belief that it, or they, are unreliable. Patterson *et al.* (1993) note several possible causes for incongruence between data sets. These include different representative species within the taxa being sampled, different methods and statistical procedures being applied to the different data sets, and different hypotheses of homology and character polarity being used by different authors. Any residual incongeunce after these have been allowed for must be due to homoplasy within the characters, as the result of convergent evolution and secondary loss. From this point, it is argued by some authors that the different data sets should not simply be combined to produce one compromise cladogram. Bull *et al.* (1993; see also Huelsenbeck *et al.* 1994) believe that in cases of strongly conflicting data sets, each set should be studied separately with a view to identifying and rectifying, or excluding the source of the incongruity. This is claimed to be more likely to lead to a realistic cladogram than would combining reliable data with data that are severely unreliable due to any of the factors noted above, other than homoplasy, whose cause can be identified.

Swofford (1991) also cautioned against too enthusiastically combining data to produce a single cladogram. The usual method for doing this is to create what is termed a consensus tree. This is a tree, or cladogram, containing all those groups present in all the separate, incongruent cladograms, but no groups or branches that are absent from any one of them (Fig. 6.6). (This is actually 'strict' consensus; there are various rather more relaxed versions such as 'majority rule' consensus, in which groups present in half or more of the separate cladograms are included.) As Swofford pointed out, however tempting it looks, this approach suffers from two disadvantages. One is that it fails to allow for the weight of support given to each separate tree. The second is that it considers only the best tree for

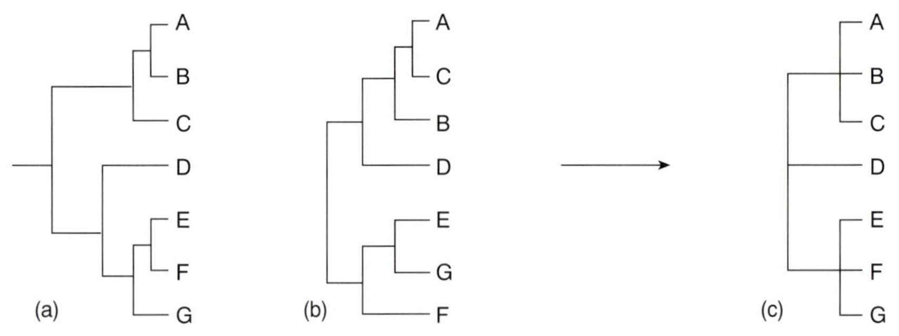


**Fig. 6.6** Consensus trees. The two cladograms (a) and (b) give the consensus tree (c), which includes only the groups common to them. (After Swofford 1991.)

each data set. There may well be a much better fit, which is to say more common groups, between slightly suboptimal trees for each of individual data sets.

The alternative argument, in favour of using all the available evidence to create the cladogram, is that any attempt to downgrade or discard data sets runs into the pattern-versus-process problem: only by knowing the phylogeny in the first place would it be possible to see which data set is accurately generating the phylogeny. All the difficulties of a priori character weighting appear (page 62). Kluge (1989), for example, argued in favour of the total evidence approach. He analysed boid snakes of the genus *Epicrates* using two data sets, one morphological and the other biochemical characters (although not actually molecular sequences). He combined the two kinds of characters together to produce a single cladogram, rather than two separate cladograms. Then he analysed the level of incongruence between the morphological and the biochemical sets of characters, and found, surprisingly, that the amount of incongruence was no more than that found within either the morphological set or the biochemical set alone. Indeed, he found most incongruence occurred within the skeletal subset of morphological characters. Therefore he concluded that there were no grounds for rejecting either character set. Although this example did not include molecular sequence data, the implication is that a comparable investigation of incongruity within data sets would show whether the combination of the two sets was acceptable or not.

To summarize this debate, combining molecular and morphological data sets avoids the circularity of a priori character weighting, but risks distortion by including a particularly misleading data set. Treating the data sets separately may allow the misleading set to be identified and rejected or improved, but risks the rejection of good data solely because it is incongruent with what is, unknowably, bad data. It is possible in principle to recognize poor data by non-circular argument, as long as it is done by a method that is independent of any reference to other data sets, or to a preconceived phylogeny. In their classic paper of 1967, Sarich and Wilson published an early, primitive molecular analysis of the relationships between the higher primates, using an immunological technique. This indicated that gorillas and chimpanzees are more closely related to humans than they are to orang-utans, *Pongo*. Furthermore, the similarity is such as to suggest a divergence of *Homo* from the African apes no more than about 5 million years ago. This conclusion spectacularly contradicted the prevailing view based on morphological evidence that *Ramapithecus*, a 15-million-year-old fossil ape, was a hominid, and therefore more closely related to *Homo* than to the other great apes (Fig. 6.7a). After much debate, and in the light of new, better preserved specimens, it eventually became clear that the characters relating *Ramapithecus* (nowadays called *Sivapithecus*) to hominids were either misinterpreted, or convergent. Thus, more careful analysis of the morphological data revealed its poorness. Meanwhile, new molecular data sets have continued to corroborate Sarich and Wilson's hypothesis, except that the divergence was probably a little earlier, between 5 and 10 million years ago.

Another example demonstrates how the converse can be true, and the molecular data sets prove to be misleading compared with morphology. Gardiner's (1982) controversial claim that mammals and birds are sister groups and share numerous morphological characters is supported by several molecular studies, such as myoglobin sequences (Fig. 6.7b). Extensive and careful analysis of the morphology shows that the hypothesis of a relationship of birds to crocodiles and other reptiles rather than to mammals is actually far more strongly supported (Kemp 1988*a*; Gauthier *et al.* 1988). Nevertheless, the molecules persistently point to the former relationship. It is now clear, however, that the high metabolic rate and body temperature that is found as a presumed homoplasic character in birds and mammals caused similar molecular substitutions to occur. This indicates that the molecular data set is misleading because of extensive convergence, and should be rejected in favour of the morphology.

These two practical examples point to the view that conflicting data sets should first be inspected by independent criteria to try and detect the cause of the non-congruence. Only if no such cause can be found should they be combined in some version of a consensus tree.

## Fossils and formal classification

If fossils are usable in phylogenetic analysis, or even if the argument is accepted that they cannot contribute significantly to the analysis but must be added onto cladograms based on living organisms, then it follows that in principle fossils can be classified along with living organisms in a formal hierarchical arrangement of monophyletic groups. Ammonites form one of the subgroups of the class Cephalapoda, in the phylum Mollusca. Unfortunately, fossils can actually prove very awkward in formal taxonomy,

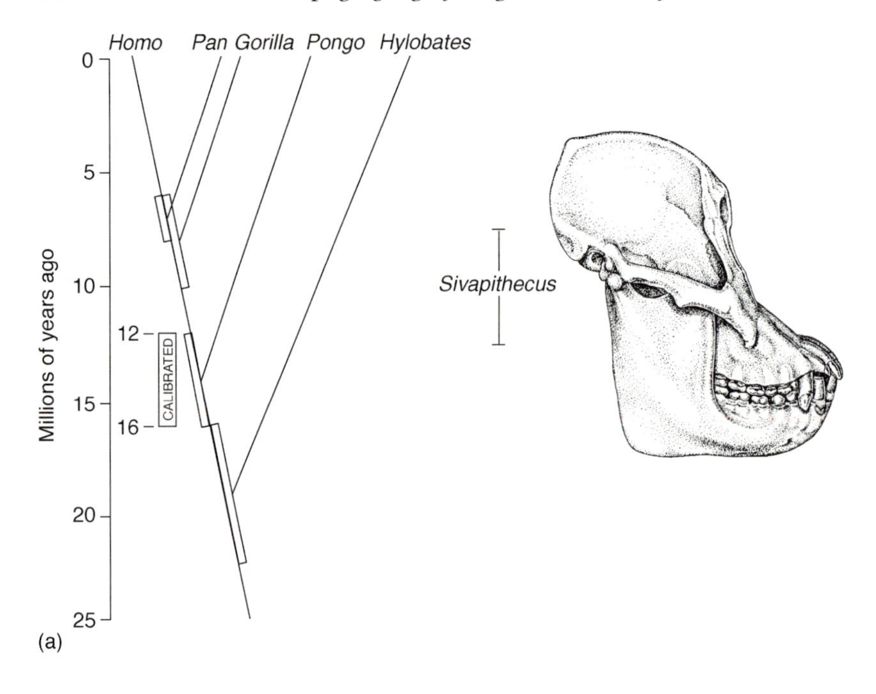


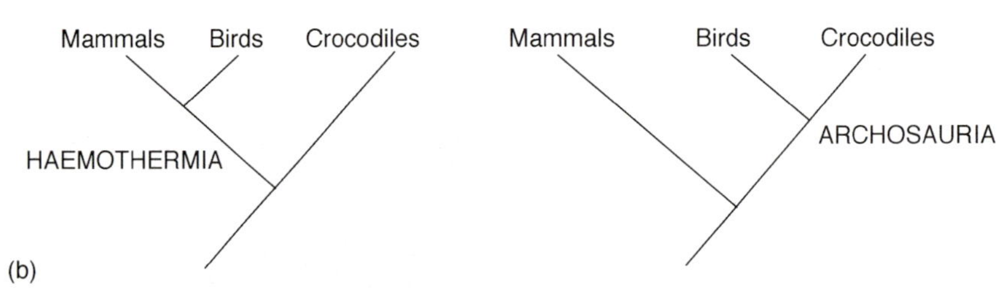


**Fig. 6.7** (a) Primate phylogeny with molecular-dated branching points, and showing the stratigraphic range of the fossil ape *Sivapithecus (Ramapithecus)*. (b) Amniote phylogeny according to molecular data (left) and morphological data (right). ((a) After Kelley 1992 and Carroll 1988.)

for two reasons. One is that in the strictest of cladistic classifications, all monophyletic groups should be named, and, furthermore, sister groups should be given the same ranking in the taxonomic hierarchy (page 54). So to repeat the example cited earlier, a single species of fish from the Jurassic called *Pholidophorus bechei* (Fig. 6.8b) is hypothesized to be the sister group of all the other 25 000 or so species of fossil and living teleost fishes put together. Formal procedure demands that *P. bechei* is given a new higher taxonomic name all for itself, and that the name refers to a taxon at the same level as the teleosts. The second complication introduced by fossils concerns the delimitations of the living groups. *Archaeopteryx* resembles modern birds in some of its characters, such as feathers, but in

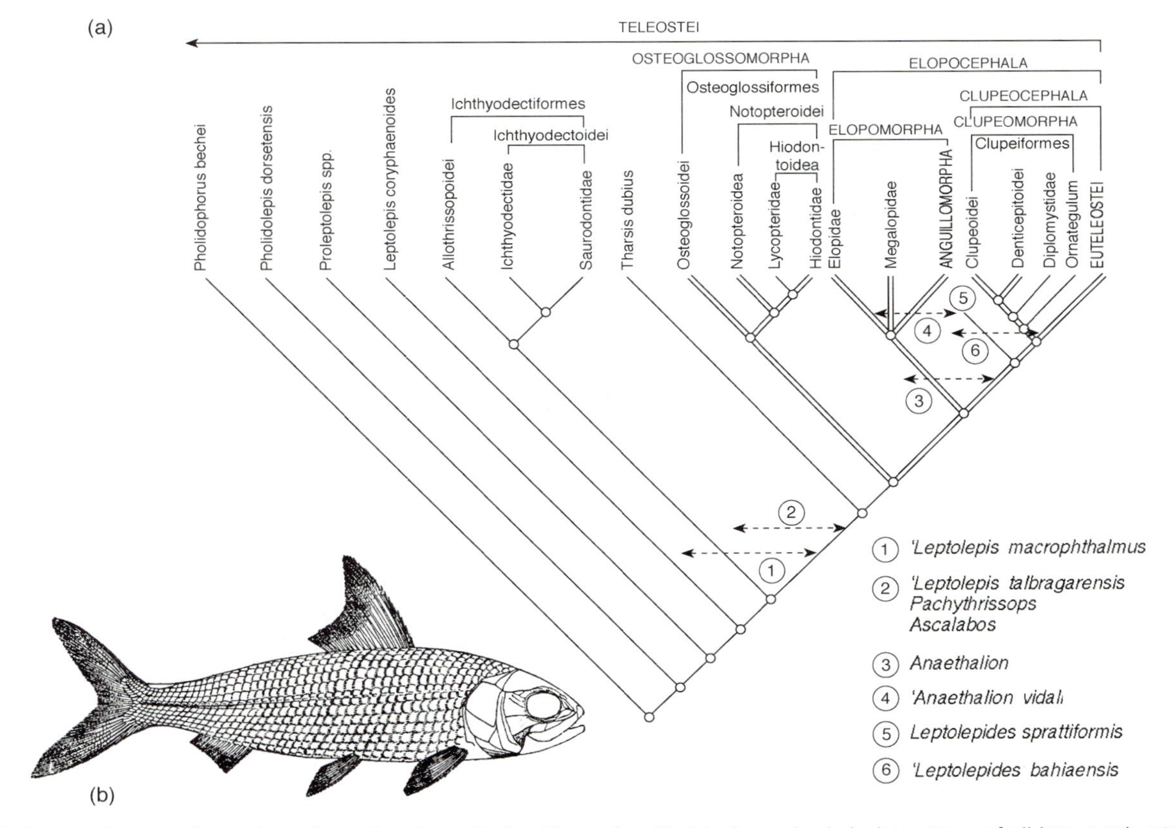


**Fig.6.8** (a) Cladogram of teleost fishes, including extinct taxa. (b) The Liassic fish *Pholidophorus bechei*, sister group of all known teleosts. ((a) After Patterson and Rosen 1977, (b) after Romer 1966.)

others, such as its teeth and long bony tail, it is like its more primitive 'reptilian' forebears. Should it be classified in the class Aves because of the first, or excluded from Aves because of the second? A subjective decision must be made about what characters are taken to define, and therefore which organisms come to constitute, the monophyetic group Aves.

One response to these formal difficulties has been to exclude fossils from classifications of living organisms altogether (for example, Crowson 1970; Løvtrup 1977). The disadvantage of doing this is that information relevant to the creation of the classification is effectively thrown out with the fossils, and, of course, it leads to different but equally irritating confusions—such as what is meant by the expression 'extinct arthropod class'. Consequently, most authors have instead tried to accommodate the fossils in ways that avoid both the clumsy excess of taxonomic names and the ambiguities of taxon boundaries. Patterson and Rosen (1977) introduced the term 'plesion' for exclusively fossil monophyletic groups. The phylogenetic positions of the plesions incorporated in some particular cladogram can be seen in Fig. 6.8a. Their positions can also be expressed in a list of taxa that starts with the most basal and adds the increasingly crownward plesions on each of the stems of the cladogram in turn (Table 6.1). Conversely, given a listing of the monophyletic taxa in this way, their respective interrelationships can be recovered completely and unambiguously. The fossils are therefore completely incorporated in the classification, while no extra names for groups need to be created. Loosely conventional ranks can be applied to the plesions for convenience, such as genus or family, but are not a necessary part. By convention, they are highlighted by a dagger symbol in the classification.

The second difficulty mentioned about fossils in formal classification concerns the effect they can have on the definitions or boundaries of groups of living organisms. This has been the subject of much more discussion because it is not just a matter of finding a convenient nomenclature to express otherwise uncontroversial relationships, but rather a matter of finding agreement on a subjective question of deciding at which point in a cladogram a particular modern group should be deemed to have arisen. For example, there is an increasingly good fossil record illustrating the emergence of the tetrapod vertebrates from their fish-like predecessors (Fig. 6.9). Expressed as a cladogram, there are several fossil groups that can be described as a series of plesions in the sense already described. Each successive plesion has at least one more character of tetrapods, which is, of course, why it is successive. The decision to be made is which particular node should be considered the beginning of the Tetrapoda; in other words, which characters should define the group.

Whichever node is taken, the Tetrapoda will be a monophyletic group: that is not the problem. The problem is that logically any one of the nodes could be used. The traditional palaeontological solution is to select a

**Table 6.1** Formal classification of teleosts based on the cladogram in Fig. 6.8(a) and including the extinct taxa as plesions

- plesion † *Pholidophorus bechei*
- plesion † *Pholidolepis dorsetensis*
- plesion † *Proleptolepis*
- plesion † *Leptolepis coryphaenoides*
- plesion † Ichthyodectiformes
    - SUBORDER † Allothrissopoidei, new
        - FAMILY † Allothrissopidae, new
    - SUBORDER † Ichthyodectoidei
        - FAMILY † Ichthyodectidae
        - FAMILY † Saurodontidae
- plesion † *Tharsis dubius*
- SUPERCOHORT Osteoglossomorpha
    - ORDER Osteoglossiformes
        - SUBORDER Osteoglossoidei
        - SUBORDER Notopteroidei
            - SUPERFAMILY Hiodonoidea
                - plesion † Lycopteridae
                - FAMILY Hiodontidae
            - SUPERFAMILY Notopteroidea
- SUPERCOHORT Elopocephala
- Elopocephala *incertae sedis* † *Anaethalion*
    - COHORT Elopomorpha
    - Elopomorpha *incertae sedis* † 'Anaethalion' vidali
        - ORDER Elopiformes
        - ORDER Megalopiformes, new
        - ORDER Anguilliformes
            - SUBORDER Albuloidei
            - SUBORDER Anguilloidei
    - COHORT Clupeocephala
    - Clupeocephala *incertae sedis* † 'Leptoleis' bahiaensis
        - plesion † *Leptolepides sprattiformes*
        - SUBCOHORT Clupeomorpha
            - ORDER Clupeiformes
                - plesion † *Ornategulum*
                - plesion † Diplomystidae
                - SUBORDER Denticipitoidei
                - SUBORDER Clupeoidei
        - SUBCOHORT Euteleostei

*Source*: after Patterson and Rosen (1977).

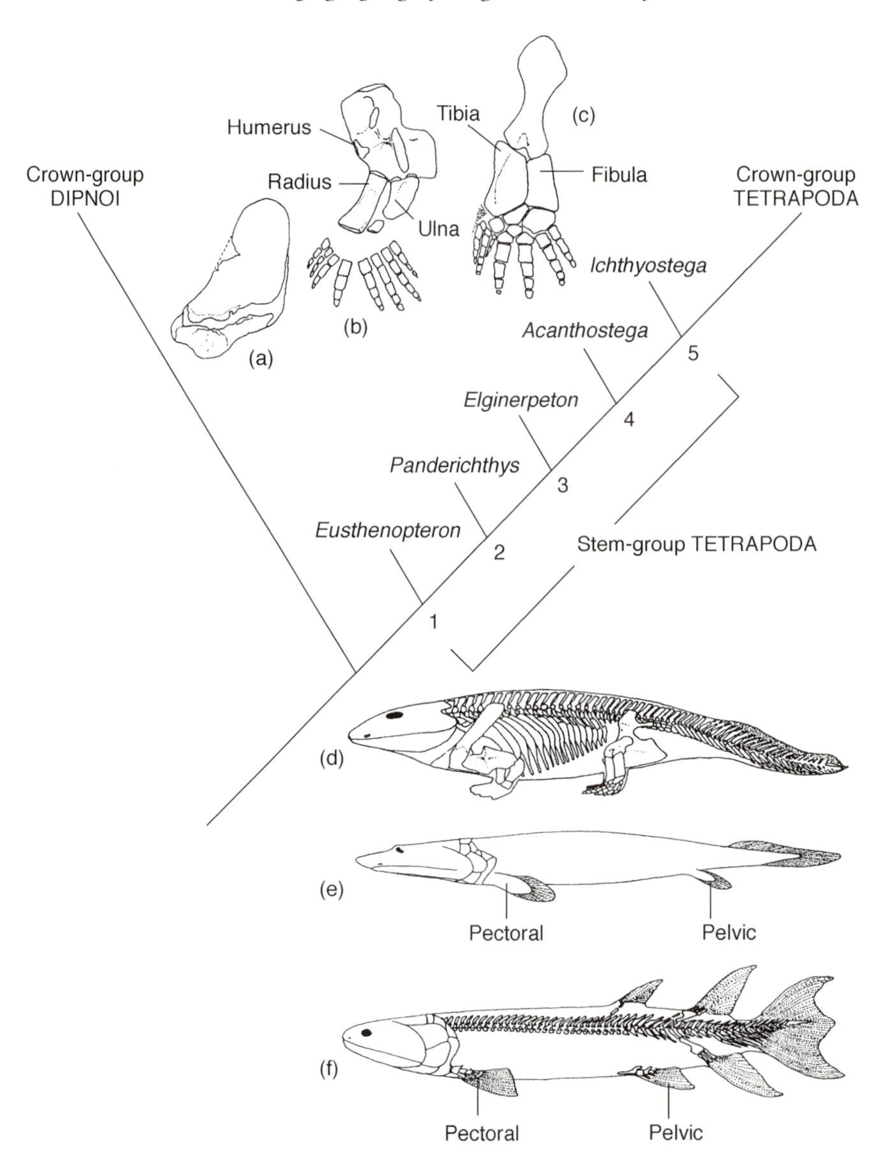


**Fig. 6.9** Relationships of the stem-group Tetrapoda. (a) *Elginerpeton* humerus, (b) *Acanthostega* limb, (c) *Ichyostega* limb, (d) *Ichthyostega*, (e) *Panderichthys*, (f) *Eusthenopteron*. (After Ahlberg and Milner 1994, reprinted with permission from *Nature*, © 1994 Macmillan Magazines Limited.)

character or set of characters believed to indicate an important new biological development, such as adoption of a new function or the invasion of a new habitat. In this case, the evolution of the tetrapod limb from the fish fin has usually been taken to indicate the invasion of land, surely a biologically significant event by any criterion. *Ichthyostega* and *Acanthostega* are therefore members of the clade Tetrapoda. But the description of *Acanthostega* with short, paddle-like eight fingered limbs, coupled with a set of fish-like gills, may lead some to doubt whether this animal lived all that different an existence from its aquatic predecessors (Ahlberg and Milner 1994). Indeed, the panderichthyids, traditionally 'fishes', do not seem so very different from *Acanthostega*, and the morphological gap between the two is already being filled in by ever more primitive tetrapodal forms, such as *Elginerpeton* (Ahlberg 1995). There is clearly less and less reason to take one particular tetrapod character, the limb say, and to make it the arbitrary basis of the definition of the group Tetrapoda.

The method now used to overcome the need to establish an arbitrary definition of the monophyletic groups with living members involves the concepts of the 'crown group' and 'stem group'. In the case of tetrapods, the crown group Tetrapoda consists of all those living organisms for which the group Tetrapoda was originally founded. It is thus defined as organisms that possess all the synapomorphic characters of the living forms. It includes logically the last hypothetical common ancestor of all the living forms, and any extinct forms that also descended from that ancestor. But it excludes any fossil taxon that has less than the full complement of modern tetrapod characters. This leaves the stem-group tetrapods as all those fossils with at least one, but not all, of the crown-group tetrapod characters. Taken as a whole, the stem group is a paraphyletic group that excludes some of the descendants of its common ancestor, and it cannot have formal cladistic status as a clade. For this reason Ax (1987), an enthusiastic proponent of this approach, calls the stem group a stem lineage. Whether group or lineage, however, what it actually consists of is a series of plesions, each one a monophyletic group in its own right. In a formal classification, the various members of the stem group would be expressed as listed plesions in the order in which they occur in the cladogram.

The advantage of defining modern groups as crown groups is that it removes the subjective element and therefore the ambiguity from the definitions of groups. It also prevents the discovery of fossils with new combinations of ancestral and derived characters from causing changes in group definitions. No longer can the absurd complaint be made that the more fossils discovered, the less it is known what a mammal, or a bird, or an echninoderm is. There are some trivial objections to the crown-group concept; such as what would happen if part of the modern group should become extinct at the hands of humans—say, for example, the monotremes—or have already disappeared, like the dodo? No doubt an arbitrary qualifying date such as 1758 or 1859 could be argued over. . . !

## Further reading

Cracraft and Eldredge's (1979) edited volume embraces all points of view about the use of fossils in taxonomy; their own book (Eldredge and Cracraft 1980) is broader in scope but narrowly cladistic in approach. Also in this vein is Ax (1987), who develops the stem-group concept. Kemp (1988*a*) considers the use of fossils in the context of a particular case, and Forey (1992) reviews the field briefly and generally. Smith (1994) is an excellent, comprehensive review of cladistic analysis and fossils.

Two recent adjacent papers in *Paleobiology*, Clyde and Fisher (1997) and Hitchin and Benton (1997), discuss the relationship between morphology and stratigraphy.

Marshall and Schopf (1996) is a good, easy introduction to molecular taxonomy, while Avise (1994) is a much more detailed source of knowledge about the techniques and applications of molecules. Patterson *et al.* (1993) review the relation between molecular and morphology in taxonomy, and there are a number of papers in Givnish and Sytsma (1997) with examples using combined sequence data and morphological data.

# 7
# Speciation: gradual, punctuated, or what?

As has been alluded to in Chapter 5, species play a central part in evolutionary theory. From one point of view, species are the distinguishably different kinds of organisms that exist—the basic diversity of the natural world that evolutionary theory purports to explain. From a second point of view, species are the interbreeding populations that share separate gene pools and thus create and disseminate the genetic variation that is available for evolutionary change. From either viewpoint, the species is the evolutionary unit, be it, respectively, the unit of pattern or the unit of process. In Chapter 3 it was also noted that there is a lot of debate about species and speciation, with two fundamental questions: is the process of the evolution of new species accountable for entirely in terms of natural selection, or does chance have a role too? And can the species as a whole be subject to a selection process independently of the selection processes acting on the individual organisms that constitute the species population?

The trouble is that the process of speciation falls right into the epistemological gap (page 21). So far as can be judged from probable dates at which known pairs of sister species first became isolated from one another, the whole process typically takes of the order of a few thousand years. This is obviously too long a time span for direct observations from the start to the completion of a speciation event. But it is too short a timescale, and therefore too high a resolution event for the details to be apparent in most fossil records. As with everything to do with the epistemological gap, severe disagreements arise as different authors choose to believe different things (Barton 1988; Otte and Endler 1989).

Before considering the role of fossils in understanding speciation, it is necessary briefly to review the evidence from studies of living organisms and their populations that bear on the process.

## The neontological perspective

Living organisms provide a great deal of information about the nature of the species that they constitute. But this information is very complex, enormously variable from species to species, and quite confusing when it is used to generate hypotheses about how new species arise. A flavour of this variability is necessary in order to understand why there are so many distinct theories of speciation, and so little agreement about the causes and course of events associated with any particular case of speciation.

***Population structure.*** As reviewed in Mayr's (1963) classic work, some species have a geographically continuous population, over a large or over a small area. Others have disjunct populations, for example where the organisms utilize a patchy food source or occupy a set of islands. The disjunctness itself may be in the form of a few large populations, many small ones, or any other arrangement in between. Different patterns of dispersion of the young and migration of the adults add a dynamic dimension to the variability in what is presumably one of the most significant characteristics of a species.

***Phenotypic variation.*** The extent of the morphological, biochemical, and other kinds of phenotypic variation within a species and between related species is itself very variable. There may be little geographical variation between the members of a species over large areas, or there may be polymorphic variation within the same area to the extent that the different morphs would be regarded as separate species, were it not known that they formed a single, freely interbreeding population. Conversely, sibling species are practically indistinguishable on the basis of morphology alone, although on grounds of mutual recognition and lack of interbreeding, they are certainly separate.

***Genetic variation.*** There are several ways of getting values representing the degree of genetic similarity between organisms, such as electrophoretic studies of proteins, DNA hybridization, and nucleotide sequences of rapidly evolving genes (Avise 1994). When the genetic variation between members of the same species is tested, it shows enormous differences in degree from case to case. There can be more variation within a species than there is typically between closely related separate species, and there can be distinct species with barely detectable genetic differences. Furthermore, there is no close relationship at all between genetic variation and morphological variation. Cases of morphologically very similar species differing genetically far more than average, and vice versa, are known. All in all, the genetic variation appears to depend on a complex of factors, including at the very least geographical setting, population structure, dispersal pattern, mating system, and local selection forces.

***Intermediate states.*** Neontological studies have also revealed a variety of anomalies in the form of populations that show some but not all the expected features of good separate species (Mayr 1963). These are particularly important because they can often be interpreted as cases of incipient, or as yet incomplete speciation events. As such they should illustrate something of the overall process of speciation. One important category is hybridization between distinct populations. If the hybrids are infertile or generally less fit than the parents, it is assumed that the two respective populations are good species, but that the speciation process has not quite been completed. There is still the evolution of biological isolation to come, which will prevent the wasteful effort of hybridizing with the wrong mate. Another category of intermediate state is subspeciation, where geographically separate but adjacent populations, the subspecies, are morphologically distinct from one another, but are still capable of successfully interbreeding at the boundaries where they meet. A special and much celebrated version of this is the ring-species, where a sequence of contiguous subspecies extends around a central inhospitable region. The two ends of the sequence may meet and overlap, but the two respective terminal subspecies are incapable of successful interbreeding. They behave as good species towards one another, but nevertheless, in theory at least, they can exchange genes via the chain of intervening subspecies. Subspecies, including ring species, suggest that sections of a geographic cline can in principle diverge genetically to the point of speciation.

Given all these ways in which species as populations or gene pools can differ from one another, there should be no surprise at the plethora of theories proposed to account for the origin of new species, or at the inability to agree on which is or are the most widespread mechanisms. Indeed there is never much agreement on what actually caused any particular example of a speciation event that happens to be being studied. There are two interrelated questions that need answering in connection with the mechanisms of speciation. The first is what the environmental circumstances are that allow reproductive isolation (in effect, genetic incompatibility) to arise between two populations that formerly formed a single, interbreeding whole. The second is what the evolutionary processes are that cause the subsequent genetic divergence between the two populations to occur.

Every standard account of speciation lists the three basic theories of allopatry, parapatry, and sympatry, which constitute three proposed answers to the first question. Allopatric theories (Fig. 7.1a) claim that geographical separation of the two populations is necessary so that genetic divergence can accumulate without the homogenizing effect of continual interbreeding. Parapatric theories (Fig. 7.1b) claim that the requirement of complete geographic isolation in every case is too restrictive to account for all the species that have evolved, but that contiguous populations can differentiate even in the presence of a limited amount of gene flow between them. Sympatric theories (Fig. 7.1c) propose even less-restrictive environmental circumstances, and suggest various ways in which a subpopulation can differentiate within the bounds of the existing species population. Some sympatric theories propose virtually instantaneous speciation, by such devices as spontaneous polyploidy or macromutation. Others suggest that appropriate combinations of differential resources, selective mate choice, and high-enough selection pressures can cause irreversible genetic divergence to occur.

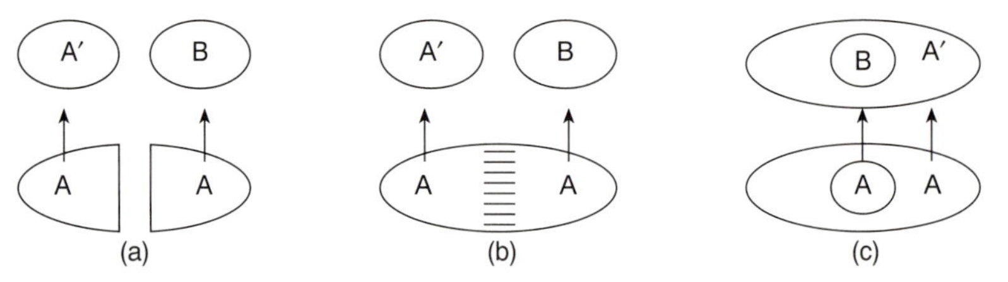


**Fig. 7.1** Speciation models in which an ancestral species A splits into two daughter species. (a) Allopatric speciation; (b) parapatric speciation; (c) sympatric speciation.

The other main question is whether natural selection alone is an adequate explanation for the genetic divergence between two populations that leads to speciation, or whether to a small, a great, or even an exclusive extent non-adaptive or chance processes are responsible. The standard neo-Darwinian response is that different selection forces occur because the two respective populations are in geographically distinct areas, and, furthermore, each population will now have different mutations and genetic combinations for selection to act on. Therefore natural selection is far the most important cause of divergence. The alternative view, developed notably by Ernst Mayr (1954, 1982*a*; Carson and Templeton 1984), is that an important part of the speciation process consists of what is termed a genetic revolution. The argument is that a well-adapted species consisting of a large gene pool is relatively resistant to evolutionary change because of the interactions between the different genes. The selective value of any particular gene is not an absolute value, but depends among other things upon the rest of the genes within the organism; that is, the genetic environment in which it occurs. These epistatic interactions between genes mean that the percentages of the different genes in the population as a whole are such that particularly favourable combinations occur with high frequency. Such a system tends to resist the spread of new genes because they are unlikely to combine so harmoniously with the existing genes in the gene pool. If this is true, then in order to have significant evolutionary change to the extent of producing new species, there must be some way of overcoming the genetic homeostasis.

According to Mayr's concept, a likely cause of such a genetic revolution is the geographical isolation of a tiny sample of the existing population. In the sample, which could be as small as a single pregnant female insect, three of four birds blown to a remote island by a freak wind, or one seed transported across an ocean, the gene pool will be very unrepresentative of the parent population from which it came. Some genes will be completely missing, and others that were rare in the original gene pool will be overrepresented. When the founder population expands within the new area, the finely tuned interactions of the genes will have been lost, freeing the population to respond to selection and also to the random losses and fixations of genes caused by genetic drift in what is still a small population. Not only will genetic divergence occur, but it will occur particularly rapidly. A genetic revolution of this kind has to be regarded as a non-adaptive accident, and not as a consequence of natural selection. An organism cannot be selected on the basis of, for example, its likelihood of being blown to a distant island, but rather it suffers from that fate by the ill-fortune of being in the very spot that the gust of wind struck. To this extent, therefore, a theory of speciation requiring a genetic revolution requires a significant non-adaptive component of the speciation process.

Other theories of non-adaptive speciation tend to involve mutations that lead to instantaneous speciation, by which is meant instantaneous reproductive isolation. They are also, therefore, theories of sympatric speciation, of the kinds briefly mentioned earlier.

To sum up the neontological perspective on speciation mechanisms, little is known with confidence because of the large variation in practically every parameter between the different species that exist. Every combination of population structure and evolutionary force imaginable has been used by someone as the basis for a theory of speciation. One way of testing these hypotheses consists of looking with ever more precision and detail at the actual population and genetic organization of particular cases. An example that illustrates the difficulties concerns hybrid zones (Hewitt 1989). These occur where two closely related but apparently good, separate species meet geographically along a line. Interbreeding between the two produces hybrids that are presumed to be less fit than the respective parental species because they do not become incorporated into either of the latter populations. The hybrid zone remains exactly that—a zone of hybrids. One explanation offered by many for the phenomenon is that it marks the secondary contact of two populations formerly geographically isolated for long enough to diverge to speciation, but not for long enough to prevent hybridization occurring. But an alternative explanation offered by others is that the zone marks the line between which two populations are in the later stages of the process of diverging parapatrically. There is no agreement on even such a straightforward question as this, and it seems likely that different processes occur in different cases.

## The palaeontological perspective

The fossil record could add further tests to hypotheses of speciation mechanisms in several ways. First, it could indicate how frequently speciation actually occurs. Whether it is a very common event or an extremely rare one would affect the judgement of whether some proposed mechanism is of the right order of probability to account for the observed frequency of speciation. Second, the fossil record could show what the rate of change of morphology is during a speciation event. Whether there is a constant rate of change, a variable rate of change, or an episodic rate of change will again have a bearing on hypotheses about the mechanism. If it emerges that either the frequency or the time-course varies in different groups of organisms, then interesting correlations might be found between particular speciation parameters and the particular attributes of different kinds of organisms. This, too, could provide an insight into the underlying process.

## The frequency of speciation

The frequency of speciation within a taxonomic group is variable both from group to group and within one group from time to time. But it is always a rare to extremely rare event as measured on the ecological time scale. The simplest way to measure speciation rate is to divide the number of species appearing by the length of time over which the frequency is being estimated. For example, MacFadden (1992) states that about 150 species of equids appeared between the earliest *Hyracotherium* species of the Eocene, and the Holocene *Equus* species, a span of about 58 million years. This gives an average frequency of one new species per 380 000 years.

To put this in context, the number of fossil species present at any one time during this period varies but lies between a minimum of 2 or 3 to a maximum of about 15. Therefore a rough calculation suggests that any one equid species can expect to undergo a speciation event roughly once in 4 million years. In another relevant observation, despite intensive study, Coope (1979) found no new species of beetle to have evolved in the 1.7 million years of the Quaternary period. Comparably, but in an entirely different context, Brett and Baird (1995) studied some Silurian to Middle Devonian marine sections of the Appalachian Basin, and found that over periods of time up to 7–8 million years, up to 80 per cent or more of the species across all taxa persisted unchanged, and no speciation events at all were detected.

Simple, rather anecdotal observations like these obscure variations in speciation rate that occur within taxa, and the record of Phanerozoic life indicates that after periods of mass extinction, a rebound effect can involve a greatly accelerated rate of speciation. They also fail to take into account rare species that are never likely to be known as fossils. In modern habitats, most species are relatively rare, leaving the biota to be dominated by a relatively small number of common species (May 1975). If the same was true of past biotas, and there is no reason to doubt it, then the rate of speciation could easily be, say, 10 times greater than it appears to be from the fossils.

Another approach to speciation probability is to measure the longevity of species in the record. If there were no change in the overall number of species present, then the average frequency of a speciation event would be the inverse of the average species longevity. Calculations for mean species longevity in different taxa have been made for many groups since Simpson's (1952) overall estimate of between 0.5 and 2.75 million years and, invariably, they are found to be significantly greater. Stanley (1979, 1985) has compiled many of these estimates (see Fig. 8.8). Even at the extremes of high rates of species turnover as found in mammals, birds, trilobites, and some ammonite families the figure for mean species duration is as high as 1–2 million years. For other macroinvertebrate, vertebrate, and higher plant groups the mean rises to respective figures between 3 and 15 million years, while for foraminiferans and diatoms it reaches over 20 million years. As in the case of the direct observations of speciation frequency, here again there are possible biases inherent in ignoring the potentially large numbers of rare species originating but never appearing in the fossil record, and which may have had significantly briefer durations.

Despite the imprecision and variability of estimates of frequency of speciation, it is beyond question that it is an extremely rare happening by the standard of the ecological time scale and human lifespan. The chances of observing a naturally occurring speciation process in the field is vanishingly small. Furthermore, the commonly used argument that some such proposed mechanism of speciation is too improbable to be acceptable, requires careful scrutiny. Speciation is sufficiently uncommon that an extremely rare, improbable process could easily be the cause. Indeed, the argument can be reversed, because a proposed mechanism that is too widespread and frequent in nature would have difficulty accounting for why speciation events are so rare.

## The time-course of speciation

All modern consideration of fossils and speciation begins with Eldredge and Gould's (1972) paper on punctuated equilibria. Indeed, it has even been suggested that this moment marks the maturing of the whole subject of palaeobiology, inasmuch as it represents the transition from passive interpretation of the fossil record in the light of the received neo-Darwinian theory to an active input from palaeontology into the development of theory itself.

The expression 'punctuated equilibrium' was used by Eldredge and Gould to label the situation where a fossil species was replaced instantaneously in the record by an apparently descendant species but with no intermediate forms between the two occurring (Fig. 7.2a). They contrasted this with the alternative, termed 'phyletic gradualism', in which a series of intermediate forms occurs between the ancestral and the descendant species (Fig. 7.2b). Their first radical proposal was that the punctuated-equilibrium pattern is much the commonest in the fossil record, and that the widely held assumption that the fossil record shows gradual evolution over long periods of time was based on the expectations of palaeontologists rather

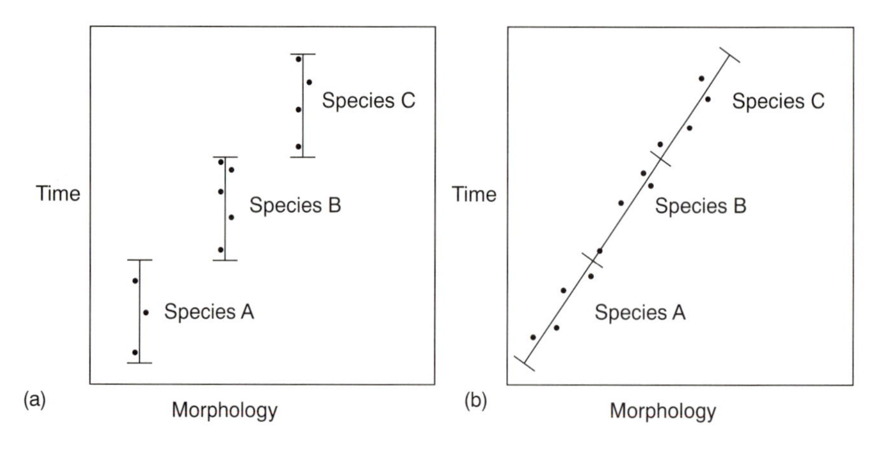


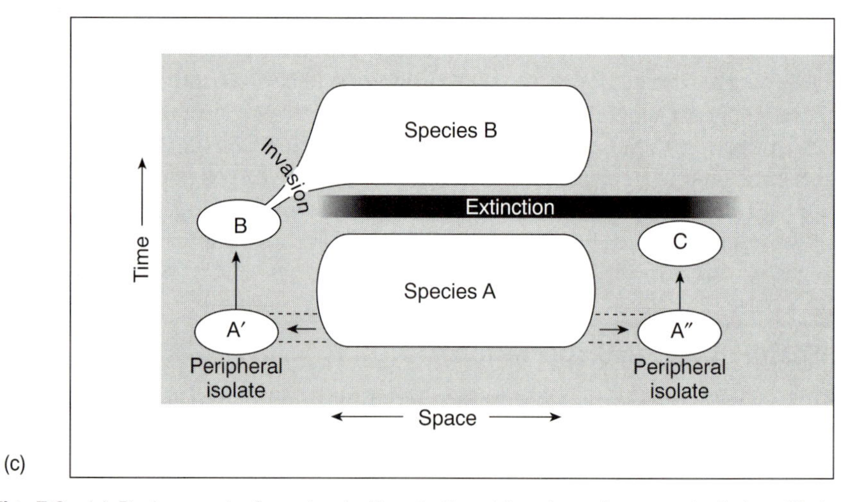


**Fig. 7.2** (a) Replacement of species in the stratigraphic column in a punctuated-equilibrium pattern. (b) Replacement of species in the stratigraphic column in a phylogenetic gradualism pattern. (c) Peripatric speciation generating a punctuated-equilibrium pattern.

than accurately documented observations. Their second radical proposal was that the punctuated-equilibrium pattern represented accurately what happened during the process of speciation (Fig. 7.2c). Referring explicitly to Ernst Mayr's ideas of rapid evolution in small, isolated populations (peripheral isolates as they came to be called), Eldredge and Gould pointed out that the intermediates between the ancestral and the descendant species were most unlikely to be found as fossils. They would have occurred as small, very rapidly evolving populations away from the main arena occupied by the ancestral species. Only after the new species had expanded its population and secondarily invaded the area of its now-extinct ancestor would its fossils be likely to be discovered. By this time, the transition from ancestor to descendant had been completed. Therefore, in any stratigraphic column, a series of distinct species replaces one another, instantaneously on the geological time-scale. The failure to discover intermediate forms is due to the nature of the processes involved, and not, as had been widely supposed, simply to the incompleteness of the fossil record.

There were several themes implicit in the original presentation of the punctuated-equilibrium view that were more explicitly developed in the context of the extensive and often passionate debate that promptly ensued (Somit and Peterson 1989). Gould and Eldredge (1977) introduced the expression 'stasis is data', which soon became a bit of a cliché. This moved the focus of punctuational thinking away from the punctuational events themselves, where one species replaced another, to the stasis or long periods of changelessness that a fossil species shows. The punctuations, consisting as they do of the absence of intermediate fossils, cannot be studied but their existence merely noted; the stasis, consisting of series of actual fossils through sediments, can be. From the perceived nature of stasis, a series of inferences can be drawn that leads eventually to a complete, internally consistent theory of the causes of macroevolution, the term applied to the pattern of origination and extinction of species through geological time. The principal inferences so made are:

1. A species spends far the greatest part of its lifespan from origin to extinction in a state of stasis, certainly 99 per cent and more. This is the normal state of affairs and speciation—or, indeed, anything other than minor intraspecific evolutionary change—is extremely rare.
2. It follows from (1) that significant evolutionary change happens only during the very rare, brief periods when speciation is occurring.
3. Over the lengths of time for which a species exists without morphological change, typically one to several million years, there must have been considerable environmental changes and therefore changing selection pressures. The failure to evolve that is shown by stasis indicates a failure to respond to these presumed selection pressures. A state akin to the genetic homeostasis of Mayr's speciation theory must exist.
4. If response to selection pressures in a fully adapted species is not possible except during speciation, then the speciation process itself cannot be caused solely by natural selection. It must be initiated by some sort of non-adaptive, accidental trigger that causes an event like a genetic revolution, which then removes the constraints on change.
5. Therefore, microevolutionary processes such as natural selection that occur within a normal interbreeding species population cannot be the complete cause of the macroevolutionary patterns of speciation over geological time. This is because a non-adaptive event is interpolated

between each ancestral species and its descendant. As the expression has it, macroevolution is decoupled from microevolution.

6. It follows from the concept of decoupling that species may behave as discrete units in macroevolution. If there are properties of the species as an entity in its own right, such as its population size, structure, or dispersal mode, which could cause different species to have different probabilities of speciation and extinction, then it is these properties that will decide the pattern of macroevolution over time. This is the process of species selection (page 42). It is distinct from organism-level selection in which it is the differences in probabilities of the births and deaths of the organisms that indirectly cause the survival or extinction of the species they constitute.
7. The possibility that organism-level selection and species-level selection can occur as independent processes leads to the idea of evolution being hierarchical, in that different levels in the hierarchy undergo their own respective evolutionary processes, with only indirect linkages between them (page 45).

## The debate and the synthesis

This perspective on speciation seems at first sight to correspond best with those neontological-based theories of speciation that most stress very rapid evolution. As has been seen already in the original formulation of punctuated equilibrium, the most obvious comparison is with the peripatric speciation theory of Mayr, involving a genetic revolution and rapid evolution in a small, isolated founder population. But any mechanism that implies more or less instantaneous reproductive isolation between the ancestral species and the incipient new species could explain the punctuated pattern. Thus various forms of sympatric speciation, such as genetic macromutation or chromosomal rearrangement, which combine instantaneousness with non-adaptiveness, are of possible relevance. These are the kinds of processes least favoured by neo-Darwinian theory, which immediately explains why there was and is still such a powerful reaction among biologists to the development of the punctuated-equilibrium thinking. This has really been something of an over-reaction due to a failure to appreciate the profound difference in scale between ecological and geological time.

It has to be permanently borne in mind that a geologically instantaneous event such as a punctuational speciation could well have occupied some thousands to tens of thousands of years. In the course of the debate, other sources of confusion have arisen. One is that the expression 'punctuated equilibrium' has been applied to both the pattern of individual fossils observed in the fossil record and also to the theoretical explanation for such a pattern. As far as rates of change are concerned, it would be perfectly possible in principle to explain an observed punctuated speciation by gradualistic selection theory. Yet another source of confusion is that there is no absolute definition of what constitutes a punctuational event as distinct from a small unit step in a gradualistic transition, or for that matter how much change is actually to be permitted in an example that is still describable as a case of stasis.

The debate has centred on three interwoven areas: whether the fossil record does show a predominantly punctuated pattern of speciation; whether a punctuated pattern needs a special explanation; and whether the implication of species selection as a distinct and significant process necessarily follows.

## The empirical pattern

Several possible causes of bias can occur when interpreting the existing fossil record, and the one most stressed by Eldredge and Gould (1972) is the nature of the prior belief of the observer about what kind of pattern to expect. There is no doubt that expectation that the fossil record would show a simple, naive version of neo-Darwinian gradual change has coloured many earlier interpretations. For example, the great classic case they quote of apparently gradual evolution is that of the Jurassic oyster genus *Gryphaea*, including one of the most reproduced of all palaeontological diagrams (Trueman 1922). It appears to show a perfectly gradual transition from ancestral to descendant species (Fig. 7.3a). But others, notably Hallam (1982), have subsequently interpreted this record rather differently, to include stasis in some lineages, and a speciation punctuated event (Fig. 7.3b). Ironically, however, the situation concerning observer bias has sometimes been reversed in the last couple of decades, as enthusiastic expectation of a punctuated pattern may have been allowed to distort perception of a more gradualistic pattern that actually exists.

Incompleteness of the fossil record as a source of bias is more difficult to allow for, because gaps in the record will tend to break up what may have been a gradually changing sequence into a series of discrete samples through time, with morphological gaps between them (Fortey 1985). The case of the transition from the foraminiferan species *Globigerina plesiotumida* to *G. tumida*, in the Deep Sea Drilling Project core 214 in the Miocene sequence of the Indian Ocean, was mentioned elsewhere (page 92). MacLeod (1991) showed that what appears as a punctuated evolutionary event fails to take into account a hiatus in the sequence, the effect of which is to throw considerable doubt on the significance of the raw observation.

A further cause of bias is the nature of the traditional classification system, in which specimens are placed into discrete species (Sheldon 1993). Collections of variable specimens from different horizons and localities tend to be placed in a single species, the description of which consists

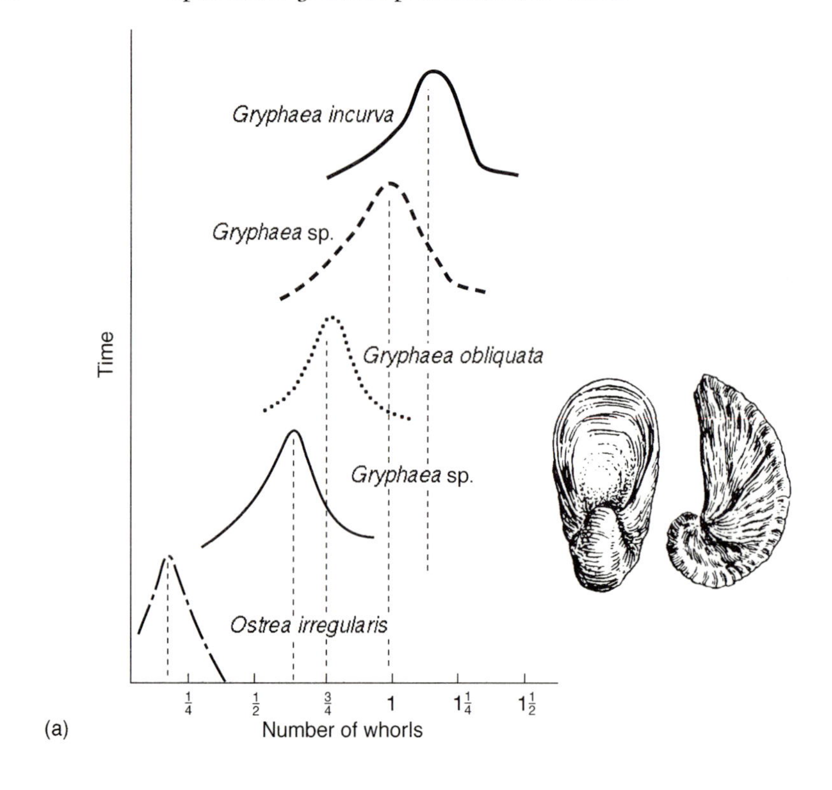


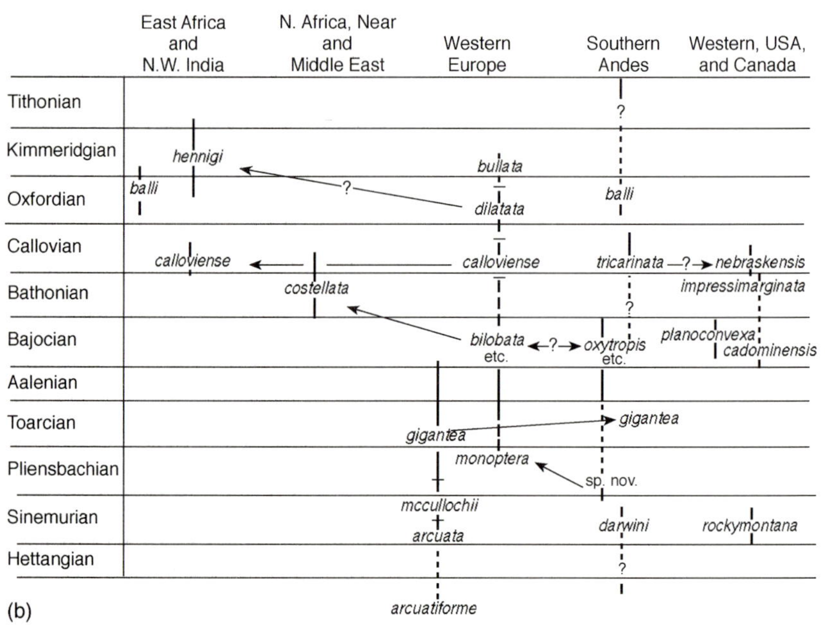


**Fig. 7.3** (a) Trueman's (1922) classic diagram of the evolution of the Jurassic oyster genus *Gryphaea*. (b) Hallam's (1982) interpretation of the evolution of *Gryphaea*, showing stasis and punctuation events as well as gradual evolution.

of the average morphology. This, and the absence of a formal nomenclature for expressing degrees of difference or of intermediateness between fossil species, tend to cause the reduction of a genuine morphological continuum into a set of separate species with perceived punctuational differences between them. Cladistic methods of taxonomy tend to aggravate the problem because of the stress on discrete alternative character states for defining groups, including species (Erwin and Anstey 1995*a*).

All in all, it is therefore not enough simply to look at existing descriptions of various fossil species in order to find out whether they show stasis and punctuational change, phyletic gradualism, or some intermediate or combined pattern. In this light, an increasing number of very carefully conducted studies have been made that use adequately dense and complete fossil records, pay attention to possible hiatuses in the stratigraphic sequence, and apply appropriate statistical tests.

Still one of the best, most revealing examples of stasis concerns the Quaternary beetles studied extensively over many years by Coope (1970, 1979). Over the entire 1.7 million years of this period of extremely fluctuating environments, he found not a single case of significant evolutionary change; every Quaternary species known is identical to a modern species.

Moving to a longer-term, marine case, Cheetham (1986) looked at the Caribbean genus of bryozoans, *Metrarabdotos*, over a period of about 4.5 million years in the Upper Miocene to Lower Pliocene (Fig. 7.4). He calculated the mean time between samples as 160 000 years, with an expected completeness of 63 per cent at that time-scale. Using a discrimination analytical method on 46 measured characters, the lineages showed virtually no change at all within species, so that the differences between successive species cannot be due to long-term intraspecific gradual evolution. One possible cause of this result could be that using so many characters actually obscures gradualistic trends affecting some of the separate characters within the species. Therefore he corrected for this by looking at single characters within the same specimens (Cheetham 1987). He found that a few of the characters did show evolutionary changes, and so the species did not exhibit complete stasis. There were, however, very few characters evolving in this way, and significantly they were not the characters that were involved in the evolution of the new species. On the face of it, therefore, the speciation events were still punctuated, and not explicable as merely the continuation of evolutionary changes within species. In this example, speciation does appear to be decoupled from microevolution within the species.

Another careful study was done by Lieberman *et al.* (1995), using two species of brachiopods from Middle Devonian rocks of New York State (Fig. 7.5a). The stratigraphic sequence was from 380 to 375 million years ago, a period of 5 million years, and had a thickness varying from place to place of between 100 and 1000 metres. They took a series of measurements

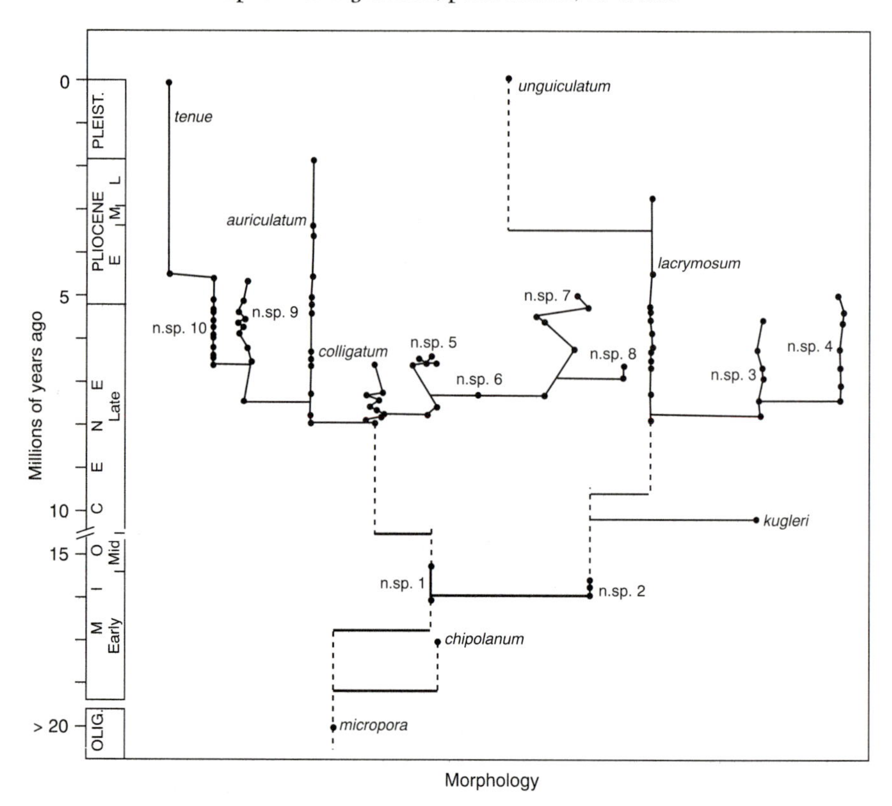


**Fig. 7.4** Evolution of the Tertiary bryozoan genus *Metrarabdotos*. (After Cheetham 1986.)

representing shell form on a very large number of specimens, and found that there was no significant difference between a sample at the bottom and a sample at the top of the sequence. This apparent case of stasis is not so simple, however, for they did find morphological differences between samples from some of the adjacent intervening horizons. Unexpectedly, these were in the form of reversible oscillations in morphology, rather than continuous trends in one direction. Furthermore, they could divide the whole stratigraphic sequence into a series of about eight different biofacies (Fig. 7.5b), on the basis of such things as depth and rate of sedimentation as well as the associated biota. When they looked at a sequence of specimens from a single biofacies through time, they found greater morphological oscillations than occurred within the species as whole through the same period of time. Several other authors have reported morphological reversals or oscillations of this kind, notably Sheldon (1993) in Ordovician trilobites and Anstey and Pachut (1995) in bryozoans, suggesting that this may be a widespread phenomenon.

Geary's (1990*a*,*b*) detailed example concerns a sequence of Miocene gastropod molluscs of the genus *Melanopsis* from a wide area of eastern

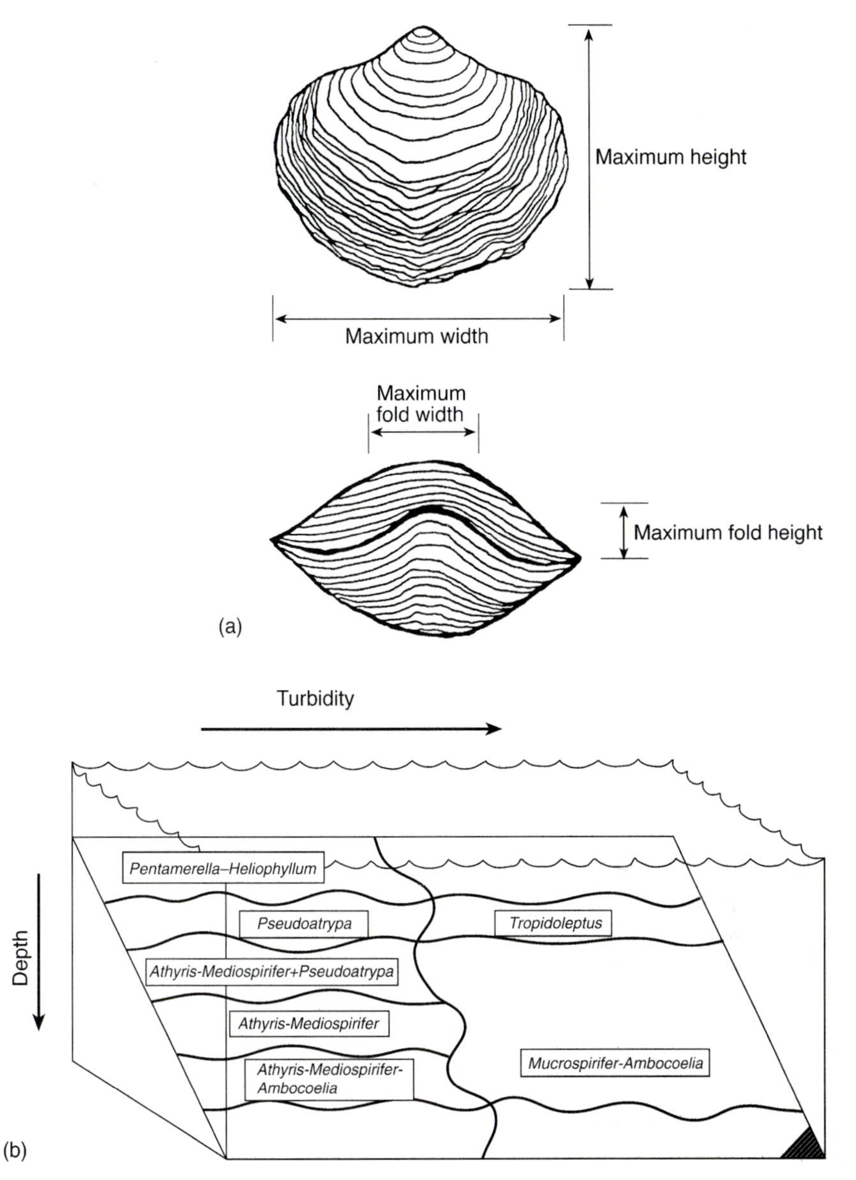


**Fig. 7.5** (a) Dorsal and anterior views of the Devonian brachiopod *Mediospirifer audaculus*. (b) The different biofacies of the Hamilton Group environment in which *M. audaculus* occurs. (After Lieberman *et al.* 1995.)

Europe and western Asia (Fig. 7.6). She made a series of measurements on 1611 specimens from 56 samples over the range and demonstrated an example of stasis in the species *M. impressa* lasting for about 7 million years in the lower part of the sequence. Then, over the course of an estimated 2 million years, she found a gradually changing series in which the

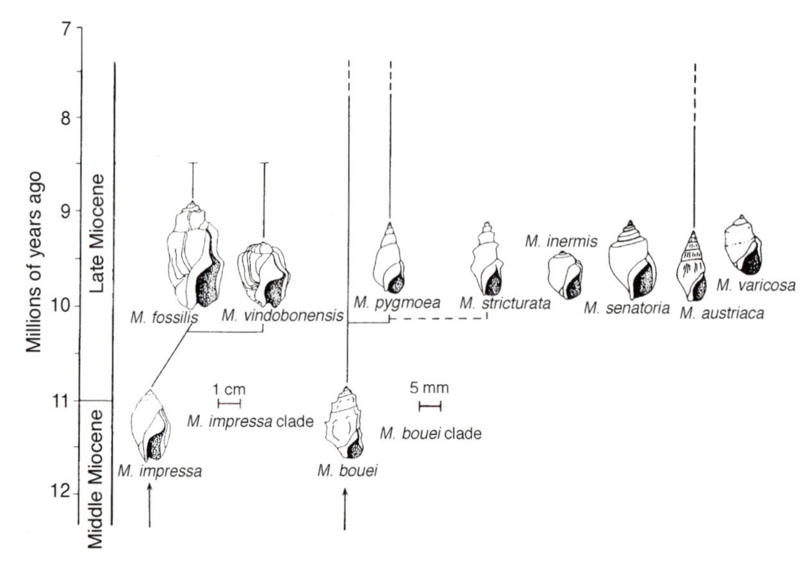


**Fig. 7.6** Evolution of the Tertiary gastropod genus *Melanopsis*. (After Geary 1990*b*.)

size increased and the shell developed a distinct shouldering, ending with what is taken to be the new species *M. fossilis*. The onset of this transition coincides with the final disappearance of marine species as the conditions altered from marine to fresh water, suggesting a change in selective forces. In contrast to the mode of origin of *M. fossilis*, however, another species *M. vindobonensis* apparently evolved from *M. impressa* very rapidly, with samples from only one locality (itself of unclear stratigraphic position) showing intermediate form. This example, therefore, shows cases of respectively stasis, phyletic gradualism, and a punctuated speciation event.

Examples showing clear-cut phyletic gradualism require particularly high-resolution stratigraphy. Some of the best observations are from within the 1.7-million-year span of the Quaternary. Lister (1993) notes several cases of mammals where enough gradual morphological change occurred to warrant the erection of separate ancestral and descendant species. For instance, the mammoth lineage in Europe shows gradual change in certain characters, including the number of enamel bands of the molar teeth, that connect the character states found in the conventionally accepted species *Mammuthus meridionalis*, *M. trogontherii*, and *M. primigenius* (Fig. 7.7). During this time he found no good evidence for any lineage splitting events, and so interprets it as a single, gradually evolving lineage.

Other examples of apparently good phyletic gradualist trends are known from the Deep Sea Drilling Project (DSDP) cores, involving planktonic and benthic protists. Malmgren and Kennet (1981) found that gradual change dominated the lineages of the late Miocene, South Pacific site DSDP 284,

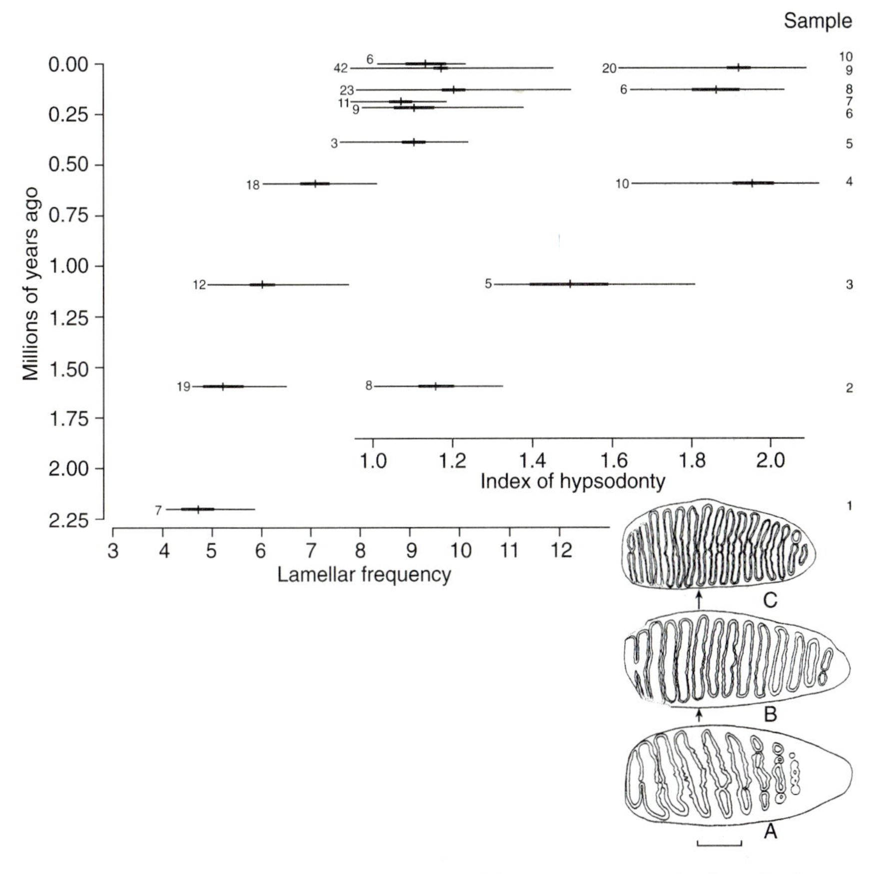


**Fig. 7.7** Evolution of the third upper molar teeth of Quaternary mammoths. Lamellar frequence is the number of enamel ridges per 10 cm of tooth crown. Index of hypsodonty is the ratio of crown height to crown width. For each sample, the mean, the range, the standard error of the mean, and the sample size is indicated. Samples 1–3 are referred to *Mammuthus meridionalis* (tooth A); sample 4 is referred to *M. trogontherii* (tooth B); and samples 5–10 are referred to *M. primigenius* (tooth C). (After Lister 1993.)

in which the sampling interval is estimated to be 100 000 years over a span of 8 million years (Fig. 7.8)

There is a growing feeling among commentators that examples such as these are pointing with increasing clarity to the conclusion that the fossil record indicates a complete range of patterns of speciation, from geologically instantaneous punctuation to gradualism at various rates. Even within the gradualistic mode, morphological change may be unidirectional or may involve random fluctuations in direction, and morphological change to an extent that would be typical of a new species can occur both with and without cladogenetic speciation. Erwin and Anstey (1995*a*) have categorized all the good, detailed studies made since 1972 that they could find

(Table 7.1). This confirms the perception of great variability in pattern, distributed amongst different combinations of stasis, gradual change, punctuational change, unbranching, limited branching, and multiple branching.

**Table 7.1** Numbers of different patterns of evolution and speciation in detailed species-level studies made between 1972 and 1995

| | Gradualism | Gradualism and stasis | Punctuation and stasis |
|---|---|---|---|
| Non-branching | 13 | 13 | 6 |
| Limited branching | 2 | 2 | 6 |
| Multiple-branching | 2 | 0 | 14 |
| Total | 17 | 15 | 26 |

*Source*: After Erwin and Anstey 1995*a*.

Given this apparent multitude of modes of speciation, the search is now on for environmental features and organism attributes that correlate with one particular pattern or another. For example, from a study of mammalian evolution, Vrba (1985) claimed that speciations are associated with physical perturbations of the environment, and hence synchronous pulses of speciation across many different lineages occur from time to time, with intervening periods of stasis. She called this the 'turnover pulse hypothesis', with the implication that stasis is associated with constancy, and

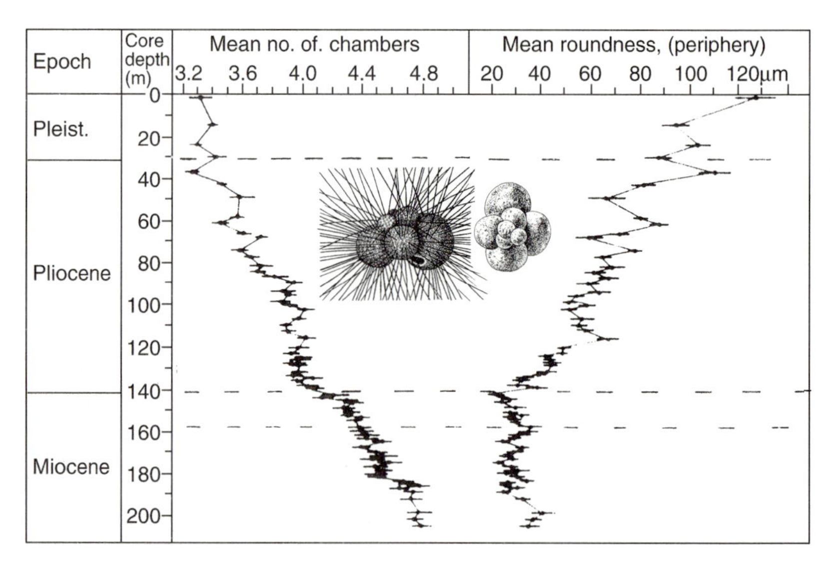


**Fig. 7.8** Evolution of the foraminiferan *Globorotalia conoidea* in the Deep Sea Drilling Project core DSDP 284. (After Malmgren and Kennett 1981.)

punctuated speciation with fluctuations of the environment. In marked contrast, from the viewpoint of marine invertebrates, Sheldon (1993, 1996) claimed that morphological stasis is a response to a widely fluctuating environment, whereas gradual evolution occurs in constant or narrowly fluctuating physical conditions. Interestingly, he believes that this accounts for the large number of cases of stasis in the fossil record, because most fossil species happen to occur in shallow water and/or temperate regions, which are the habitats most associated with large fluctuations in environment.

From initial tentative proposals like these, a debate has arisen about the circumstances associated with stasis of species within communities. It centres on the concept of 'co-ordinated stasis' introduced by Brett and Baird (1995). They studied Middle Palaeozoic facies of the Hamilton Formation of the Appalachian Basin, and found that they could recognize ecological–evolutionary (e–e) subunits of 3–7 million years in duration. Within such e–e subunits, there is a very high level of stability. Typically up to 80 per cent and more of the species persist unchanged, at most about 10 per cent become extinct, and immigration and emigration of species are low. The inference drawn from these findings is that stasis of species is co-ordinated by some actual process amongst the species of the community. Their explanation is that stasis relates to an overall ecological resilience in the face of environmental change, and punctuated speciation presumably depends on a breakdown of that stability. Others have failed to find a clear-cut pattern of co-ordinated stasis (Miller 1997), and it is possible that even if such a pattern does exist, the explanation lies in patterns of immigrating and emigrating species, rather than extinctions and speciations in phases at the boundaries between the e–e subunits.

At any event, much more empirical evidence is likely to emerge in the coming years about the particular environmental and ecological circumstances associated with different patterns of speciation as seen in the fossil record.

## The theoretical interpretation

It is now indisputable that stasis, or perhaps morphological oscillation about a mean, occurs in many and probably a majority of cases of fossil species where the record is adequate to discern it at all. A species does, typically, survive with no or very little morphological change for one to several million years. Equally it seems beyond dispute that speciation usually occurs so rapidly in terms of the geological time-scale that the process is below the resolution of the fossil record. At best, as is shown by studies of high-resolution fossil records, this means a period of time of no more than a few thousands to tens of thousands of years, although it still allows for far longer periods in most other cases.

The explanation that is referred to as the punctuated-equilibrium theory has already been described (page 137). Briefly, to reiterate, the theory proposes that stasis of the species is caused by a form of genetic homeostasis that prevents a large interbreeding population from responding to selection pressures. A punctuation is caused by a genetic revolution of some kind that frees a part of the species population to evolve. The genetic revolution could be a founder effect, or perhaps a macromutation significantly altering the organism. The consequence of the genetic revolution is very rapid evolution to the level of a new species, by natural selection and genetic drift. The main evidence for the theory is stasis itself, with the claim that this occurs over too long a period of time to be accounted for by the absence of any environmental changes causing altered selection forces, to which the population would respond if it could; therefore it cannot. The major weakness of the theory is the continued failure to identify convincingly either the mechanism underlying the implied genetic homeostasis, or the nature of the assumed genetic revolution.

The main alternative explanation is the neo-Darwinian one, that the punctuated-equilibrium pattern is caused by different rates of change under different selection regimes. Stasis is due to stabilizing selection, where the selection pressures favour the existing morphology. The net result of this phase of natural selection is a zero rate of evolution. The punctuated events are explained as responses to new selection pressures that are strong enough to promote rapid directional change in the morphology, too rapid for the intermediate stages to be resolvable in the fossil record. The evidence for this neo-Darwinian explanation is, first, that the rates of change proposed are well within the rates of change known to occur in nature, and indeed well below the maximum rates attainable in laboratory investigations of response to intense artificial selection. This is an argument that the punctuated-equilibrium pattern is compatible with neo-Darwinian mechanisms. It is coupled with the philosophical view that because natural selection is known to occur, there is no need to invoke any other mechanisms, and that the scientific principle of simplicity of hypothesis therefore forbids such invocation. In short, neo-Darwinian microevolutionary processes are enough to explain the whole macroevolutionary pattern. The main weakness of this explanation is its inability to explain satisfactorily stasis over lengths of time when large environmental perturbations must have occurred.

Different theoretical explanations for an observed punctuated fossil record are well illustrated by the discussions following publication of Williamson's (1981) example of the Plio-Pleistocene mollusc fauna of the Lake Turkana deposits of East Africa (Fig. 7.9). Williamson described 13 lineages of molluscs showing stasis over a period of about 4 million years. At certain times in the record, many of them simultaneously show a virtually instantaneous shift in morphology, which Williamson attributed to punctuated

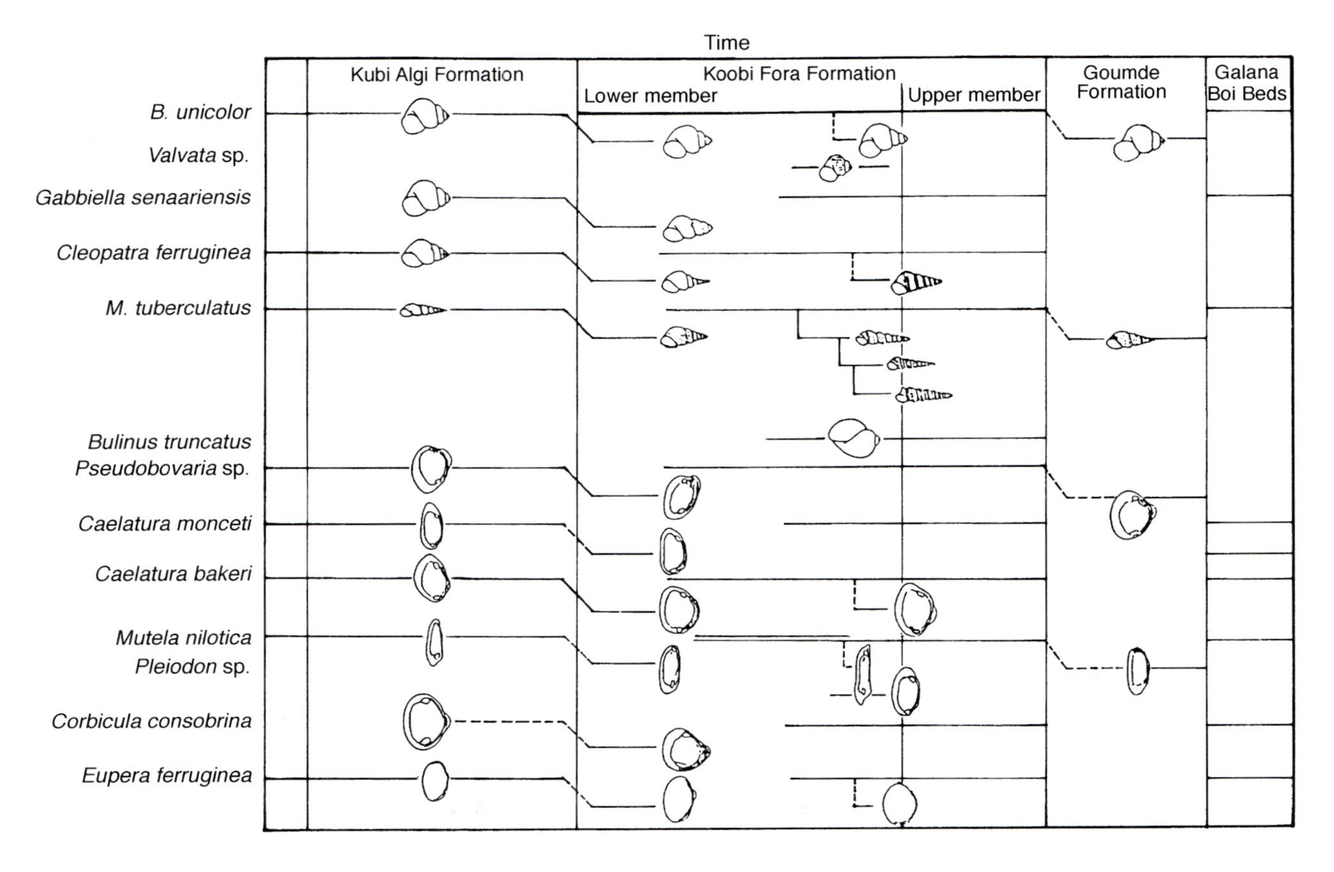


**Fig. 7.9** The pattern of evolution of Lake Turkana Plio-Pleistocene molluscs. (After Williamson 1981.)

speciation events. He also discerned a brief period of increased morphological variability in the population of each species, lasting an estimated 5000–50 000 years, which he took to be an indication of a period of disturbed embryological development associated with a genetic revolution.

Four quite distinct explanations have been offered for this punctuated-equilibrium pattern, one of the most-detailed and high-resolution fossil studies available (Kemp 1989). Williamson's own is based directly on the punctuated equilibrium theory. The 4-million-year stasis is attributed to developmental and genetic homeostasis, without specifying exactly what this means biologically. A genetic revolution is triggered in all the species simultaneously by an environmental deterioration—namely change in the level of the lake—and after the brief period of unstable development, it leads rapidly to the descendant species morphology. In each case the ancestral species must have persisted elsewhere until extinction of the descendant species, at which time it re-invaded the area. Conversely Charlesworth and Lande (1982) found no difficulty in ascribing the pattern to different rates of natural selection, from stabilizing to rapidly directional. The other two explanations offered both imply that no evolution actually occurred at all. Fryer *et al.* (1983) argued strongly that the whole pattern is no more than an ecophenotypic effect, where changes in the environment directly affected the developmental pathways of the organisms, causing an altered adult morphology. They were impressed by the simultaneous changes in so many different lineages, and by the subsequent return of the ancestral morphologies, and they argued that the degree of morphological change was entirely in keeping with the known ecophenotypic effects in modern molluscs. Cohen and Schwartz (1983) denied that there had even been any change in morphology at all, but that what was seen represented the invasion of the area by different species from elsewhere, in conjunction with the environmental changes.

The point to draw from this example is how impossible it is to decide which is the best explanation without a good deal more information. Without it, the hypothesis preferred depends largely on preconceived notions of the mechanism of speciation. With suitable *ad hoc* assumptions, any of these four explanations can be made to fit the limited observations of the fossils and their stratigraphic and palaeoecological setting.

The necessary additional information is beginning to accrue, and in particular this focuses on the key question of stasis over long periods of time. An enormous amount is now known about the extent of environmental change over geologically significant time periods, and there is no doubt it is of an amplitude and frequency that would be expected to induce changing selection forces and therefore evolutionary change in species populations over the lifespan of species generally. Measurements of stable isotope ratios (page 95) in marine deposits have revealed the Milankovitch cycles of climatic change that are believed to be due primarily to a series of physical changes in the Earth's rotation (Webb and Bartlein 1992). The eccentricity of the orbit (100 000 and 400 000 years), the tilt of the axis (41 000 years), and the precession of the rotation (23 000 and 19 000 years) all have predictable cyclic variations with the periodicities indicated. These are implicated in the very severe climatic fluctuations marked by the Quaternary ice-ages, but they must have been a factor affecting the environment from the very origin of the Earth itself. Far less is known about whether there are longer term regular periodic variations. Over the order of millions of years, however, tectonic events affecting continental disposition and orogenic events certainly cause continual, often large climatic alterations.

The question of why species show stasis under these circumstances is a real one indeed. One possibility, touched on at the end of the last section (page 147) is whether the stasis of a species is part of a process of co-ordinated stasis of many species in a community or habitat. Do they effectively buffer one another against environmental change in a way that is not yet clear, or does each species independently react to the environment in the same way as all the others (Miller 1997)? Sheldon (1996) suggested that species living in a widely fluctuating environment only survive by their immunity to the fluctuations in the first place. Species of this nature, therefore, will not experience a changing environment as changing selection pressures. Hence the stasis, and hence the expression *plus ça change (plus c'est la même chose)* model for his explanation of stasis. Lieberman and Dudgeon (1996) suggested an alternative explanation for stasis in a fluctuating environment. If the species occurs in several different environments, it will be subject to different selection pressures in different parts of its population. There will be no net directional change, because these differences will tend to cancel out. As the environment changes over time, this same buffering against overall directional change will occur. Their model is in accord with the observations of oscillating morphological change, and the differences in the pattern of oscillations found in different parts of a species range, mentioned earlier (Lieberman *et al.* 1995).

Less consideration has yet been given to the possible causes of rapid speciation when it does occur. Indeed, if the explanation for stasis is along the lines of these recent suggestions, there is no longer a problem with assuming that under certain environmental conditions natural selection will drive a species to the point of speciation. The conditions may be the stable or narrowly fluctuating environment as envisaged by Sheldon (1996), or simply a small population occupying a single biofacies as would be implied by Lieberman and Dudgeon's theory. If, as co-ordinated stasis theory has it, stasis is due to interactions between species in a habitat, then breakup of the habitat would presumably lead to a combination of species extinctions and speciation events. There is still much to be learnt.

## The implications for species selection

If it is accepted that the punctuated-equilibrium pattern, when seen in the fossil record, can be accounted for solely in terms of the neo-Darwinian processes that occur within interbreeding populations, then there are obviously no profound implications beyond the questions of why the rate of change is sometimes very fast and sometimes very slow, and why particular morphological trends observed within species cannot be shown to extend beyond species boundaries and be expressed as trends between successive species. If, on the other hand, it is accepted that perceived punctuated-equilibrium patterns cannot be explained solely in those terms, but that they imply a non-adaptive, chance process at least initiating each speciation event if not actually being the sole cause of speciation, then there is a profound implication. It is that the species could in principle act as an evolutionary unit in its own right, and that a pattern of speciation through geological time could be due to the process termed species selection, analogous to but distinct from the organism selection of standard neo-Darwinian theory (page 42).

Vrba (1980, 1984) introduced an extraordinarily convincing and now classic example of apparent species selection in fossil antelopes of southern Africa (Fig. 7.10). The phylogeny of these particular bovid species of the last 5 million years consists of a pair of sister groups. One, the Aepycerotini, is represented only by the modern impala, and no more than one or two precursors that may be separate species. The other group is the Alcelaphini, which consists of over 30 modern and Plio-Pleistocene species, including the

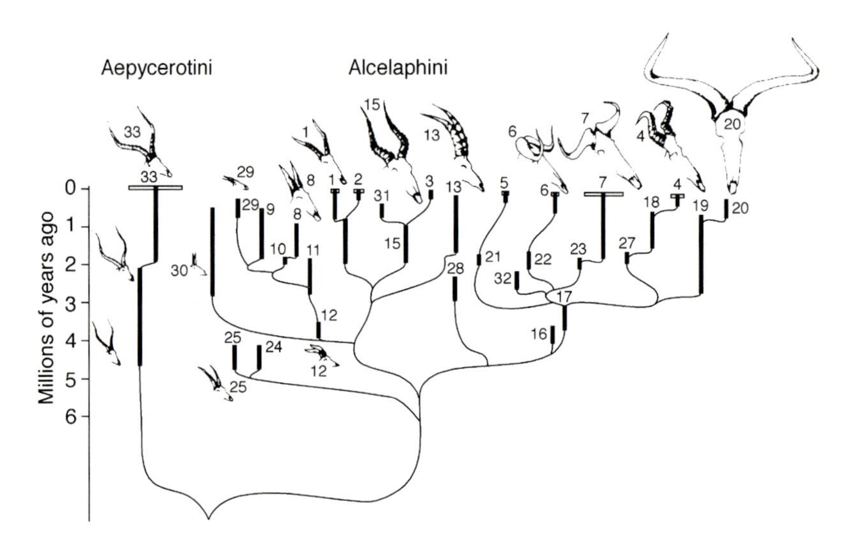


**Fig. 7.10** The phylogeny of Plio-Pleistocene bovids of Africa. Each number refers to one species. (After Vrba 1980.)

wildebeests and hartebeests among others. Why is one group so species-poor and the other so species-rich? Vrba's explanation is that the impala is a generalist relying on geographically wide-ranging resources. The Alcelaphini, on the other hand consists of specialist species, each one tending to rely on dispersed resources. These ecological differences cause differences in population structure, the aepycerotines being widely distributed but the alcelaphines having patchily distributed populations. This in turn affects the probability of isolated parts of the population remaining isolated to the point of diverging into separate species; the probability of speciation of an aepycerotine species is much less than that of an alcelaphine species.

If this is what actually did happen in these animals, then the process involved is true species selection, for the characteristic that determined whether species splitting was more probable or less probable was population structure—a species-level and not an organism-level character (page 42). In this light, it is difficult to explain the evolutionary pattern solely by neo-Darwinian selection acting on individual organisms. While natural selection has to be invoked to account for the adaptations of the organisms constituting each species, there is no obvious sense in which natural selection of organisms can favour speciation of the species to which the organisms belongs. Despite the attraction of this explanation, however, all that the fossil record actually reveals is a correlation, in this case between specialist trophic biology and high speciation rate. The explanation offered by Vrba assumes that the former is the cause of the latter, acting via the emergent species-level property of patchy population structure.

An alternative explanation could be based on the assumption that the cause and effect were the other way around: the high speciation rate was the cause of the specialist trophic biology and its associated population structure. Suppose, for instance, that specialist species face greater intraspecific competition because of the more limited resources available in any one area. Then selection might be imagined to favour those individual organisms that evolved slightly different resource requirements, such as diet. Therefore divergent evolution might be initiated more readily, and consequently the probability of speciation enhanced. As speciation events proceeded in this way, they would cause ever more specialist species to arise. Such a process would be perfectly in accord with neo-Darwinian theory, because in this case the character that determined the probability of speciation was the organism-level character of dietary preference. Nevertheless, it would account for the greater speciosity of the alcelaphines.

This general problem of how to test between species selection and organism-level selection as alternative explanations for a macroevolutionary pattern has been addressed by Lieberman (1995) in connection with another example. It has long been known that in certain taxa of marine gastropod molluscs such as the Turitellidae, in which some species have a planktotrophic larva and others a non-planktotrophic larva, there are two to three times less planktotrophic species than non-planktotrophic. The explanation normally given is that planktotrophic larvae, due to their long life in the plankton, have a high dispersal ability and therefore there is a reduced probability of geographic isolation of a subpopulation occurring. On the other hand, the non-planktotrophic species lay sessile eggs and have very poor dispersal, and therefore isolation is more likely to be maintained (Jablonski and Lutz 1983).

Lieberman accepted this possibility of a species-selection process, but pointed out possible organism-level mechanisms. For instance, a non-planktotrophic larva has lost its feeding adaptations and is perhaps unlikely to be able to reverse this and evolve into a planktotrophic form with well-developed feeding structures. A planktotrophic form, however, could presumably evolve equally easily into either type of species. This asymmetry implies that, everything else being equal, non-planktotrophic species will accumulate at a greater rate than planktotrophic species due to the organism-level character of feeding structures or their absence. Leiberman argued that in principle a cladistic analysis showing the number of times each of the two larval types evolved into the alternative would allow a test to be applied distinguishing between the two kinds of explanation. The species-selection process predicts one, or at least very few transitions from a planktotrophic to a non-planktotrophic species. On the other hand, the proposed organism-selection process implies frequent transitions from planktotrophic to non-planktotrophic species. He used a 16S ribosomal RNA gene from the mitochondria of some extant species to derive a cladogram, which unfortunately seems to show that neither of the explanations alone is adequate. The cladogram is not very robust, however, and the importance of his study is more in the idea than in the particular instance. It indicates that in principle tests can be found for deciding whether the best explanation for a given macroevolutionary pattern is a species-selection or an organism-selection process.

Most commentators accept the logical possibility of species selection, but many continue to doubt whether it is a significant process in the overall phylogenetic picture. Evolutionary biologists with a neontological background tend to feel that species selection would be too slow and weak a force to offset the action of natural selection acting on species, and also that it cannot directly generate adaptations of organisms, which they tend to feel should be the focus of evolutionary study (e.g. Dawkins 1996). To many, though by no means all, palaeobiologists this view misses the very important point that species selection will affect different but equally real aspects of the total evolutionary picture. The numbers of species in clades, the rates of species turnover, and the history of the increasing expansion into the Earth's potential ecospace are all features of phylogeny that are both long term and not necessarily connected directly with organism adaptation. The one argument that should not be levelled against the idea of species selection is that it is not interesting.

At any event, its importance is not only in the strict context of speciation mechanisms, but also in any consideration of patterns of species turnover and higher levels of phylogeny, which matters occupy the following chapters.

## Further reading

This is a field with a vast literature, and the following are guides into it, without any attempt to be comprehensive. The benchmark for the biological species concept and allopatric speciation is still Mayr's (1963) classic. Otte and Endler's (1989) edited volume contains useful reviews of modern thinking, such as Templeton's (1989) overview and Barton's (1989) thoughts about the role of founder effects. Ereshefsky (1992) has brought together a collection of the most important papers of the previous 20 years or so about species concepts.

The origin and evolution of punctuated equilibrium can be followed from its inception by Eldredge and Gould (1972), through their further thoughts about stasis in Gould and Eldredge (1977), to Gould's (1983) polemic and Eldredge's (1985) development of hierarchical thinking. Subsequently Eldredge has been developing his naturalistic views; Eldredge (1996) is a recent statement and Eldredge (1995*b*) a more popular version. A typical critique of punctuated equilibrium theory of the time is Hecht and Hoffman (1986), while Hoffman's (1989*a*) book is an extended essay criticizing all punctuated evolution and species-selection thought. Somit and Peterson (1989) offer an outsider's overview of the debates.

Kemp (1989) used the Lake Turkana mollusc record as a text for thinking about how to interpret a real fossil record of speciation, but Erwin and Anstey's (1995*b*) edited volume contains papers that are by now deeply embedded in the habit of much more careful analysis that marks contemporary investigations; their own review essay is an excellent statement of the position today.

Miller (1997) reviews the idea of co-ordinated stasis.

Most volumes of the journal *Paleobiology* have contributions bearing on species and speciation in the fossil record.

# 8
# Rules and laws of taxonomic turnover: are there any?

The last chapter considered the question of speciation—the splitting of one ancestral species into two or more species—and what relatively little light the fossil record can actually throw upon the evolutionary mechanisms involved. But the one thing perhaps above all else that the fossil record can demonstrate is that the history of life has consisted of a hugely complex interwoven sequence of ever-changing patterns of taxonomic diversity. For the whole 3500 million (3.5 billion) years of the fossil record, species and clades of species have continually arisen, flourished, and duly disappeared to be replaced by new but equally ephemeral taxa. The timescale of this taxonomic turnover is greatly variable. The longevity of even a single species (or, more precisely, of a morphotype that is taken to be a species) can be as much as 2000 million years in the case of certain single-celled micro-organisms still extant, or a period too brief to be measurable in the fossil record: from virtually infinite to virtually zero. As seen earlier, typical figures for a wide range of species are of the general order of 1–10 million years, with a mean of around 4 million years.

Because a lot of real information is available from the fossil record at this scale of events, an enormous amount of effort has been spent in trying to discover non-random patterns within the overall taxonomic turnover, with a view to discovering laws, rules, or at least statistically significant regularities that are amenable to explanations in terms of evolutionary processes. First, there is interest in the total biodiversity of the Earth in the past, and whether such things as the number of species, and the average rates of speciation and extinction have remained constant or changed in a regular fashion. Second, there is the question of what causes the diversity changes within individual clades. Third, there has been much interest about the processes involved in inter-clade relationships, particularly when one clade of organisms ends up replacing another clade of organisms within the same general habitat.

## The units of taxonomic turnover

The fossil record is very incomplete at the level of the species, with perhaps only 1 per cent of all the one-time existing species ending up preserved, discovered, and described on the basis of at least one specimen (Raup 1995). Therefore plots of numbers of species against time are poor indicators of actual diversities. In theory, increasingly higher-ranked taxa (genera, families, and so on) will be represented in the fossil record by increasingly high percentages of the actual number of taxa at that level that existed. For this reason plots of these taxa against time will be increasingly complete illustrations of the changing diversity as measured in terms of these respective higher taxonomic levels. There are also practical reasons for utilizing higher taxa: first, because of the enormous numbers of species actually described from many horizons; second, because distinguishing fossil species from one another tends to be a subjective exercise, affected more by whether a particular palaeontologist is a 'lumper' or a 'splitter' than by careful morphometric assessments of populations of specimens. The higher taxa are probably more consistently distinguished from one another by different people. As a consequence of these theoretical and practical considerations, the units of diversity used in most investigations have been taxa above the species level, principally families or genera but occasionally even higher groups right up to phyla.

Given this, it is necessary to ask what exactly it means to speak of the origination and the extinction of such taxa. Origination of a supraspecific taxon results from a sequence of speciations, and therefore the actual unit that is subject to the process is the species: origination of the taxon is the summation of a series of speciation events. Extinction of a taxon is the result of the disappearance of each of the constituent species, so, again, appears to be a species-level process. Indeed, it could be argued that the extinction of a species is actually the sum of the death of all its constituent organisms, in which case strictly it could be regarded as an organism-level process.

There are two possible conditions under which the use of higher taxa can be justified for plotting diversity, and diversity changes. First, if there are around the same number of species in each of the higher taxa, then the taxa can stand proxy for species-level diversity. But if this condition does not hold, then the plot of the taxa through time will be a distortion of the species plot that it is supposed to be representing. If four families each had, say, 50 species, while a fifth had only 5 species, then the extinction of the latter would show as a 20 per cent reduction in family diversity but only an insignificant drop of 2.4 per cent in species diversity. Exactly the same bias arises if the origination of a low-diversity taxon is regarded as quantitatively equal to the origination of a high-diversity taxon at the same taxonomic level. Signor (1985) compared the numbers of described taxa at several ranks and found a poor match between them. From the end of the Ordovician onwards, the number of orders of marine animals remains roughly constant at around 100. Numbers of families, genera, and species also do not alter greatly until after the end of the Permian.

From this point, all three ranks increase but at very different rates. In particular, species diversity increases much more rapidly, which suggests that the numbers of species per genus increased severalfold. Despite the reservations implied by Signor's figures, computer simulations by other palaeobiologists do show a reasonably acceptable correlation between diversities of species and higher taxa in terms of altering rates of change of diversity through time (Sepkoski and Kendrick 1993).

The more intriguing justification for using higher taxa is that they might actually be legitimate units of diversity in their own right, because each one consists of a series of species adapted to a set of related niches in the habitat. If this implied division of the habitat itself into sets of niches represents a division into real ecological entities, then each taxon, as the occupant of each respective division, does itself have an ecological meaning. The five families proposed above now represent five different general ways of life, or kinds of adaptations. Loss of any one of them, including the one with few species, represents an equivalent 20 per cent reduction in the occupation of kinds of niches. It is certainly not at all clear yet whether this is a biologically reasonable interpretation of higher taxa (Hoffman 1989), although several palaeobiologists do defend it (e.g. Stanley 1990; Eldredge 1996).

Patterson and Smith (1987) raised a further question about the use of higher taxa as units of diversity change, having noted that many of the published compilations include non-monophyletic taxa. They argued from a cladistic point of view that because only strictly monophyletic taxa are real, objectively definable entities (see page 52), then only they should be used. Study of changing diversity through time, however, is concerned only with changing numbers of species through time, not with the sequence of phylogenetic branching and its representation in the form of a hierarchy of monophyletic groups as such. If taxa are taken to be proxies for a given number of species, it does not matter in the least whether the taxa are monophyletic, paraphyletic, or even polyphyletic, just so long as each one contains roughly the appropriate number of species. Furthermore, if the alternative view is taken, that taxa represent sets of related niches in the ecosystem, then all that matters is that the species within such a taxon are sufficiently similar to one another to be reasonably thought of as ecologically so related. Again it is irrelevant whether the taxa are monophyletic or not. That well-known paraphyletic group 'fishes' has an ecological consistency that the strictly monophyletic group Sarcopterygii containing lungfish, coelacanths, and all the tetrapod vertebrates certainly does not. For example, the statement that birds are living dinosaurs and therefore the Dinosauria did not suffer extinction at the end of the Cretaceous may be cladistically pure, but it hardly detracts from the view that *something* biologically meaningful surely fell to zero! Sepkoski and Kendrick (1993) studied the effect of using non-monophyletic groups in computer-generated phylogenies, and showed that it does not significantly reduce the information about diversity changes if paraphyletic groups are included.

To conclude this section, the use of higher taxa as the units for measuring diversity change over time is still necessary at present, even though it can be justified only weakly, and therefore the underlying quantitative biological meaning of what follows is often tentative. Nevertheless, the qualitative impression is still one of dramatic taxonomic turnover at the species level.

## The total diversity of the Earth's biota

It would say much about the nature of life on Earth and its evolution if the change in the total number of species present could be discovered for the whole of evolutionary time. Has this number been rising gradually throughout recorded time, has it reached a constant number at some earlier stage and remained constant thereafter, or has it fluctuated extensively about some mean? It would also be interesting to discover whether, in addition to the total diversity, the mean rates of origination and extinction have varied at different times. Attention must be restricted initially to the marine habitat, because terrestrial plants and animals did not appear until relatively late in evolutionary history. Subsequently, it will be instructive to look at the changing diversity of terrestrial species to see if it matches the pattern found in the sea.

The first organisms to appear as fossils are the Archaean stromatolites—multiple-layered structures built by mats of prokaryotic cyanophyte bacteria, which date from about 3500 million years ago (Fig. 8.1b) Generally, the chances of small, single-celled organisms other than stromatolite-formers being preserved is presumably far too small to make any estimate of the number of species meaningful. The Cyanobacteria themselves are the dominant forms, and within various families of these, Schopf (1995) recognized that 20–40 per cent of what he carefully called species-level morphotypes are indistinguishable from modern species (Fig. 8.1a). Thus it appears that this early phase of evolutionary history consisted of extremely low rates of species turnover, and presumably extremely low levels of total biotic diversity by Phanerozoic standards.

Single-celled eukaryotic organisms (Fig. 8.1c) are likewise expected to be hopelessly under-represented as fossils, although several hundred Precambrian and Lower Cambrian localities have in fact yielded specimens, so some very rough estimates of diversity and species turnover rates can be made. Knoll (1995) did just that, and found that there were apparently significant changes. Until about 1000 million years ago, protist diversity and turnover rate were both low. At about that time, he found a significant increase in both figures, an increase in what George Gaylord Simpson (1944) classically termed 'evolutionary tempo'. Various new taxa appeared,

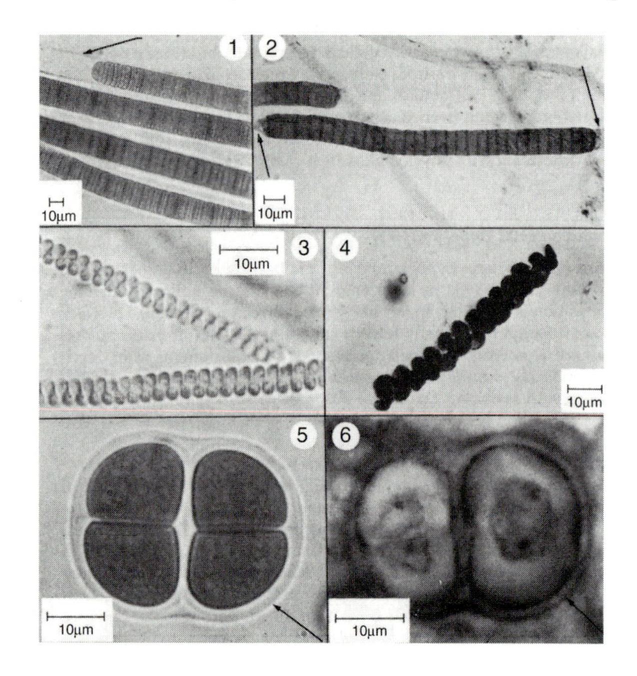


(a)

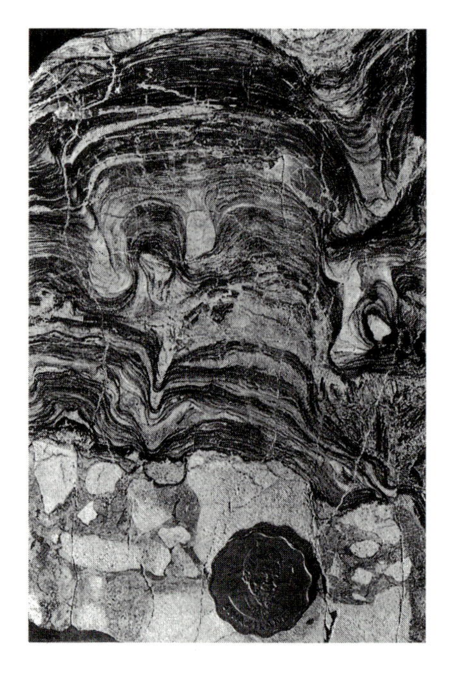

(b)

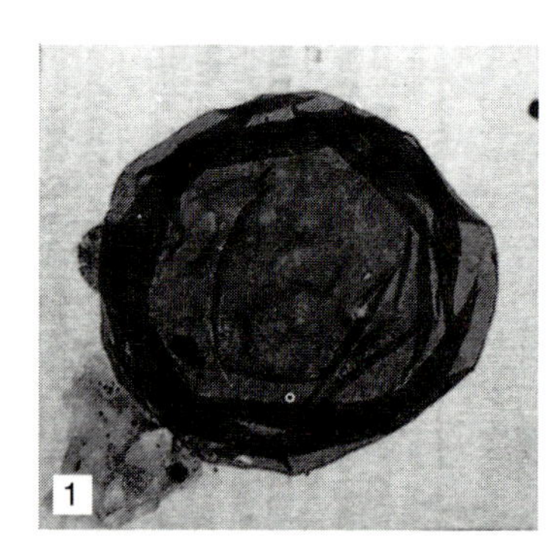


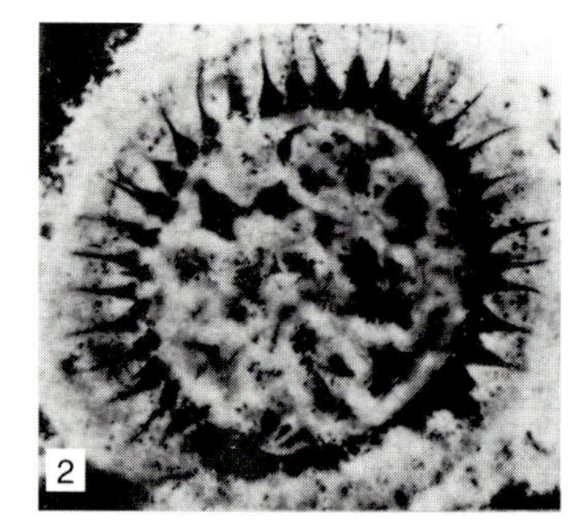


(c)

notably the red, the green, and the chromophyte Algae. Furthermore, molecular evidence from modern organisms demonstrates that several other more advanced eukaryotic groups such as the ciliates and the fungi probably had their origins at around this time.

The next phase in the fossil record is the Ediacaran fauna (see Fig. 9.5a), named after the original locality in Australia but now known from many places in the world (Glaessner 1984). These localities vary in date between about 620 and 550 million years ago, all of which is late Precambrian, and the fauna itself consists of some 100 species in the form of impressions of organisms and trace fossils such as burrows and trackways. No attempt to estimate the diversity world-wide has been made from such limited information, but it certainly would seem to have been very low.

All detailed work on diversity levels and changes has concerned the Phanerozoic, from the Cambrian to the Holocene. The rise of the many groups of fossils with hard skeletons makes assessments of actual numbers of species a great deal more realistic. Because only a small percentage even of potentially fossilizable species have actually been described, however, a series of more indirect methods of calculating species-level diversity has had to be devised, as reviewed by Signor (1985). Most accounts have been based on the number of known higher taxa, particularly families or genera. In a seminal piece of work, Sepkoski *et al.* (1981) used various methods for estimating species diversity through time, at least partially independent of each other (Fig. 8.2a). They counted the diversity of trace fossils, the number of described species, the number of genera, the number of families, and also the richness of species within a single habitat. Together, these led to the conclusion that there has been a marked increase in the number of species in existence over Phanerozoic time, with a particularly marked rise since the end of the Cretaceous (Fig. 8.2b).

An ingenious alternative approach by Signor (1985) was to attempt to get rid of the biases inherent in the fossil data by applying appropriate corrections to them. The main sampling biases preventing the use of the number of known fossil species being taken directly as a measure of the actual diversity are threefold (Fig. 8.2c): (1) the volume of sediment of that particular age that is exposed, (2) the intensity of collecting in those areas, and (3) what he termed the degree of palaeontological interest devoted. The relative value of each of these can be estimated for each geological period, along with the number of known fossil species. But in order to use these three values as correction factors, they have to be calibrated by reference to a

**Fig. 8.1** (opposite) (a) Three examples (right) of Archaean species of Cyanobacteria (2, *Palaeolyngbya*, 950 million years old; 4, *Heloconema*, 850 million years old; 6, *Gloeodiniopsis*, 1500 million years old) that are virtually indistinguishable from respective modern species (1, *Lyngbya*; 3, *Spirulina*; 5, *Gloeocapsa*). (b) A fossil stromatolite in section. (c) Two species of Proterozoic acritarchs: 1, *Leiosphaeridia*; 2, *Tanarium*. ((a) After Schopf 1995; (c) after Golubic and Knoll 1993.)

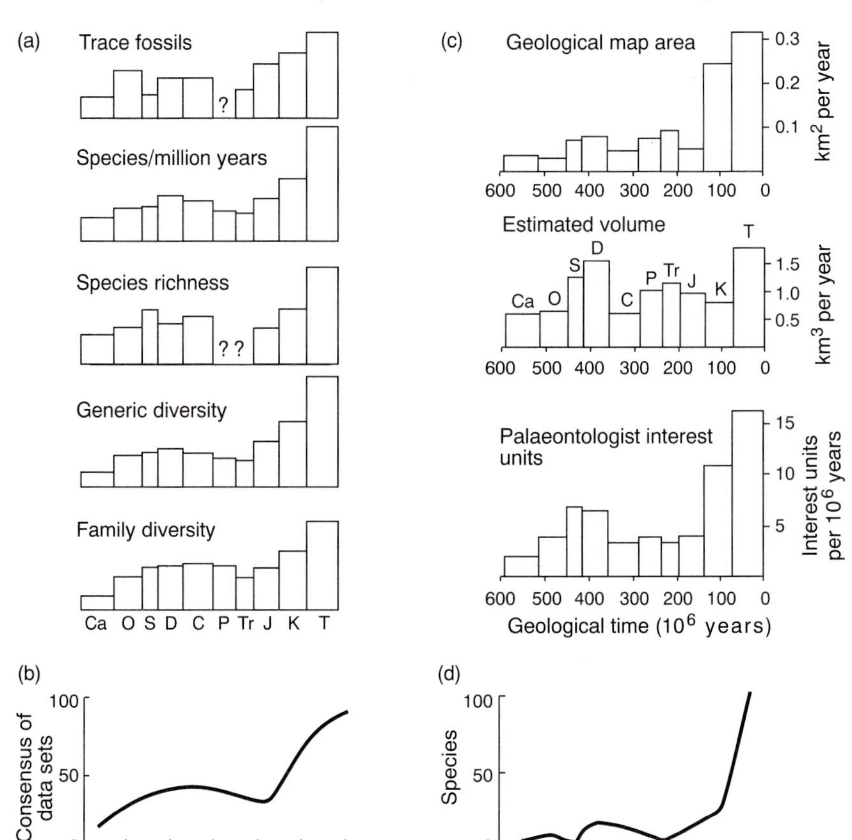


**Fig. 8.2** (a) Five empirical plots bearing on global species diversity during the Phanerozoic. (b) The consensus curve for all five data sets, expressed as the percentage of modern species diversity against Phanerozoic time. (c) Correction factors for estimating actual species diversity from apparent species diversity. (d) Corrected species diversity curve expressed as the percentage of modern species against Phanerozoic time. Note the marked increase in the number of species since the end of the Cretaceous period (K). ((a) After Sepkoski 1981; (b–d) after Signor 1985.)

period when both the number of fossil species described and the actual number of species are known. These can be estimated for the Cenozoic era from three measurements:

(1) the number of potentially fossilizable modern marine species (namely those with suitable hard skeletons); this is around 100 000–150 000;
(2) the pattern of change in number of species throughout the Cenozoic; and
(3) the mean duration of Cenozoic species from origination to extinction (both (2) and (3) can be found from the fossil record).

Together these produce an estimated total Cenozoic diversity of about 605 000 species, compared with the total number of fossil species described from this era of about 46 000 (Raup 1976). These figures of inferred actual compared with known numbers of fossil species for the Cenozoic can be combined with the three sampling-bias values to give an estimated correction factor applicable to earlier geological times. A final curve of diversity against time for the Phanerozoic can now be drawn (Fig. 8.2d).

Signor's method produced a somewhat similar result to the earlier method of Sepkoski *et al.* (1981), and confirms that there were indeed low numbers of species during the Palaeozoic, a small, gradual increase during the Mesozoic, and a large increase in the Cenozoic. But Signor's calculation suggests that the latter rise was very much greater than that calculated by Sepkoski *et al.*

Whichever method is used, the conclusion is circumscribed by the many assumptions that have to be made in order to arrive at it, but a general increase in species-level diversity in the marine environment does seem to have occurred. There are several explanations on offer for why this pattern should have occurred, based on Sepkoski's (1982) huge, comprehensive compilations of the described fossil taxa. He discerned three successive evolutionary faunas that displaced one another through time (Fig. 8.3):

1. The *Cambrian fauna*, with trilobites as the dominant group but also containing among other groups inarticulate brachiopods and monoplacophoran molluscs. It began in the Early Cambrian, increased in diversity, and then declined through the remainder of the Palaeozoic.
2. The *Palaeozoic fauna*, dominated by the articulate brachiopods, crinoids, corals, and several other groups. Although it began in the Cambrian, the Palaeozoic fauna reached its maximum diversity in the Ordovician, and maintained it until a precipitous decline at the end of the Permian, at the time of the great end-Permian mass extinction (page 209).
3. The *modern fauna*, with gastropod and bivalve molluscs, echinoid echinoderms, and the vertebrates among the dominant groups. Although the origins of the modern fauna may be detected in the early Palaeozoic, it increased dramatically in diversity only after the end of the Permian, and is therefore associated primarily with the Mesozoic and Cenozoic eras.

Sepkoski's (1981; Sepkoski and Miller 1985)) explanation for this pattern was that each of the three evolutionary faunas obeyed a simple logistic relationship, such that the rate of increase in diversity fell with increasing diversity level, and therefore the number of species approached an equilibrium value. Superimposed on the model is the idea that each successive fauna competed with and largely replaced the previous one, due to the evolution of increasingly advanced adaptations by the member species (Fig. 8.3d).

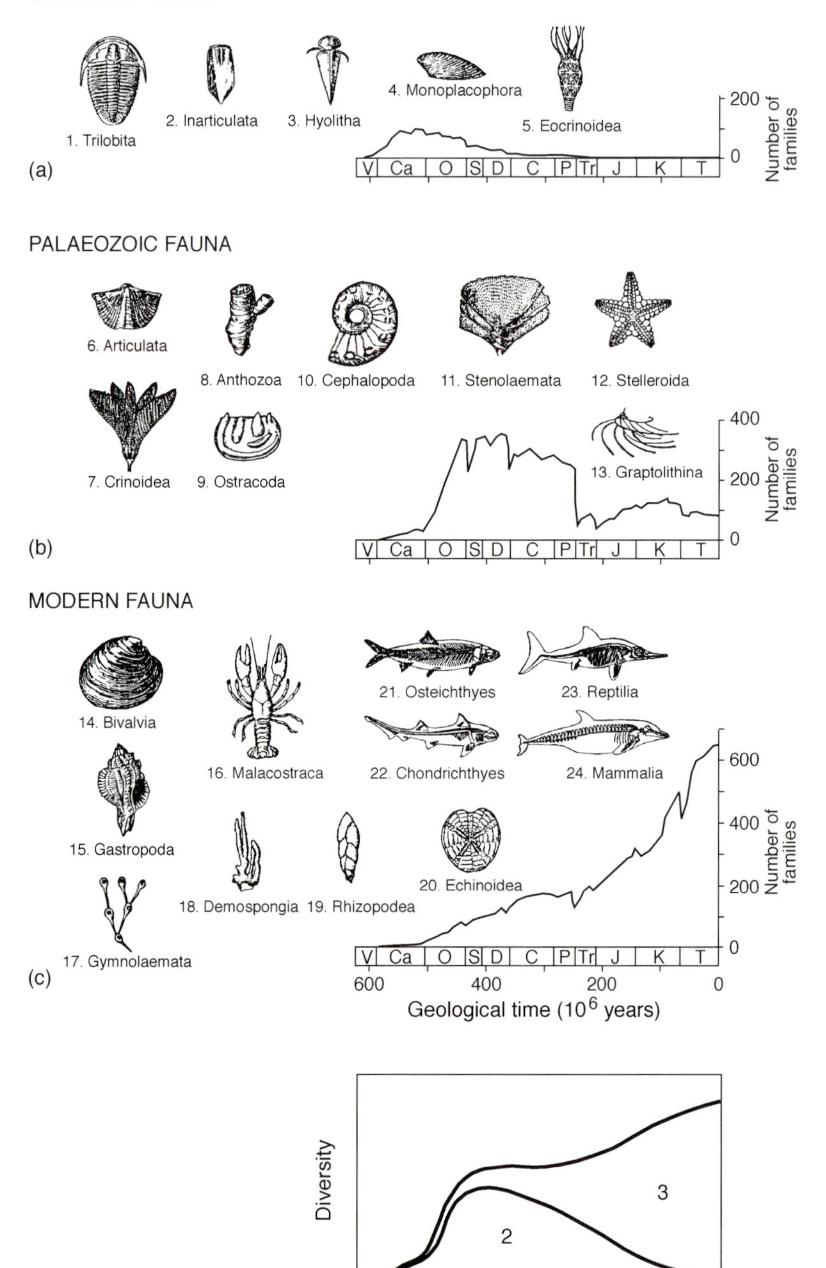
CAMBRIAN FAUNA
1. Trilobita
2. Inarticulata
3. Hyolitha
4. Monoplacophora
5. Eocrinoidea
(a)
200
0
Number of families
V Ca O S D C P Tr J K T
PALAEOZOIC FAUNA
6. Articulata
7. Crinoidea
8. Anthozoa
9. Ostracoda
10. Cephalopoda
11. Stenolaemata
12. Stelleroida
13. Graptolithina
(b)
400
200
0
Number of families
MODERN FAUNA
14. Bivalvia
15. Gastropoda
16. Malacostraca
17. Gymnolaemata
18. Demospongia
19. Rhizopodea
20. Echinoidea
21. Osteichthyes
22. Chondrichthyes
23. Reptilia
24. Mammalia
(c)
600
400
200
0
Number of families
600
400
200
0
Geological time ($10^6$ years)
Diversity
1
2
3
(d)
Time

Others have disagreed with Sepkoski's prime assumption that there is indeed a theoretical equilibrium level at which the rate of speciation equals the rate of extinction, causing the level of diversity to remain roughly constant. Kitchell and Carr (1985), in particular, were able to simulate the actual pattern of changing marine diversity by assuming only that occasional perturbations of the environment caused high levels of extinction, along with occasional evolutionary innovations that led to the expansion of particular groups. Together, these two processes prevent the simulated diversity generated by their model from ever approaching any particular equilibrium value. Hoffman (1989a) was even more sceptical of any such simple, high-level explanation as Sepkoski's, on the grounds that the use of families as the units implies that higher taxa as such have individual evolutionary properties, which, he argues, they do not. He proposed that the probabilities and therefore the rates of speciation and of species extinction varied randomly and independently of each other over time, and that the overall pattern of changing diversity that resulted amounted to the summation of myriads of individual, independent species-level events. He likened the effect to a statistician's double random walk, one of speciation and the other of species extinction.

Whatever view is taken of the underlying cause of the global pattern of species turnover, it is still fairly certain that the marine biosphere changed over time in such a way that it accommodated more and more species. There is no shortage of speculation about the ecological nature of such change. The simplest possibility is that more places have appeared geographically that can accommodate different species. There is indeed a correlation between the level of provinciality—the number of geographically distinct areas or provinces capable of supporting a biological community—and the global diversity (Valentine *et al.* 1978; Signor 1990). Continental drift alone seems to be a major part of any explanation for the increased number of species. In addition, there seems clearly to have been a general increase in the numbers of adaptive kinds of organisms, occupying an increasing volume of ecospace.

Bambach (1983, 1985) argued that the documented increase in species numbers has been a result of the evolution of new ways of making a living since the early Palaeozoic; there has been an increase in the number of what he termed 'guilds' or modes of life (Fig. 8.4). Most of this evolution has concerned the relatively low taxonomic levels of orders, families, and genera, rather than the appearance of radically new major body plans that would be reflected, conventionally, by recognizing new phyla or classes. Ausich and Bottjer (1982, 1985) introduced the more

**Fig. 8.3** (Opposite) Sepkoski's three evolutionary faunas, with characteristic members: (a) Cambrian fauna; (b) Palaeozoic fauna; (c) modern fauna. (d) Model of the three superimposed faunas based on three logistic equations of population growth. (After Sepkoski 1984.)

CAMBRIAN FAUNA

| | Suspension | Herbivore | Carnivore |
|---|---|---|---|
| Pelagic | Trilobita (agnostids) | | |

| Epifauna | Suspension | Deposit | Herbivore | Carnivore |
|---|---|---|---|---|
| Mobile | | Trilobita<br>Ostracoda<br>Monoplacophora | Monoplacophora<br>Ostracoda | |
| Attached low | Inarticulata<br>Articulata | | | |
| Attached erect | Eocrinoidea | | | |
| Reclining | ?Hyolitha | | | |

| Infauna | Suspension | Deposit | Carnivore |
|---|---|---|---|
| Shallow passive | | | |
| Shallow active | Inarticulata | Trilobita<br>'Polychaeta' | 'Polychaeta' |
| Deep passive | | | |
| Deep active | | | |

(a)

PALAEOZOIC FAUNA

| | Suspension | Herbivore | Carnivore |
|---|---|---|---|
| Pelagic | Conodontophorida<br>Graptolithina<br>?Cricoconarida | | Cephalopoda<br>Placodermi<br>Merostomata<br>Chondrichthyes |

| Epifauna | Suspension | Deposit | Herbivore | Carnivore |
|---|---|---|---|---|
| Mobile | Bivalvia | Agnatha<br>Monoplacophora<br>Gastropoda<br>Ostracoda | Echinoidea<br>Gastropoda<br>Ostracoda<br>Malacostraca<br>Monoplacophora | Cephalopoda<br>Malacostraca<br>Stelleroidea<br>Merostomata |
| Attached low | Articulata<br>Edrioasteroida<br>Bivalvia<br>Inarticulata<br>Anthozoa<br>Stenolaemata<br>Sclerospongia | | | |
| Attached erect | Crinoidea<br>Anthozoa<br>Stenolaemata<br>Demospongia<br>Blastoidea<br>Cystoidea<br>Hexactinellida | | | |
| Reclining | Articulata<br>Hyolitha<br>Anthozoa<br>Stelleroidea<br>Cricoconarida | | | |

| Infauna | Suspension | Deposit | Carnivore |
|---|---|---|---|
| Shallow passive | Bivalvia<br>Rostroconchia | | |
| Shallow active | Bivalvia<br>Inarticulata | Trilobita<br>Conodontophorida<br>Bivalvia<br>Polychaeta | Merostomata<br>Polychaeta |
| Deep passive | | | |
| Deep active | | Bivalvia | |

(b)

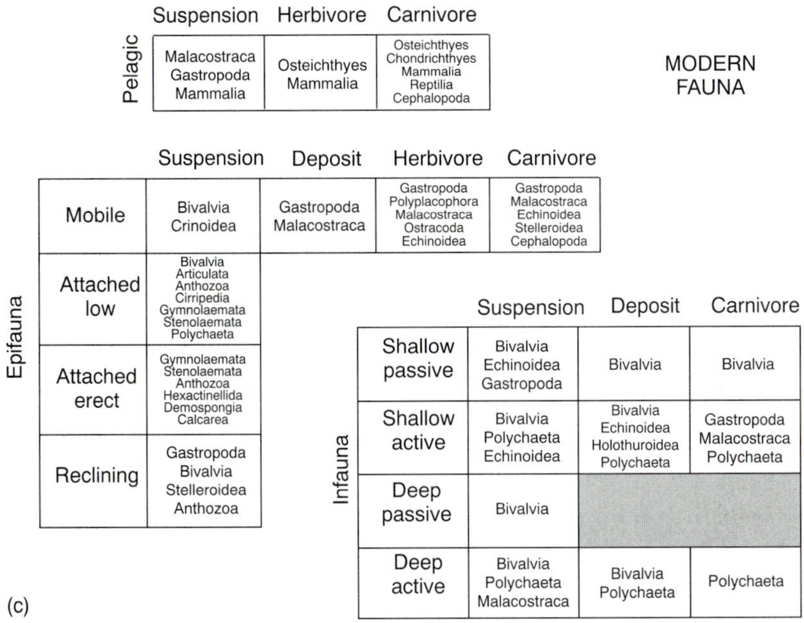

MODERN FAUNA

| Pelagic | Suspension | Herbivore | Carnivore |
|---|---|---|---|
| | Malacostraca Gastropoda Mammalia | Osteichthyes Mammalia | Osteichthyes Chondrichthyes Mammalia Reptilia Cephalopoda |

| Epifauna | Suspension | Deposit | Herbivore | Carnivore |
|---|---|---|---|---|
| Mobile | Bivalvia Crinoidea | Gastropoda Malacostraca | Gastropoda Polyplacophora Malacostraca Ostracoda Echinoidea | Gastropoda Malacostraca Echinoidea Stelleroidea Cephalopoda |
| Attached low | Bivalvia Articulata Anthozoa Cirripedia Gymnolaemata Stenolaemata Polychaeta | | | |
| Attached erect | Gymnolaemata Stenolaemata Anthozoa Hexactinellida Demospongia Calcarea | | | |
| Reclining | Gastropoda Bivalvia Stelleroidea Anthozoa | | | |

| Infauna | Suspension | Deposit | Carnivore |
|---|---|---|---|
| Shallow passive | Bivalvia Echinoidea Gastropoda | Bivalvia | Bivalvia |
| Shallow active | Bivalvia Polychaeta Echinoidea | Bivalvia Echinoidea Holothuroidea Polychaeta | Gastropoda Malacostraca Polychaeta |
| Deep passive | Bivalvia | | |
| Deep active | Bivalvia Polychaeta Malacostraca | Bivalvia Polychaeta | Polychaeta |

(c)

**Fig. 8.4** Increasing number of modes of life (guilds) of organisms through time. (a) In the Cambrian fauna, less than half the potential guilds had evolved; (b) in the Palaeozoic fauna, almost three-quarters were filled; (c) in the modern fauna, all 20 are occupied. (After Bambach 1985.)

detailed idea of increasing diversity by 'tiering', which refers to the vertical extent to which the habitat is utilized. Most of the Palaeozoic species were low-level epifaunal, living and feeding on or close to the surface of the mud. In the Mesozoic, the fauna had become a mixture of epifaunal and infaunal species, the latter living within the mud. During the Cenozoic there has been a great increase in the extent of infaunal organisms, extending to greater depths within the substratum.

Looking above the substratum of the sea bed, the evolving relationship between the planktonic and the benthic faunas illustrates tiering rather well (Signor and Vermeij 1994). During the Cambrian, the benthos contained few suspension feeders, and those that were present tended to be either passive or feeders on very small, bacterial-sized prey, because the plankton was poor in both adult and larval metazoans. During the Late Cambrian and the Ordovician, the benthic fauna started to include many active suspension feeders capable of creating their own water currents, and many species became elevated well above the substratum. Simultaneously, metazoans such as cephalapods and graptoloids, and also no doubt metazoan planktotrophic larvae, became increasingly abundant in the plankton. What seems to have been evolving was an increasingly complex interrelationship between the substrate-based community of the benthos and the free-living planktonic community, in the course of which many new niches evolved, and therefore many more species could be accommodated.

Yet another detailed proposal for a mechanism behind increasing diversity has been described at length by Vermeij (1987) as 'escalation'. He envisaged an arms-race situation, in which new adaptations for predation in carnivores correlate with new protective and escape devices in the prey species. There is ample documentation that something along these lines has happened at the descriptive level: there are indeed many more ways of killing or evading being killed now than in the Palaeozoic. The concept can be criticized as excessively over-simple, however, for it assumes that species react in a very one-dimensional manner towards one another in an evolutionary theatre of a fixed environment. Kitchell (1990) points out that, to start with, organisms tend to evolve trade-offs between conflicting requirements. In such a case as prey evolution in the face of a predator, escalation points to the evolution of, say, a thicker shell. But this may well increase the time taken to reach sexual maturity, and the fraction of the resources devoted to reproduction. An equally good response could be the evolution of a thinner shell, which allows more rapid growth and therefore earlier and more fecund reproduction. A further complication is that any evolutionary change in one species affects the nature of the selective forces acting on the second; but an evolutionary response by the second in turn alters the environment of the first: organisms do not simply solve problems in fixed environments. All in all, the evolutionary track followed by a lineage will depend on a complex of subtly integrated factors, including the nature of the evolving organism itself. Of course, even if simple escalation is rejected as a useful general concept, the idea of a more complex, multidimensional form of niche evolution can certainly be part of the explanation of increasing diversity.

Up to this point, only the marine biota has been considered. The invasion of land began in the Silurian, and studies have been made to see if the pattern of changing diversity of terrestrial plants and vertebrates was similar to that of the marine fauna. The general conclusion is that both land plants and tetrapod vertebrates do indeed show a comparable pattern of increase in diversity levels through time, punctuated by periods of mass extinction. Niklas *et al.* (1985) reviewed the diversity of land plants since the first appearance of the group some 420 million years ago in the Late Silurian, and they recognized four successive floras (Fig. 8.5a) comparable to Sepkoski's marine evolutionary faunas (Fig. 8.3). The earliest flora consisted of simple vascular plants such as the Rhyniophyta. It was dominant in the Late Silurian and Early Devonian, by which time the elements of the second flora had appeared. This was characterized by pteridophytes ('ferns'), including heterosporic and tree-like forms. There was a fourfold

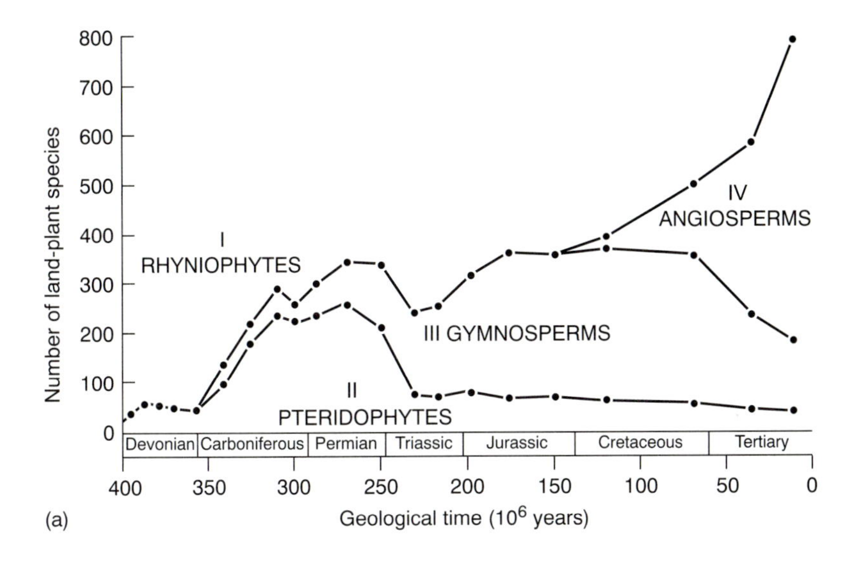


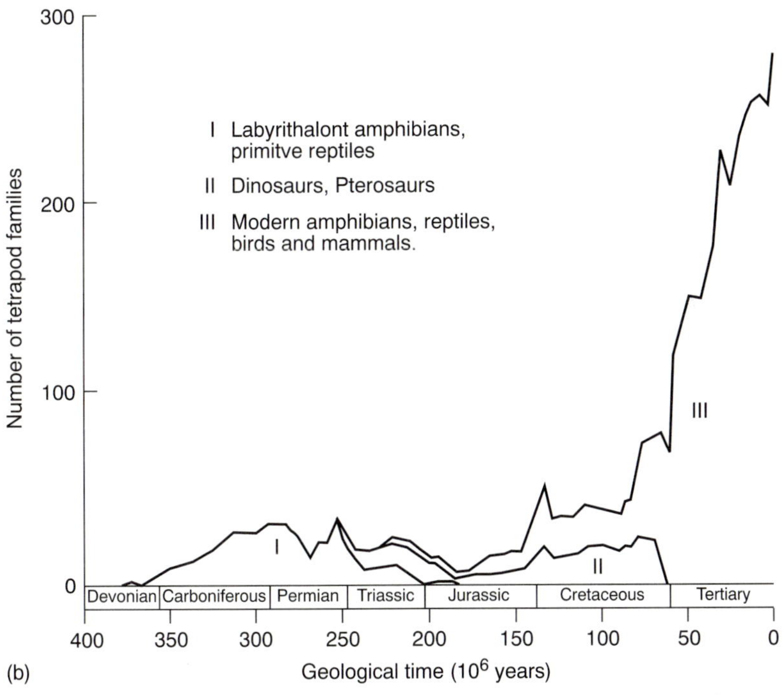


**Fig. 8.5** Pattern of changing diversity of (a) species of land plants, interpreted as four apparent evolutionary floras, and (b) tetrapod vertebrate families, interpreted as three apparent evolutionary faunas. ((a) After Niklas 1986; (b) after Benton 1989.)

increase in species numbers and this flora remained dominant until the end of the Carboniferous. The third flora consisted of seed plants, arising in the Devonian but not culminating until the gymnosperm radiation of the Mesozoic. Finally, the fourth, or Tertiary Flora, underwent yet another severalfold increase in species numbers as the angiosperm radiation proceeded to dominate from the Late Cretaceous onwards.

Niklas *et al.* (1985; Niklas 1986) attributed the increasing overall number of plant species seen in the fossil record both to the extension of the resources of the terrestrial habitat available, and to the subdivision of existing niches. As far as the succession of one flora by the next is concerned, they carefully state that: 'This is consistent with the speculation that a macroevolutionary analog to competitive displacement is operative in the succession of evolutionary floras.' It is left unclear just what kind of mechanism might reasonably be proposed to underlie the analogue unless it be the dubious concept of clade competition (page 43).

Benton (1985, 1989) has plotted the diversity of terrestrial tetrapod families against time and has also shown a continual increase punctuated from time to time by mass-extinction events. He proposed that there are three successive major radiations, presumably to be thought of as equivalent to evolutionary faunas (Fig. 8.5b). The first consists of labyrinthodont amphibians, primitive reptile groups, and mammal-like reptiles; the second largely of dinosaurs and pterosaurs; and the third of the 'modern' groups of frogs, salamanders, lizards, snakes, turtles, crocodiles, birds, and mammals. But each of these three respective faunas, particularly the last one, contains such a disparate collection of kinds of tetrapods that they cannot realistically be regarded as in any sense unified faunas. All there is by way of comparison with the marine invertebrate and the land-plant patterns is the increasing number of taxa, which by inference indicates an increasing number of species of land vertebrates. On seeking explanations for this, Benton (1990) finds some evidence of increasing endemicity in the greater climatic gradients and the breakup of Pangaea since the Permian. But he regards adaptive expansion, as witnessed by an increasing range of habitats occupied and food resources utilized by vertebrates, as more significant. Interestingly, he also detects a trend of increasing numbers of terrestrial vertebrate species preserved in *Lagerstätten* (page 72), from a mean of about 20 in the Carboniferous to around 50 in the late Tertiary. If these are reasonably comparable samples, they imply an increasing subdivision of niches within habitats. Unlike Niklas *et al.*'s suggestions regarding plants, Benton (1996) looked for but failed to find convincing evidence of any kind of interfaunal or interclade competitive process involved with faunal replacements.

# Diversity changes within taxa

One possible explanation for the changing diversity through time of the total Earth's biota is that it is the simple summation of the diversity changes within the individual clades that make it up. It is appropriate, therefore, to consider the latter, to see if there are general rules underlying patterns of change at this lower level. In principle, any clade begins as a single species and therefore its diversity can only increase at first, although in practice it may have diversified at a rate too high for the initial expansion stage to be detectable in the fossil record. In this case, even at its first appearance, it will exist at high diversity. Equally obviously, the diversity must eventually decline to zero at extinction, although taxa with modern representatives have not yet reached that point. Within these two self-evident limits, every pattern of change of diversity imaginable can be found, as is shown by the spindle diagrams of diversity against time from origin to extinction of the clade (for example, Sepkoski and Hulver 1985) (Fig. 8.6). The commonest pattern is an increase to the maximum diversity, followed by a decline to final extinction. Yet, even here, the maximum may occur early, late, or at the midpoint, and clades may also show multiple maxima, particularly higher-level clades that have survived periods of calamitous decline at mass-extinction events. Yet another pattern is virtually parallel-sided indicating a very long period at constant diversity.

One general rule about clade shape has been proposed. Gould *et al.* (1987) defined the 'centre of gravity' of a clade, and found that clades appearing early in the history of a higher group had on average a significantly lower (that is, earlier) centre of gravity compared with the equivalent clades that first appeared later, whose centre of gravity on average lay at the midpoint of the clades' history. The difference was small. In the case of 703 families of marine invertebrates, the mean centre of gravity of the genera appearing in the Cambrian and Ordovician occurred at 0.482 units, compared with 0.499 units for the later-appearing genera. Gould *et al.* (1987) found a similar result for clades of Tertiary mammals, but, in this case at least, Uhen (1996) found just the opposite. His study describes the earlier-appearing clades as being slightly 'top-heavy', and the later ones having a mean around 0.5.

Apart from this claim and counterclaim, no general rules relating certain kinds of organisms to certain shapes of spindle diagrams have been discerned. Each clade appears to have its own unique shape, which suggests that the underlying cause consists of a unique combination of the various factors that must affect speciation and extinction rates. There is not even any apparent correlation between the speciation and the extinction rates themselves, because few clades have a constant diversity for significantly longer than the average longevity of the contained species. Stanley (1990)

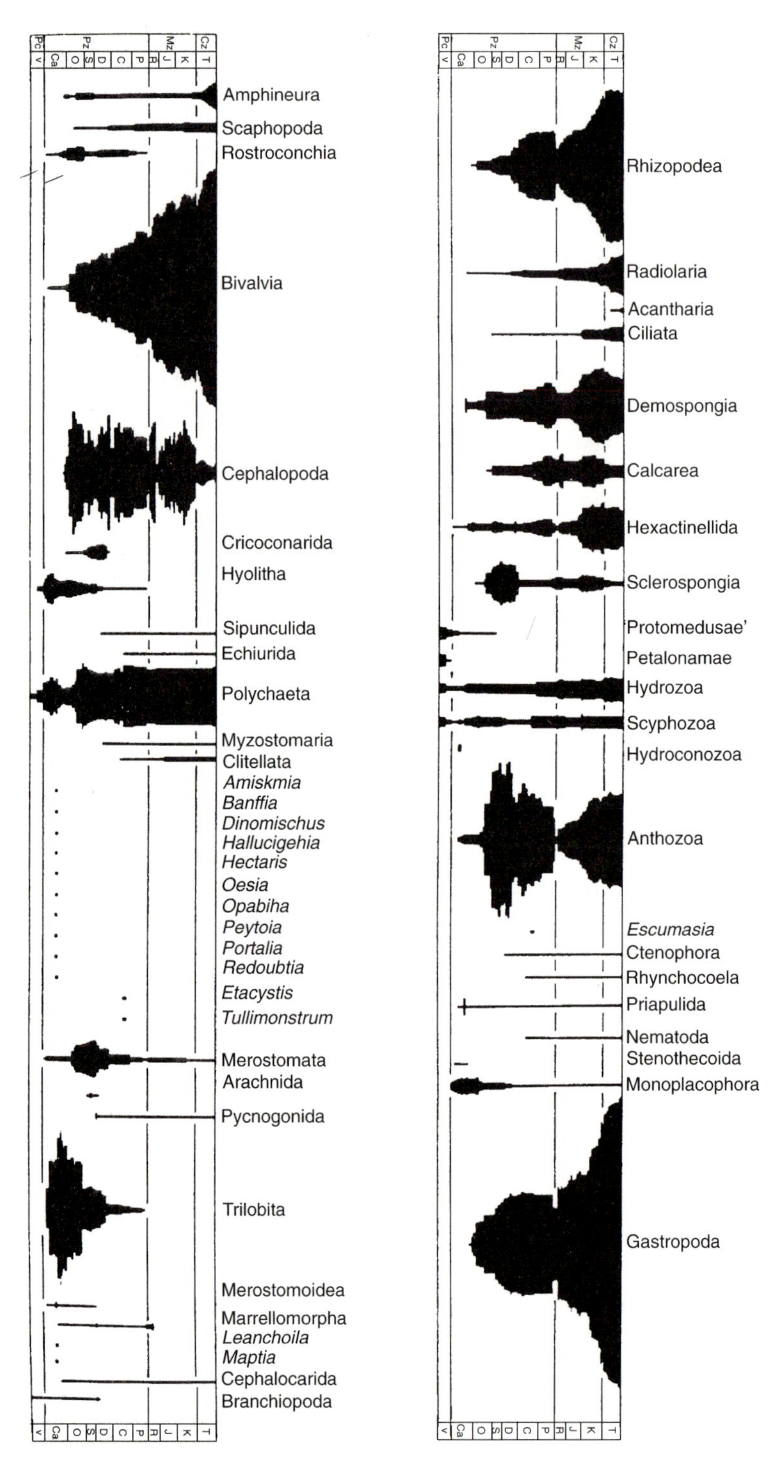

**Fig. 8.6** A variety of spindle diagrams depicting numbers of families within orders through time. Several patterns are evident: bottom-heavy (trilobites, etc.), top-heavy (gastropods, etc.), parallel-sided (polychaetes, etc.), and multiphased (anthozoa, etc.). (After Sepkowski and Hulver 1985.)

proposed that there is a weak correlation between these two parameters, but not for the simple ecological reason that extinction of one species creates space for another species to occupy. He argued that there are several ways in which probability of speciation is affected independently of, but in the reciprocal direction to the probability of extinction. Species with a narrow, specialized niche may, he suggested, be more likely to speciate because of the greater chance of isolation of populations. They may, however, also be more likely to go extinct because of loss of the niche. Similarly, species with geographically restricted populations, or which consist of more complex organisms, may have higher speciation and, independently, higher extinction rates. If these positive but weak correlations are true, they would account on the one hand for the high degree of variation in the changes shown by the spindle diagrams, but on the other for the fact that clades can nevertheless be very long-lived despite the species turnover within them.

The details of any particular case may well be superimposed on certain general rules about clade history, and a number of potentially important suggestions of causes of the patterns have been made.

## Clade diversity and species turnover as stochastic

Several computer models have been constructed based on predetermined probabilities of a species persisting, speciating, or going extinct at each successive interval of time (see, for example, Raup *et al.* 1973; Raup 1977; and Fig. 3.6). These can generate remarkably realistic-looking changes in diversity of a hypothetical clade and, taken at face value, suggest that the null model of clade dynamics is a stochastic pattern. Unsurprisingly, palaeontologists have not been happy to accept the idea that their perceived patterns have no discoverable causes, and at least two reasons for not simply accepting that diversity changes within real taxa are random have been proposed. One is that the probabilities that have to be used in the models lie within definable limits, and the limits themselves would presumably have causes in the real world. Second, Stanley *et al.* (1981) pointed out that the models have to incorporate unrealistic assumptions about such things as the starting diversity of the clades. Empirical arguments explicitly set against the idea that randomness is the explanation for diversity changes are based on the discovery of evidently non-random patterns. For example, Flessa and Levinton (1975) showed that more than the expected random number of real clades appear simultaneously in the fossil record. Gilinsky and Bambach (1986) performed a statistical bootstrap analysis on a series of real clades, and showed that these clades have a more rapid early diversification period than expected by chance alone.

In an appropriate warning on this question of randomness, Kitchell and Carr (1985) pointed out that, as well as a simulated stochastic process being able to appear to be deterministic, the reverse is also true. It is quite possible for a fully deterministic simulation, in which precise relationships are assumed to exist between events such as extinction, speciation, and perturbations of the environment, to generate patterns that appear to be random. Given the relatively small sample of cases studied in the real fossil record, great care is needed to avoid over-interpretation when inferring particular processes from either apparently non-random, or apparently random patterns of taxonomic turnover.

## Constant probability of extinction

The Red Queen hypothesis of Van Valen (1973) has become an icon of modern palaeobiology because it both addresses a simple, explicit question and is potentially testable from the fossil record. It has also become an example of how nothing is ever as simple as it seems at first sight in palaeobiology. Van Valen (1973, 1985) studied the survivorship of a whole variety of different taxa, including various planktonic foraminiferans, brachiopods, molluscs, actinopterygian fishes, and mammals. The taxonomic level studied varied from species to orders in different cases. When he plotted semi-log survivorship curves showing the proportion of the included taxa that survived to each age against that age, he obtained a generally linear curve (Fig. 8.7a). What this means is that the probability of a contained taxon such as a species becoming extinct does not change with the time since its origin: newly formed species are neither more nor less likely to go extinct than old, long-established species. (The clearest analogy is with radioactive decay of atoms: the probability of any individual atom decaying is entirely independent of how long that atom has been in the sample.)

Now this seemed surprising in the light of the general belief that evolution by natural selection is a continually 'improving' process: older species (or genera, etc.) should be better adapted than younger ones, and therefore less liable to suffer extinction. But Van Valen produced an ingenious explanation for the paradox. If the environment is continually deteriorating as far as the organisms are concerned, they will continually suffer from an adaptive lag, and will never achieve full adaptation. In the words of the Red Queen in *Alice through the looking glass*, they will forever have to keep on running simply in order to stay in the same place. In his view, the only aspect of an organism's environment that will show this phenomenon of continual, reasonably constant change will be the biotic one. The other organisms with which it interacts—predators, prey, parasites, and competitors—will all be evolving adaptations to cope better. The focal species will therefore have to continue do the same, throughout its existence. Although physical aspects of a species' environment change, this is unlikely to be anything like continual. More likely, it will be episodic, with periods of no

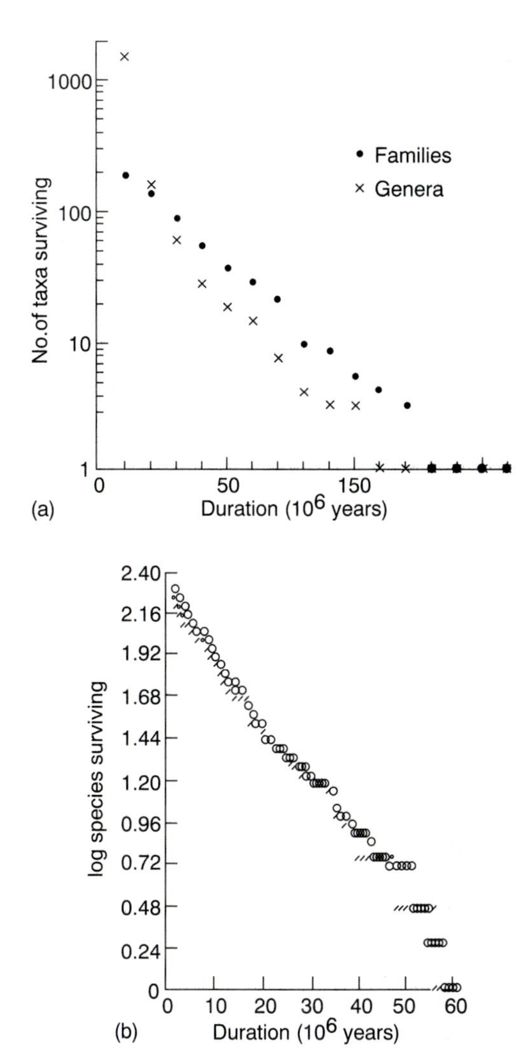


**Fig. 8.7** The Red Queen hypothesis. Semi-log plots of (a) diversity of families and genera, respectively, of ammonoids over time; and (b) species diversity of coccoliths over time. ((a) After Van Valen 1973; (b) after Hoffman and Kitchell 1984.)

change alternating with irregular periods of change of greatly varying intensity. The question that the Red Queen hypothesis therefore answers is: which is the more important driver of adaptive evolution, the abiotic or the biotic environment? To Van Valen the answer was unambiguously the latter.

Much interest was generated by the Red Queen hypothesis, and a major misunderstanding soon arose (McCune 1982). The linear survivorship curves of Van Valen say no more than that the probability of a taxon within a group going extinct is independent of the age of that taxon. It does not imply that the probability of extinction of the taxon is constant throughout its life, or that the extinction rate within the group as a whole is constant through time.

Some observations of microfossils from Deep Sea Drilling Project cores (page 89) support the Red Queen hypothesis quite well (Hoffman and Kitchell 1984) (Fig. 8.7b). Others, however, obtained results more consistent with what came to be termed the Stationery Model (Stenseth and Maynard Smith 1984), in which extinction does not occur unless there is change in the physical environment (Pearson 1992). In this case, taxa of different ages may be found to have different probabilities of extinction, depending on the historical accident of their occurrences at times of environmental deterioration (Wei and Kennet 1983).

Even given the constant probability of extinction, there are other possible explanations than the Red Queen Hypothesis. McCune (1982) pointed out that according to the punctuated-equilibrium explanation of speciation (page 137), once a species has arisen and evolved to a state of genetic homeostasis it is no longer capable of responding to selective forces, and is therefore incapable of evolving increasing resistance to extinction. All species once fully evolved will retain the same probability of extinction as one another, irrespective of their individual ages.

## Biogeographic effects on clade histories

Jablonski *et al.* (1983; Jablonski and Bottjer 1990*a,b*) noted that amongst post-Palaeozoic marine groups, the higher taxa such as orders most frequently make their first appearance as fossils in onshore environments. In contrast, there is no such bias apparent in the origin of the lower-level taxa such as the families and genera. These are as likely to appear in offshore as onshore settings, and in fact their origin seems to correlate only with the existing levels of diversity: a new genus is likely to make its first appearance in an area where many genera already occur. Other palaeobiologists have supported this general rule; for example, Skelton *et al.* (1990) for bivalve molluscs. Fortey and Owens (1990) claimed that 8 out of 10 major trilobite clades appeared in onshore habitats and subsequently spread to the offshore areas. On the other hand, Smith (1990) noted that early Palaeozoic echinoderms may be an exception to the rule, on the grounds that these forms were unattached, passive suspension feeders that could not have coped with the higher currents associated with onshore habitats.

Several explanations have been proposed for the rule, none of which is overwhelmingly convincing. Skelton *et al.* (1990), for example suggest a rather vague ecological reason, to the effect that onshore species face a more stressful, disturbed environment and are correspondingly more physiologically tolerant. Therefore they tend to have lower extinction rates. The consequence of this is that an evolutionary novelty that could be the basis of a new higher taxon is more likely to persist for long enough to allow diversification of species which possess it to reach the stage where a higher taxon such as an order is recognized. This presumably contrasts with the lower-level taxa, families and genera, where a much shorter existence of the respective subclade is necessary for it to be recognized as such. In the latter case, the probability of a suitable novelty arising can be a simple function of the number of pre-existing lower-level taxa.

Skelton *et al.* (1990) also hinted at a rule that is comparable to the onshore–offshore rule, but related to latitude. In the case of the inoceramid bivalve molluscs of the Cretaceous, they note that the diversity of the group is greatest in higher latitudes during the earlier stages of its evolution. Subsequently the group spreads equatorwards, with the diversity increasing in the mid- to lower latitudes. They propose that, as in the case of onshore habitats, the higher latitudes are associated with greater physiological stress.

## Species longevities and turnover rates

A spindle diagram shows the change in diversity of a clade through time, indicating the difference between the rate of speciation and of extinction of the contained lower taxa such as species for each increment of time. As such, it represents only the absolute diversity of the whole clade at each instant in time. It does not reflect directly the actual rates of speciation and extinction, and therefore a given spindle diagram may represent a group in which the two rates can have any value between very high and very low. Simpson (1944) recognized that these rates are highly variable in different groups of organisms, an observation for which he coined the terms bradytelic for a group with low rates and tachytelic for a group with high rates of taxonomic evolution or turnover.

Stanley (1979) published a table of the different characteristic species longevities of different groups, which approximate to the reciprocal of species turnover rates when diversity levels are roughly constant (Fig. 8.8). At one extreme reef corals have average species durations of 20 million years and foraminiferans even more, whereas at the other extreme mammals, trilobites, and ammonites have species longevities of no more than around 1–2 million years. Other invertebrate groups and fishes lie between these values. There is an evident correlation between the general level of complexity of the organisms in a taxon and rates of species turnover, with the more-complex kinds having the higher rates. Schopf *et al.* (1975) believed that this was no more than an artefact arising from the greater number of taxonomic characters, and therefore the greater taxonomic discrimination that is possible in the more complex organisms. Stanley (1979)

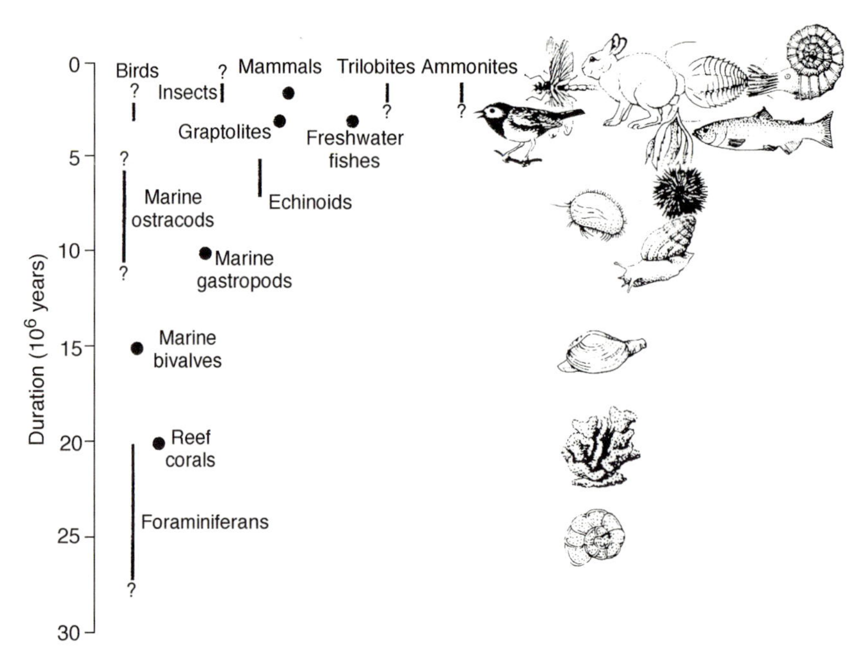


**Fig. 8.8** League table of average species longevities. (After Stanley 1979.)

countered this argument by noting that if a set of modern organisms is classified using a similar ranges of characters to those available for fossils, a largely correct species classification still results.

Stanley's (1979) own explanation for the relationship is that higher rates of speciation and extinction in certain groups may be due to one of two main reasons. One is dispersal pattern; the greater the dispersal ability, the less likely is speciation to occur because of the lower likelihood of an isolated population remaining isolated for long enough. The second possible reason is that behavioural complexity affects both the probability of speciation and of extinction. The more complex the mating behaviour is, the greater is the probability of the evolution of new behavioural isolating mechanisms and therefore of incipient speciation. Furthermore, the more complex the general behaviour, the narrower is the niche of the species, and the greater is the potential competition between species. This could cause a greater probability of extinction. Interesting as this pair of suggestions is—the one based on a species-level property of pattern of dispersal and the other on the organism-level property of behavioural responses—they are both highly speculative.

# Clade interactions

So far the discussion has centred on what happens during the history of a single clade. There are numerous clear cases in the fossil record, however, where the increase in diversity of one taxon coincides in time with the decline of a second one that consists of broadly comparable kinds of organisms. This often occurs to the point of complete extinction of the latter, a pattern that is sometimes referred to as ecological replacement.

Until relatively recently, all such cases were almost invariably explained by competition, a simple use of extrapolation of a process understood in ecological time to a pattern perceived on a geological time-scale. There are two severe difficulties to this interpretation. The first is that it implies that whole taxa can behave in some way as interacting units, but it is not at all clear whether taxa in this context have appropriate properties to be regarded as such integrated units. There would need to be some characteristic of the taxon as a whole, rather than of the individual organisms or species that constitute it, and which was the demonstrable cause of the extinction or survival of the taxon (page 43). There are no candidates for such kinds of characteristics. Failing this condition, then it has to be assumed that each individual species within the taxon shares some characteristic with all the other species such that, in every case of species-to-species competition, the species belonging to the first taxon outcompetes the corresponding species belonging to the second, irrespective of the presumed range of niches involved. Even if a characteristic like this did exist, there is no way to test whether any particular fossilized character really was the cause of the differential effect leading to the observed replacement.

The second difficulty concerning extrapolation from ecological competition concerns the time-course of the replacement. If it is sufficiently long as to be apparent in the fossil record as a coincidental decline in one taxon and increase in diversity in the other, then it is a process that must have taken of the order of millions of years from commencement to completion. Therefore the difference in average competitive ability between the species of one taxon compared with the corresponding species of the other must have been so slight that the survival or extinction of species by chance would be expected to have had a far more significant role than would deterministic competition in the overall replacement.

Gould and Calloway (1980) offered an early criticism of simple extrapolation from ecological processes to explain a major diversity change (Fig. 8.9a). During the Palaeozoic, brachiopods were the dominant two-valved shelly invertebrates, whilst bivalve molluscs occurred at much lower diversity. After the Permian, as part of the great shake up caused by the end-Permian mass extinction (page 209), bivalves had become dominant while

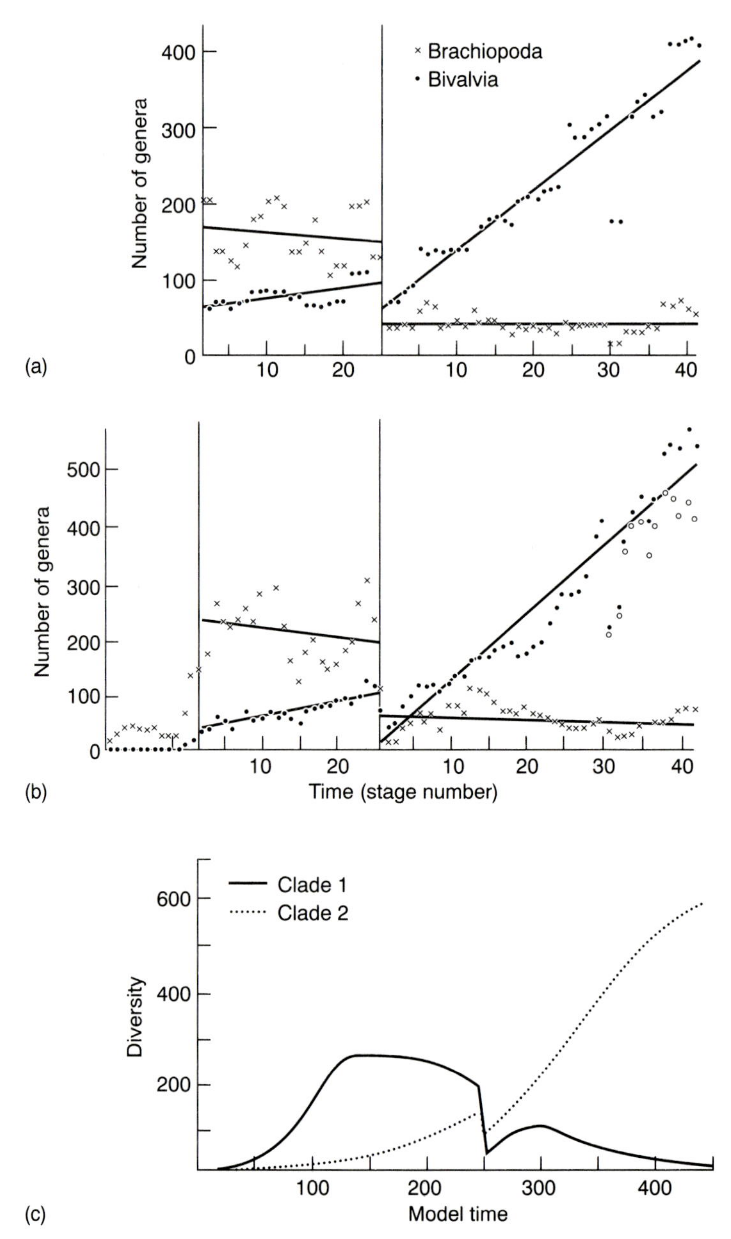


**Fig. 8.9** Diversity changes of brachiopods and bivalve molluscs across the Permian–Triassic boundary. (a) Gould and Calloway's origin version. (b) Sepkoski's modified version incorporating additional data and an extension backwards in time. (c) Sepkoski's interpretation modelled on two logistic curves compatible with competition as the cause of the change. ((a) After Gould and Calloway 1980; (b), (c) after Sepkoski 1996*a*.)

brachiopods never recovered their former prominence. The usual explanation for this had always been that bivalves had evolved competitive superiority over brachiopods, for example because they evolved a fused mantle and siphon that enhanced burrowing ability and therefore predator avoidance (Stanley 1977). Gould and Calloway, however, claimed that this kind of explanation is not at all in keeping with the actual pattern seen in the fossil record. Both groups show a massive decline in diversity, no doubt in response to some catastrophic environmental perturbation, and during which brachiopods suffered much more. After the event, bivalves began their radiation from a significantly greater starting diversity than brachiopods. Thus a much more feasible explanation of the change in fortunes between the two groups is that by chance brachiopods suffered a greater decline in the face of the imposed environmental deterioration, and that bivalves simply had a head-start during the post-Palaeozoic recovery.

Benton (1987) carefully reviewed the use of the ecological concept of competition as an explanation for clade replacement over the geological time-scale, and recognized the inappropriateness of simple extrapolation from one time-scale to the other. He proposed a test to distinguish between two possible causes of clade replacement that involved looking at the precise time-course of the pattern found in the fossil record (Fig. 8.10a). Competitive replacement should be associated with a double-wedge pattern (type 1), in which the earlier taxon declines gradually and coincidentally with the gradual increase in diversity of the later taxon. Alternatively, what Benton terms opportunistic replacement should be revealed by the rapid, more or less instantaneous decline of the earlier taxon, to be followed by a rapid increase in diversity of the later one (type 5). In this case, the cause of the replacement is the extinction of one group due to an environmental change, followed by the expansion of the second taxon into the vacated ecospace. This is not because it is competitively superior, but because it consists of species that were better able to survive the environmental perturbation or the new, post-perturbation conditions.

An additional prediction arising from the concept of opportunistic replacement is that there will be indications of the presumed environmental perturbation, such as a change in the flora. Benton was unable to find a single convincing case of a double-wedge pattern implying competitive replacement, with the possible exception of the sequence of major radiations of land plants. Subsequently, he reviewed a wide range of vertebrate clade replacements often attributed to competition, and showed that at least the large majority conform much more closely to an opportunistic pattern (Benton 1996). For example, the various explanations offered for the replacement of the mammal-like reptiles by the dinosaurs during the later part of the Triassic—such as supposedly more advanced temperature regulation or locomotory ability of the dinosaurs—fail on the grounds that the mammal-like reptiles actually declined prior to, rather than coincident

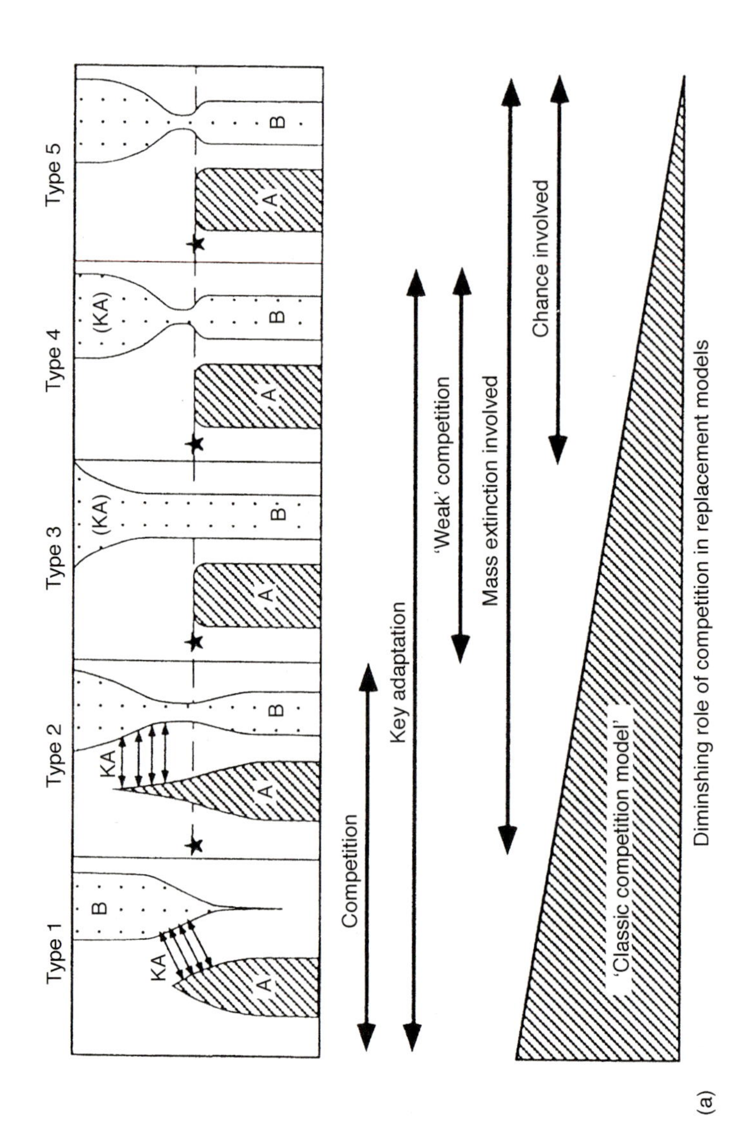
Type 1
Type 2
Type 3
Type 4
Type 5
KA
A
B
(KA)
Competition
Key adaptation
'Weak' competition
Mass extinction involved
Chance involved
'Classic competition model'
Diminishing role of competition in replacement models

(a)

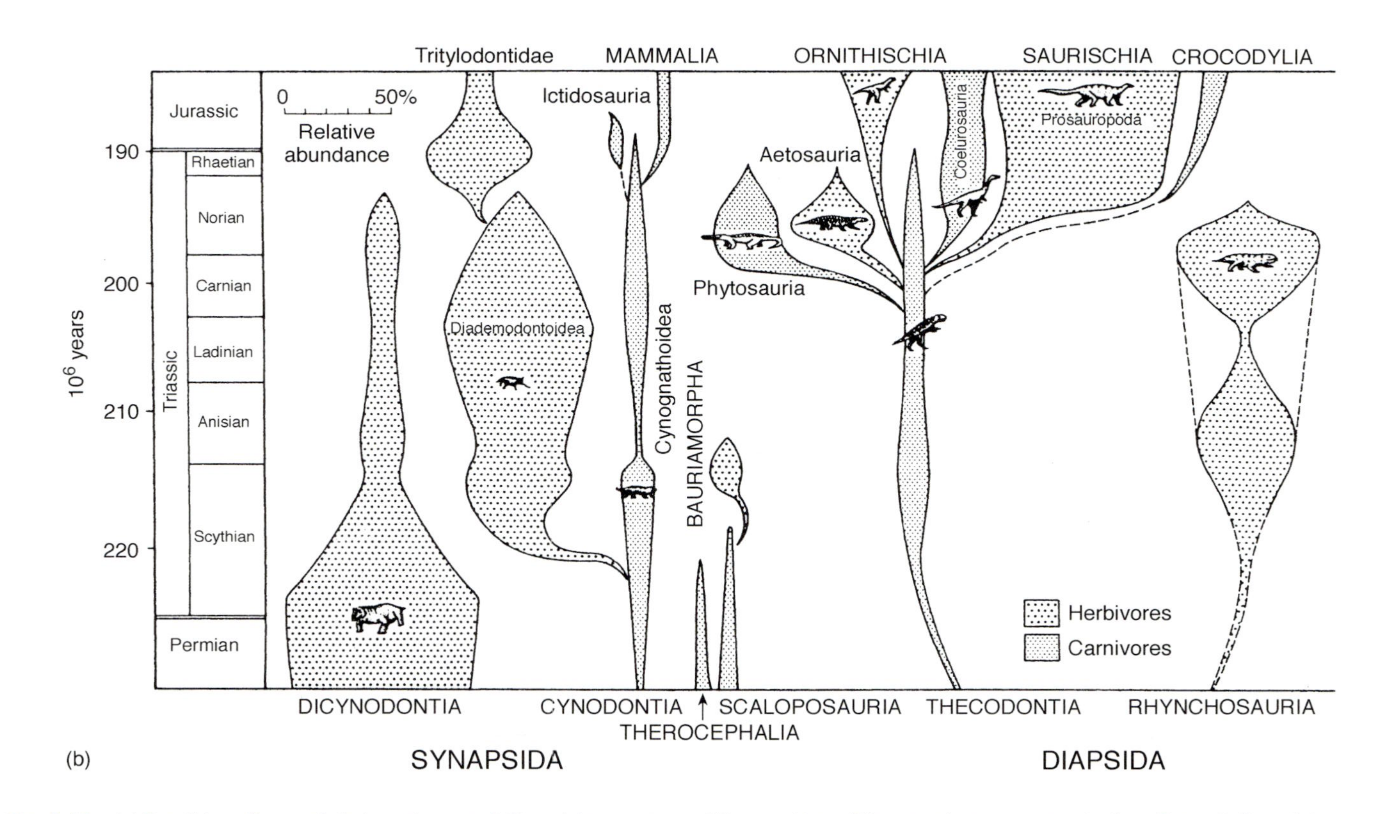


**Fig. 8.10** (a) Possible patterns of clade replacement. Type 1 is pure competition, and type 5 is pure chance or opportunism. Types 2, 3, and 4 represent degrees of combination of the two. KA = key adaptation. (b) Pattern of replacement of mammal-like reptiles (Synapsida) by dinosaurs (Ornithischia and Saurischia). (After Benton 1983, 1996.)

with, the rise in diversity of the dinosaurs (Fig. 8.10b). Hallam (1990) found the same absence of evidence for competition in the extensive Mesozoic taxonomic turnovers of ammonites and other molluscs.

It is clear that Benton's potentially testable alternative modes of explanation for clade replacement are not really independent of one another at all, because a change in the abiotic environment could presumably act by altering the competitive difference between species belonging to two co-existing taxa; equally a change in the biota due to a competitive replacement of a few species is itself a change in the environment that could conceivably result in a phase of opportunistic replacement. With these reservations in mind, more recent investigations of particular cases include taking into account additional features such as functional interpretations of the putatively competing fossil organisms, biogeographic distributions of the separate taxa, and the associations with other contemporary taxa.

Maas *et al.* (1988) studied the decline of the primitive plesiadapid primates during the Palaeocene (Fig. 8.11). Two of the families of plesiadapids, together called the 'non-paromomyid' groups, suffered the greatest decline, and their pattern of species reduction shows an inverse relationship with the pattern of species increase of rodents that is consistent with competitive replacement. But, furthermore, when the numbers of individual specimens in these plesiadapid and rodent groups, respectively, are counted, there is also an inverse relationship, as predicted by competition. The authors go on to investigate various aspects of the biology of the non-paromomyids. Estimations of body size were made from molar tooth size, diet from form and wear patterns of the teeth, nocturnality from the relative size of the orbits (Fig. 8.11c), and arboreality from the structure of the postcranial skeleton. These animals were closely comparable to contemporary rodents in all these biological features, indicating the likelihood of serious overlap of resources. Finally, they point to evidence for an Asian origin of rodents, with immigration into North America immediately before the non-paromomyids started declining. Taken together, this represents a well-corroborated hypothesis that a true competitive process underlay the replacement. Furthermore, the other generally comparable mammals at that time were the adapid and omomyid plesiadapiforms, and they do not show the same detailed coincidence in pattern of diversity change, or degree of morphological overlap with either non-paromomyids or rodents.

Other equally careful studies, however, such as that by Lidgard *et al.* (1993) on the replacement of cyclostome bryozoans by cheilostome bryozoans since the Palaeozoic, find that competitive replacement probably had only a minor role in the overall process.

The pattern of clade replacement in the fossil record certainly persists in looking like the sort of pattern that some form of competitive process would produce, and it is hardly surprising, therefore, that many palaeobiologists continue to look for evidence of dynamic clade interaction rather

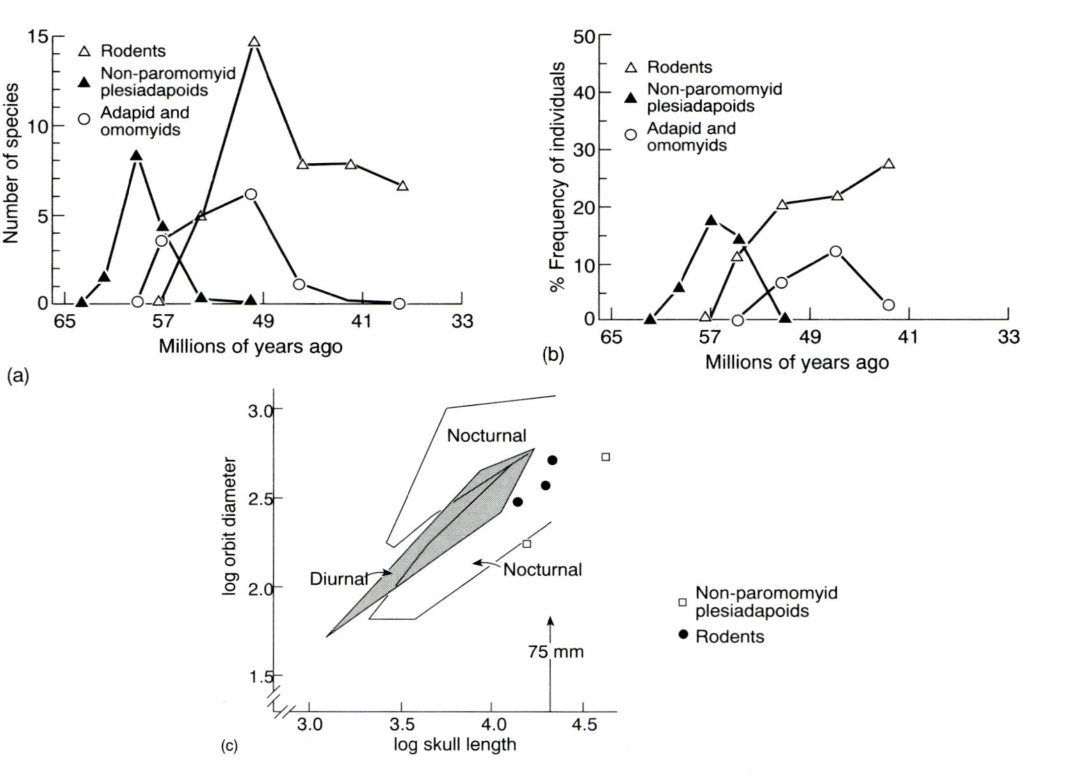


**Fig. 8.11** Pattern of diversity changes in plesiadapid primates and rodents in the Palaeogene. (a) The increase in number of rodent species coincides well with the decrease in number of non-paromomyid species, but not with the decrease in number of adapid and omomyid species. (b) The same is true of the respective numbers of individual organisms in each group. (c) The non-paromomyid and the rodent species show morphological overlap in size of the orbit, suggesting both types were small, small-eyed nocturnal animals. (After Maas *et al.* 1988.)

than passive replacement. The main stumbling block remains the difficulty of explaining the long time-course that clade replacements characteristically follow. Where the time-course is shorter, the replacement is instantaneous in the fossil record and therefore cannot be studied. Where the time-course is long enough to be preserved, then chance should predominate over the implied minuteness of the average competitive advantage of the species in one clade over those of another clade. Addressing this problem, Rozenzweig and McCord (1991) proposed a mechanism they called 'incumbent replacement'. They envisaged an incumbent clade whose species occupy a range of niches, and because of their successful existence these species are immune from competitive replacement. A 'key innovation' evolves in the ancestor of what is to become a replacement clade of comparable species. The effect of the key innovation is to increase the rate of speciation of that clade, and potentially to give these species a competitive advantage over the incumbent species. But only if an incumbent species becomes extinct from some other cause, can a comparable species of the replacement clade take over that niche, either because it is now empty or by successful competition on level terms with any remaining species of the incumbent clade. Therefore the overall rate of species replacement relies on the rate of extinction of incumbent species, which could be very low. By this mechanism, the long time-course of the replacement is explained, as also is the eventual complete replacement of all the incumbent species, and the increased rate of replacement associated with the very high extinction rates during mass-extinction episodes.

Sepkoski (1996*b*) reviewed possible ways, including this one, by which a competitive process can still explain the slow time-course. He remains convinced that relatively simple logistic growth curves underlie long-term clade replacements, indicating active interaction between them. If the two respective clades have different values for intrinsic rate of speciation, for maximum possible (equilibrium) diversity, and for starting diversity, then predictions about the pattern of changing diversities over time can be made, and shown to correspond to the real cases. For example, he reconsidered Gould and Calloway's case of the brachiopods and bivalves in the light of additional new data (Fig. 8.9b), and matched the diversity changes in the two respective groups very well to a pair of coupled logistic equations with suitable parameter values (Fig. 8.9c). By altering the parameters at different times, it was possible to model the time course across the mass-extinction and the post mass-extinction radiations quite realistically.

The question of whether clade replacements really do involve a true competitive process or merely give the appearance of one has not yet been resolved. The phenomenon of multiple, simultaneous clade replacements known as mass extinction may throw further light on it.

# Further reading

This is an area of palaeobiology that was prominent in the mid-1980s, and the volume edited by Valentine (1985) summarizes much of the work to that point. Levinton (1988) pulled a lot of the material together in his textbook, and Signor (1990) published a review paper. Recently Jablonski *et al.* (1996) edited a volume of essays in honour of James Valentine, in which several authors offer updated views; Bambach and Bennington (1966) on community evolution, and Benton (1996) and Sepkoski (1996*b*), respectively, on mechanisms of clade replacement are particularly relevant. It is indicative of current thinking that despite its title of *Evolutionary paleobiology*, a significant percentage of the material concerns modern biotas. McKinney (1997) has reviewed what little is known about which kinds of species are most vulnerable to extinction.

The journal *Paleobiology* continues to be a source of examples and developments in analytical techniques.

# 9
# Mass extinctions: resetting the evolutionary clock

From the very dawning of the appreciation of fossils and stratigraphy there was a growing awareness that from time to time many different kinds of organisms would disappear instantaneously and apparently simultaneously from the record. This observation underlay George Cuvier's concept of catastrophism, with its proposed sequence of global biotic destructions and re-creations, and in due course the palaeontological discontinuities came to form the practical basis for the division of geological time into eras and periods.

There has been a continual, lively debate about these mass extinctions: were they really instantaneous or merely artefacts due to the incompleteness of the fossil record; did they affect species randomly or were some kinds of organisms more susceptible than others; were they caused by terrestrial environmental changes, such as global cooling or falls in sea level, or by interference with the Earth's environment by one sort of extraterrestrial event or another?

The investigation of mass extinctions entered an exciting new phase in the early 1980s due to two independent studies. Alvarez *et al.* (1980) described an anomalously high level of iridium in a clay layer (1–2-cm thick) right at the boundary between the Cretaceous and the Tertiary at Gubbio in the Italian Apennines. The reason why they had measured the iridium level in the first place was that they were looking for an independent measurement of how long it took for this unfossiliferous clay section to be formed. It is generally believed that most of the iridium on the Earth's surface is derived from cosmic dust and micrometeorites from space, and that the rate of bombardment by this material is roughly constant. Therefore the amount of iridium present should be proportional to the length of time represented by the section. But Alvarez *et al.* discovered a thirtyfold increase in the level compared with the limestone rock immediately above and below the sample, which is far too high to explain by a low rate of sedimentation alone in such a deposit. It was clear that some event had introduced a very large amount of the metal over a very short period of time. Their explanation was that exactly at the moment represented by the boundary between the Cretaceous and the Tertiary (K–T), a large meteorite with a diameter of around 10 km had collided with the Earth, and that this was the actual cause of the mass extinction that resulted in the loss of the dinosaurs amongst many other taxa. At about the same time, Sepkoski (1982) had been compiling an enormous data base from the literature of fossil marine families for the entire Phanerozoic, and by 1984 Raup and Sepkoski were using this information to propose that mass extinctions had occurred with a precise periodicity of one every 26 million years, implying a clock-like extraterrestrial cause.

Such bold, unexpected, but at that time necessarily rather tentative conclusions stimulated huge interdisciplinary interest—from astrophysicists searching for possible detailed causes of meteoric or cometary bombardment, through geochemists attempting to identify more precise signs of catastrophic events at the K–T and other boundaries, to palaeobiologists trying to test whether the exact time-course of the extinction events correspond to catastrophic, stepped, or gradualistic patterns (Raup 1986). In no other endeavour than seeking the cause of mass extinctions has the profoundly interdisciplinary nature of the earth sciences been so manifest.

## The epistemology of mass extinctions

A mass extinction is loosely defined as the disappearance of a relatively large percentage of the existing species from a wide range of taxa over a geologically brief period of time and on a global scale. Typically a loss of an estimated 30 per cent or more of the species present, in what appears to be a geological instant, would count as one.

The low resolution of the fossil record (page 89) creates difficulties in developing a more precise definition on two counts, first, concerning the measurement of the percentage of species lost and, second, recognizing over what length of real time the extinction occurred while still appearing to have been instantaneous.

### Taxonomic units

The general concern about which are the appropriate taxonomic units to record when considering diversity changes with time has already been considered in the last chapter (page 157). Briefly to reiterate, as far as extinction is concerned, the point was made that it is a process that occurs at the species, or possibly even the organism level, and therefore extinction of a whole taxon is actually only the sum of the extinction of all its constituent species. Because of the very small percentage of the world-wide total of the original species that is likely to be preserved, discovered, and described as fossil species, however, a more complete picture of the extent of extinction is given by counting the number of taxa, usually either families or genera, that disappear. To proceed from this to an estimate of the actual number of species lost involves applying a correction factor derived from calculations of the average number of species per taxon. This must usually be found by reference to modern biota. If, but only if, each taxon contained roughly the same number of species as all the others, would the number of taxa going extinct even approximate to an acceptable proxy for the number of species going extinct. Furthermore, even if this condition of equal numbers of species in each taxon holds, there is still the distortion due to the fact that a taxon that loses almost all its constituent species will be counted as surviving exactly as if it had lost few or even none at all. The alternative view was mentioned that a taxon may actually represent a true unit of extinction because all its constituent species occupy a related set of niches. But this approach suffers both from the basic dubiousness of the proposal, and from the same statistical distortion of counting lightly and heavily affected taxa equally.

Whatever the theoretical shortcomings of using families and genera, the global extent of mass extinctions necessarily must in practice be described and evaluated by the use of compilations taken from the huge pre-existing palaeontological literature. For example, Sepkoski's (1996*a*) most recent diagrams are based on 3461 families, of which 2349 are extinct, and no less than 28 856 generic extinctions (Fig. 9.1). Given the vast numbers of species constituting all these genera, the patchy geographical coverage of palaeontological work, and the biases of individual authors at the level of species taxonomy, it is very much easier and far more consistent to count families and maybe genera rather than species. This gives a reasonable big picture. On the other hand, any attempt to explain the cause of a mass extinction is likely to require a thoroughly detailed study at the species level of particular local stratigraphic sections that span the event in question. Research in mass extinctions requires consideration of both these levels together.

## Time-course

Having established the taxonomic extent of a mass extinction, the question of its exact time-course must be addressed. Proposed explanations will be constrained by whether the event was catastrophic or more gradual, whether it was a single event or a succession of closely spaced events, and whether it was globally synchronous or diachronous.

The general difficulties of temporal resolution of, and correlation between different stratigraphic sections apply in this area of palaeobiology as much as in any other, as do the methods available to overcome them (McGhee 1996). An additional difficulty when studying mass extinction is that of rare species. It is well known that in any ecosystem a relatively small number of species occur in disproportionately large numbers, while the majority are rare (e.g. May 1975). In reviewing the implications of this phenomenon for fossil collections, Sepkoski and Koch (1996) showed that

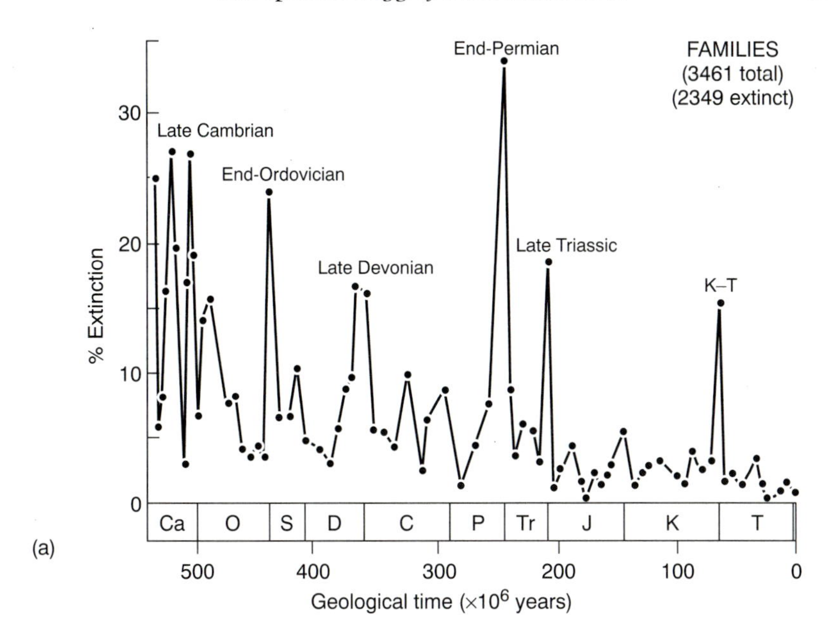


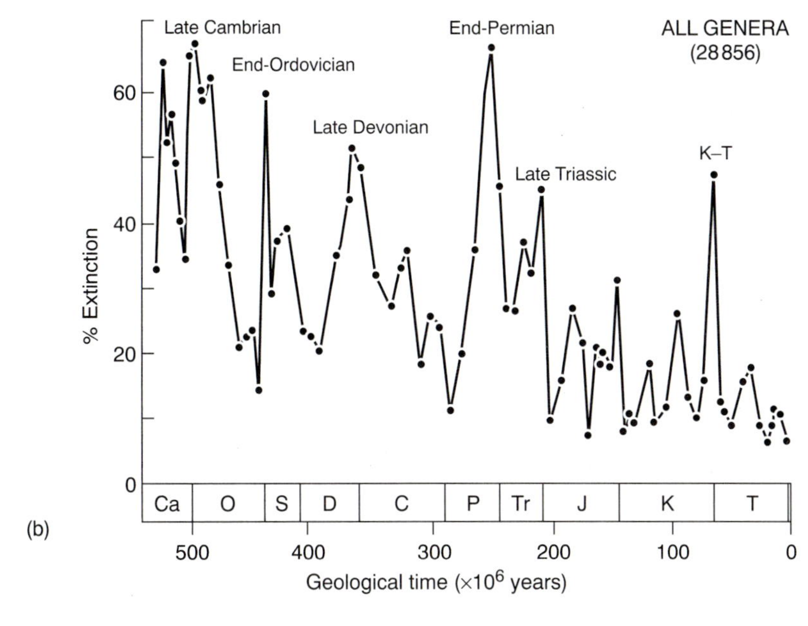


**Fig. 9.1** Rate of extinction through the Phanerozoic, as percentages of existing taxa: (a) families; (b) genera. The major mass extinctions are indicated. (After Sepkoski 1996*a*.)

a high proportion of the fossil species at any locality occur in only a small number of collections made from that area. Typically, more than 50 per cent of the species known occur as specimens in fewer than 1 per cent of the collections made. Therefore, the probability that the last known occurrence of a species in a section corresponds to the time of its actual extinction is extremely small for most species. This is termed the Signor–Lipps effect, an illusion of smearing backwards in time of the extinction event (Signor and Lipps 1982).

Raup (1989) illustrated the effect using the fossil record of ammonites from a late Cretaceous section at Zumaya in Spain (Fig. 9.2a). First, he imagined that there had been an instantaneous mass extinction at the 100-metre level, but because of the sporadic rather than continuous occurrences of the various species through the section, the last known occurrences of the respective species occurred at different times before the 'event', and the appearance is one of a gradual decline to extinction of the group (Fig. 9.2b). Conversely, Raup then imagined that there had been a period of non-deposition (and therefore no preservation of fossils) beginning at the 125-metre level (Fig. 9.2c). In this case, because of the coincidence of the 'hiatus' with the fortuitous occurrence of specimens of several species, the appearance is one of simultaneous extinction of about one-third of the species. Therefore, due entirely to the vagaries of preservation, a true mass extinction could appear to be a gradual decline, while a mass extinction could appear to have occurred when no such biological crisis had actually taken place at all.

There are methods of allowing for this kind of sampling bias if the right information is available and appropriate assumptions made. For example, Koch (1991) re-analysed a sample of molluscs from immediately below the K–T boundary at Prairie Bluff in Alabama. Of the total of 115 species recorded, 28 per cent last occurred in the lower part of the section, 14 per cent in the middle part, and 58 per cent in the upper part that marks the actual K–T boundary. The appearance is one of a stepped extinction, with three phases. However, the sizes of the samples differ in each case with 2144 in the lower, but only 558 in the middle and 680 in the upper parts. Koch assumed that the actual species abundances fitted the log-normal curve found in modern biotas; in other words that it was typical in having a few common species but that most of the species were rare.

**Fig. 9.2** The effect of preservational bias on the interpretation of mass extinctions. (a) The record of ammonites preserved in the latest Cretaceous of Zumaya, Spain. The lines connect the first and last occurrences of 21 species. The cross-bars indicate the exact levels at which specimens of each species have been recovered. (b) The record as it would be had there been a real mass-extinction event at the 100-metre level, giving an appearance of a gradual decline in species diversity. (c) The record as it would be had there been a period of non-deposition of sediments and fossils between the 125-metre and 25-metre levels, giving an appearance of the simultaneous extinction of seven species. (After Raup 1989.)

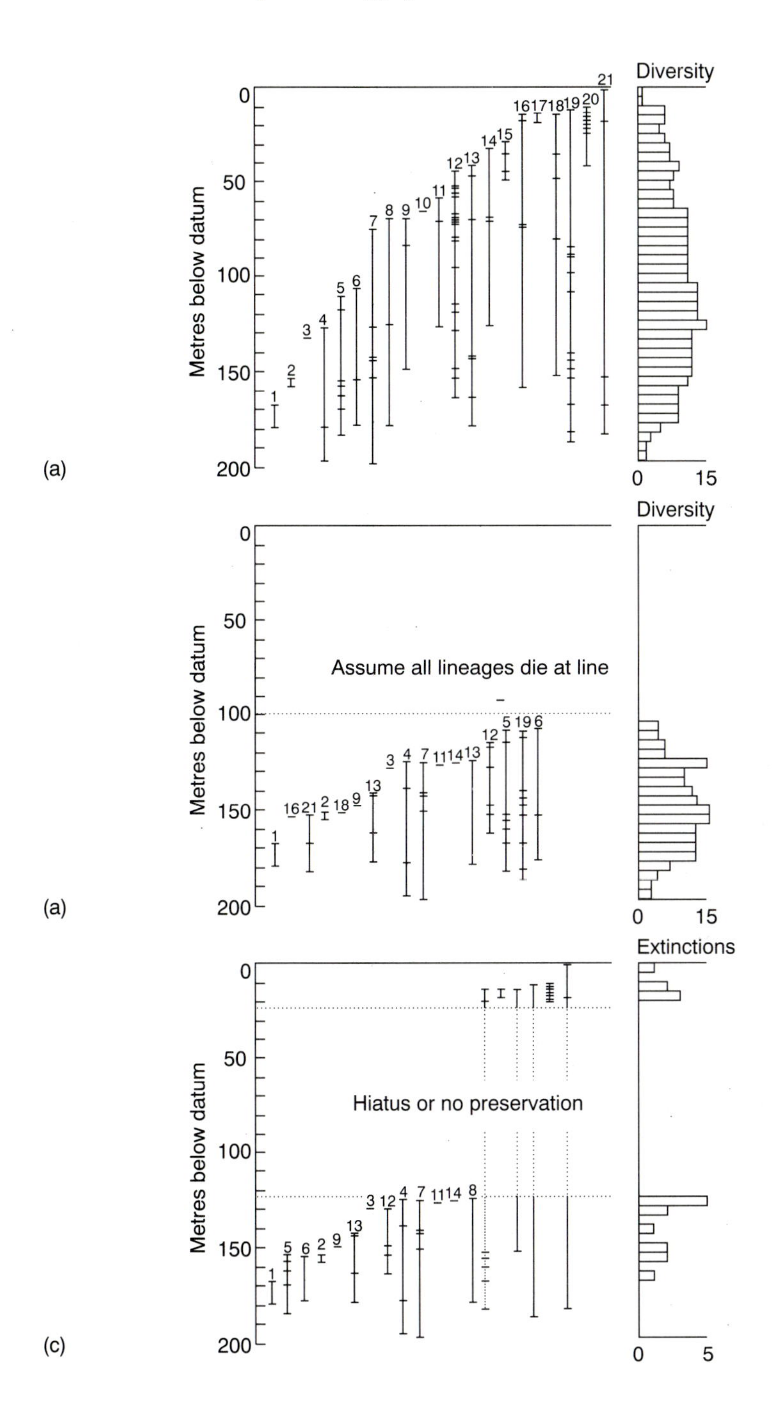
Diversity
Metres below datum
0
50
100
150
200
15
(a)
Assume all lineages die at line
(a)
Extinctions
Hiatus or no preservation
5
(c)

He was then able to calculate a corrected pattern of extinction, assuming that the sample sizes had been equal, and thus estimated figures of 8 per cent extinction in the lower, 14 per cent in the middle, and a full 78 per cent in the upper. What this means is that, because of the sampling bias, the stepped extinction pattern as recorded masked a clear-cut, single mass-extinction event. As Koch pointed out, some of the claims that particular mass extinctions were actually a closely spaced sequence of smaller events are very reminiscent of this example. Clearly, accurate information on sample sizes and relative abundances of the separate species are required before true stepped or gradual extinction patterns can be substantiated. To date, such detailed data are rare.

## Abiotic correlations of mass extinctions

About the only point of universal agreement concerning mass extinctions is that the immediate cause of a large, rapid decline of diversity must be an intolerably large change in the environment, and that this may well leave one or more recognizable abiotic signatures in the stratigraphic record (Fig. 9.3). This latter category of evidence can suffer from exactly the same problems of resolution and correlation as do the biotic data, and there can be considerable ambiguity about exactly what kind of environmental change a certain physical or chemical feature within the rocks actually demonstrates. Nevertheless, increasingly detailed investigations are revealing numerous highly suggestive possible correlations between signs of major events in the environment and biotic crises.

### *Sea-level changes*

Mapping ancient shorelines indicates that the level of the sea is more or less constantly changing. Often this is a localized effect due to small-scale tectonic changes, but rises (transgressions) and falls (regressions) on a global scale occur with considerable frequency (Fig. 6.3). There are many possible causes of such cycles of transgressions and regressions, including alterations in the size of polar ice caps, and the filling and emptying of marine basins. One of the major causes on a longer time scale is probably change in the volume of the mid-oceanic ridges and associated continental spreading (Berger *et al.* 1984).

The first detailed correlation between mass extinctions and sea-level change was made by Newell (1967), who realized that regression of the sea accompanied at least several of the mass-extinction events. Subsequently, Hallam (1989) showed that out of 13 mass-extinction events he considered, no less than 10 and possibly 11 are associated with a regression, and this includes all the major ones. On the other hand, not all marine regressions are associated with a mass extinction, and there were spectacular sea-level falls in the Cenozoic, which appear to have been unaccompanied by any significant biotic crises.

The environmental effect of a marine regression would include simple loss of continental shelf, which is the habitat for a large percentage of fossilizable organisms. It may also signal a fall in global temperature, if the cause had been the development of ice caps, while alterations in patterns of sea currents, prevailing winds, albedo, and continentality of climate would all have potentially profound and complex climatic outcomes.

### *Black shales*

Hallam (1989) also reviewed the occurrence of black shales in association with mass extinctions. These deposits are high in carbon, are believed to form under anoxic conditions, and tend to be associated with marine transgressions. One simple cause of the anoxia might be warming of sea water after the melting of ice caps that also caused the transgression; this would reduce oxygen solubility. For much of the Phanerozoic, deeper sea water has probably been poorly oxygenated, and therefore another possible result of a transgression is that this oxygen-poor water spreads over the continental shelves, destroying the habitat. A more severe effect is termed oceanic overturn, and might occur if the surface water cools to such an extent that it exceeds the density of the underlying water layers, and consequently sinks (Wilde and Berry 1984). Of his 13 mass extinctions, Hallam found that six of them certainly and four more doubtfully were associated with such black shales.

### *Shifts in stable isotope ratios*

The use of stable isotope ratios for assessing various palaeoenvironmental features has already been mentioned (page 95). Numerous shifts in the ratios of various elements occur coincidentally with many of the mass extinctions, although it must be added that many shifts do not appear to correlate with any particular major biotic event at all (Holser *et al.* 1996). As an example of strong isotope ratio signals coinciding with a major event (Fig. 9.3), the mass extinction at the end of the Permian is accompanied by a marked, global decrease in the $\delta^{13}C$ value. This particular effect, which occurs at several other event horizons as well, has been termed a 'Strangelove Ocean' (Hsu and McKenzie 1990), and the most obvious explanation for it is a massive, catastrophic fall in the level of phytoplanktonic photosynthesis. There is also a rise in the $\delta^{34}S$ value in deposits close to the end-Permian boundary, which is related to the level of activity of sulphur bacteria, and therefore implies anaerobic conditions. Third, there is a fall in the $^{87}Sr/^{86}Sr$ ratio. This is related to components of old igneous rock richer in $^{87}Sr$ relative to new, lighter basalts with less $^{87}Sr$ that are derived from beneath mid-oceanic ridges. The significance of these various shifts in stable isotope ratios is not yet very clear, but they do indicate a rather dramatic alteration to the oceanic chemistry of the time.

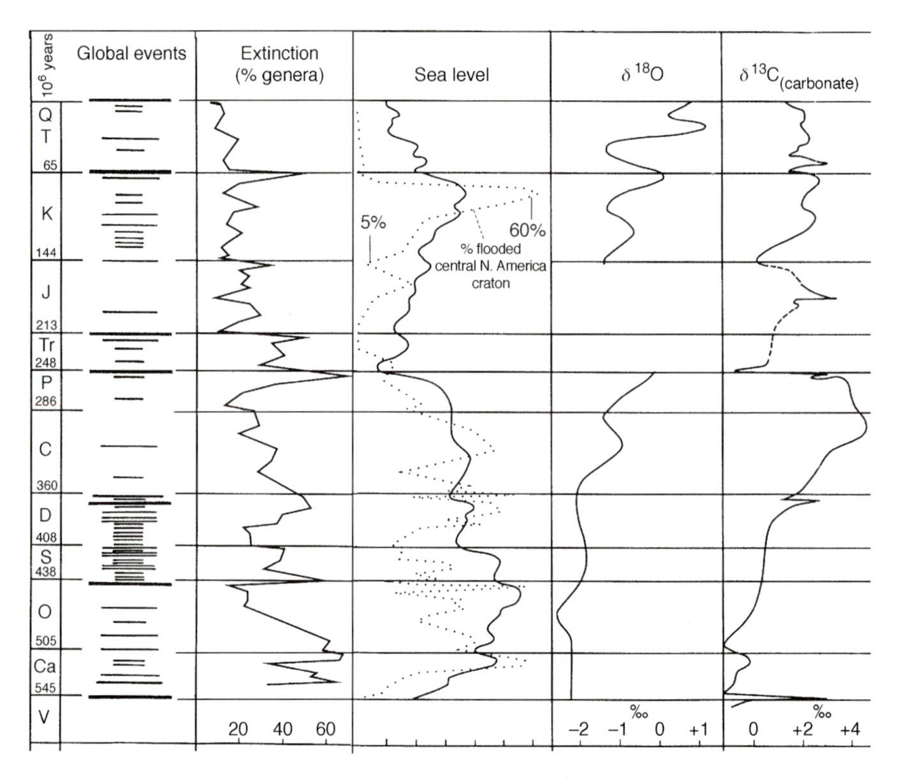


**Fig. 9.3** The Phanerozoic record of extinction events and some of the associated abiotic signals. The bars under the column 'global events' are approximately proportional to the magnitude of the extinction event, with the big six Phanerozoic mass extinctions in bold. (After Morrow *et al.* 1996.)

Against this example of the end-Permian, another case is the end-Triassic extinction. This was also a major mass extinction but it has not yet been associated with any significantly large shifts in stable isotope ratios. To complete the confusion surrounding this category of evidence, an example of a strong shift in $\delta^{13}$C occurs in the middle of the Jurassic, when no detectable biotic crisis at all appears to have been in progress.

## *Element abundances*

The enormous effect on the study of mass extinctions caused by the discovery of the enhanced iridium levels at the end of the Cretaceous has been commented upon (Alvarez *et al.* 1980). Since then every possible horizon associated with a mass extinction has been searched for enhanced levels of iridium and other elements, using techniques such as neutron activation analysis that can measure as little as five parts per $10^{12}$ (Orth 1989).

Two groups of elements are of interest. The first are referred to as siderophile because they are associated with meteoritic or cometary materials,

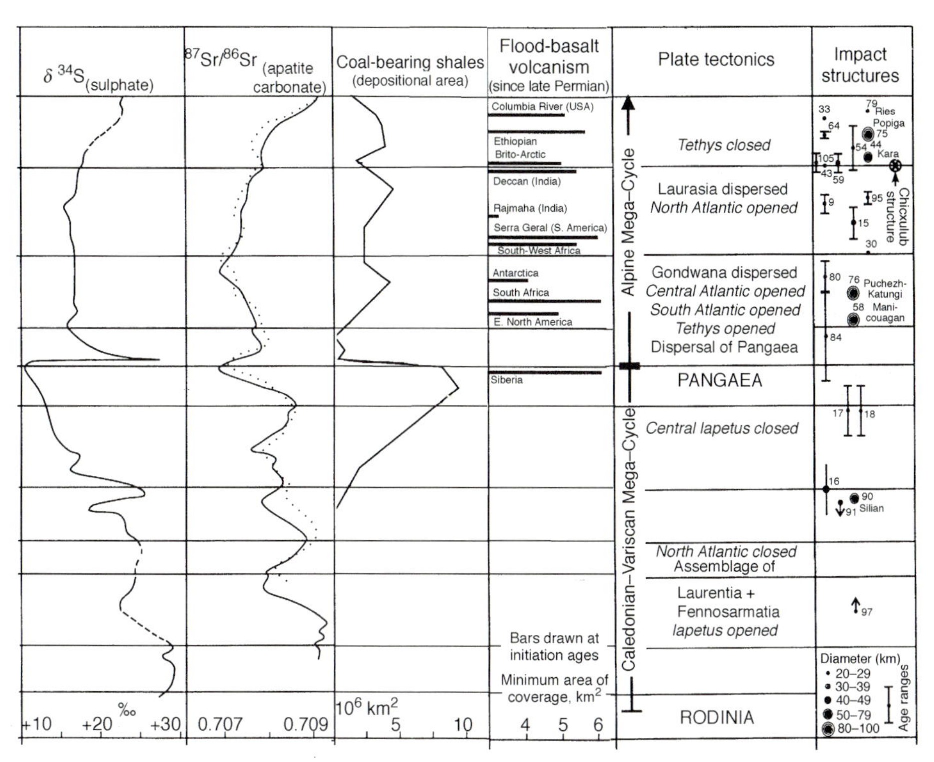


and the most important of these are the platinum-group elements. The easiest of them to measure is iridium, but rubidium, rhenium, palladium, osmium, and platinum are included. The presence of these elements is most readily explained by impact of an extraterrestrial body, a bolide, with the Earth's surface. The possibility that they can also occur as a result of extensive volcanic activity has been argued (Officer and Drake 1985), and even that they may be concentrated by bacterial activity (McGhee 1996). Therefore, interpretation of anomalous levels of iridium and the other siderophile elements depends on assessments of the period of time over which they accumulated, and on their relative proportions in the sediments, compared with these values in known volcanic products and in bolides. The consensus at present is that at least the larger of the known anomalies are best accounted for by impacts.

Chalcophile elements are those that are fairly certainly associated with volcanic rocks and include arsenic, molybdenum, and selenium. In conjunction with the volcanic traps discussed next, they are taken as evidence for significant volcanic activity.

## *Volcanic activity*

The most-direct evidence for extensive, possibly global periods of volcanic activity is the occurrence of flood basalt rocks, and two of the major mass extinctions at least approximately coincide with them (Fig. 9.3). The Siberian Traps cover about 2.5 million square kilometres to a depth of 3000 metres, and they date from the boundary between the Permian and the Triassic (Erwin 1996). This is the time of the most-severe mass extinction of all. Almost equally impressive, the Deccan Traps of central India were produced at the end of the Cretaceous although the activity started about 2 million years before the actual K–T event, and continued after it. On the other hand, examples are known where a volcanic event did occur, but not at the same time as a mass-extinction event. Sometime close to the end of the Triassic, volcanic traps in eastern North America were produced but they are somewhat younger than the end-Triassic mass extinction. And possibly the largest volcanic eruption of the Phanerozoic is indicated by mid-Ordovician rocks in the Baltic and eastern North America, but which had very little biotic effect (Huff *et al.* 1992). At the times of the other major mass extinctions, there is no evidence of elevated levels of volcanic activity at all (Morrow *et al.* 1996).

As to the likely effects of massive vulcanicity, these are severalfold. Even volcanic eruptions observed in recent centuries, all exceedingly modest compared with those from the past, have caused some weeks of much-reduced light and temperature levels. For a very large eruption the levels of nitrogen and sulphur oxides could result in extremes of acid rain carried world-wide. Even simple, global wildfires would be a possible cause of a huge environmental catastrophe.

### *Tectonic activity*

Since the continental land masses have been continually congealing and breaking up, mass-extinction events inevitably coincide sometimes with large tectonic changes, which equally inevitably must have been associated with climatic changes. New dispositions of the land masses would alter oceanic currents and the patterns of prevailing winds. For example, there is growing evidence for a single late Precambrian supercontinent that broke up around the beginning of the Cambrian, and which may coincide with the extinction of most of the Ediacaran fauna (Brasier 1989; 1996). Unfortunately the temporal correlation of the tectonic and the biotic events is still poor.

### *Glaciation*

Sedimentary rocks showing clear signs of glaciation occur from time to time throughout the Phanerozoic and occasionally coincide with mass extinctions. The inference that temperature falls may have been involved in the cause of the biotic events in such cases is obvious.

### *Extraterrestrial impact structures*

Geochemical signs of possible extraterrestrial impacts have been mentioned. There are also several other kinds of direct evidence. One thing that a bolide, be it meteoritic or cometary, colliding with the surface of the Earth is sure to cause is a crater, even if it is submarine. Indeed a large number of craters are known (Fig. 9.3)—more than 100 with diameters estimated at over 3 km (Greive and Robertson 1987; Greive 1995). Shoemaker (1984) made some calculations based on the cratering patterns of the Moon and the Earth, and on estimates of the numbers of meteors and comets in the Solar System that might be expected to reach the Earth's atmosphere. His conclusion was that an average of something like three to six asteroids with a diameter of greater than 1 km collide with the Earth every million years, with a somewhat lower figure for comets, and about six bolides with a diameter of over 10 km have probably collided during the Phanerozoic. But attempts to correlate particular craters with particular mass extinctions have not been successful, with one exception. With a date of 66 million years, the massive Chicxulub Crater discovered at the Yucatan Peninsula of Mexico coincides exactly with the K–T boundary and is widely accepted as the likely source of extraterrestrial iridium associated with this time (MacLeod 1996). But there are also equally large craters that do not coincide at all with any particular mass-extinction events. The huge Manicouagan Crater in Quebec, which was once thought to occur at the end of the Triassic and therefore to coincide with the major mass extinction at that time, is now believed to be about 12 million years older than the biotic event (Hallam 1996; Hallam and Wignall 1997).

Energy impacts can create shocked quartz, a very dense form of the material that forms only under extreme pressure. Such grains appear in association with the K–T event, further supporting the impact theory for that particular mass extinction. Yet another indication of impacts, again well represented at the K–T boundary, are microtektites. These are glass-like, silicate spherules up to about 1 mm in diameter. They form when molten rock at the point of a violent impact is blasted into the atmosphere. Here the spherules solidify in the aerodynamic forms they had adopted while in transit, before falling back to Earth.

Numerous scenarios to describe the likely effects of a large bolide-strike have been constructed, often using computer models of the Earth's climate. In an early estimate, Jones and Kodis (1982) calculated that the supposed 10-km diameter bolide that struck the Earth at the K–T boundary would have released about 2.5 million megatonnes of energy, equivalent to around 50 000 large nuclear bombs. This would be enough to cause immediate shock-wave-induced earthquakes, monstrous tsunamis, and wildfires, none of which would be expected to have much lasting effect on the global biota. The introduction of light-blocking debris in the atmosphere that would be spread widely by the atmospheric circulation, however, would be expected to have a far longer-lasting effect, and therefore cause permanent changes. Were the bolide to land in the ocean, then levels of water vapour in the atmosphere would lead to a massive, potentially catastrophic greenhouse effect.

## Biotic correlations of mass extinctions

A lot of effort has been made trying to establish whether mass extinctions affect species randomly, or whether certain taxa or particular kinds of species are less susceptible than others. All has been to surprisingly little avail so far (Raup 1995). A mass extinction consists of the loss of anything from an estimated, say, 30 per cent of species to the possibly 96 per cent species loss at the end-Permian event. Does the percentage of species lost by some taxa differ significantly from the overall average percentage extinction of the particular event in question?

Completely random extinction of species across all the taxa then extant cannot account for the loss or even the severe reduction of a taxon that consists of many species. Statistically it is virtually impossible that the extinction of, for example, the tabulate corals at the end-Permian event, most of the ammonoids at the end-Triassic event, or all the then existing dinosaurs at the end-Cretaceous event were due to random loss of species over the entire biota. If the argument is accepted that extinction is ultimately an organism-level process, consisting of the sum of the deaths of all the constituent organisms (page 189), then the only possible cause of such non-random extinction is that all these organisms possessed some characteristic that made each of them ultimately inviable in the face of the environmental cause of the mass extinction.

There have been endless speculations about what particular inadequacy of the organisms led to the demise of this particular group or that. For example, the large body size of dinosaurs has often been taken as the cause of their extinction, a view bolstered perhaps by the survival of the birds, which are strictly speaking miniaturized, flying dinosaurs. In all explanations such as this, the problem is one of testing the hypothesis and so they are usually quite unconvincing. Nevertheless, one or two generalizations have been made along these lines. One is exactly this, that taxa whose constituent organisms have large body size have an increased susceptibility to mass extinction. Perhaps this is because large organisms tend to have lower reproductive rates, or are more specialized because of their more critical mechanical design requirements. Another proposed general rule is that taxa living in and therefore adapted to tropical conditions fare worse than those from higher latitudes, perhaps because tropical forms lack the option of moving equatorwards to track suitable conditions as temperatures fall (Stanley 1987; McGhee 1996). Raup and Jablonski (1993), however, failed to substantiate this relationship for end-Cretaceous molluscs.

If the argument is accepted that extinction is actually a species-level process (page 189), then species-level characters such as population size, pattern of dispersal, and geographic extent could affect susceptibility to mass extinction. Jablonski (1986, 1989) noted one such correlation. He found that during normal times, when only the low rate of background extinction was occurring, a taxon that consisted of numerous individually widespread species was less susceptible to extinction. But at times of mass extinction, these taxa lost any such advantage. Instead, taxa such as genera with a wide geographic distribution as a whole suffered less extinction. For example, he found that at the end-Cretaceous mass extinction 55 per cent of bivalve genera with a widespread distribution survived, compared with a mere 9 per cent of those endemic to the area he was looking at. There are comparable figures for gastropod molluscs at that time, and Jablonski also quotes examples from other taxa and other mass-extinctions.

The biotic correlations are thought so weak that, overall, it has to be concluded that mass extinctions show little taxonomic selectivity, and that the differential effect on taxa must be very much a matter of particular responses by particular groups during particular events. Indeed, it has been argued that mass extinction should be seen as non-Darwinian, if not even anti-Darwinian evolution. Gould (1985) asked whether succumbing to whatever environmental perturbation that caused a mass extinction was due to 'bad luck' rather than 'bad genes' (Raup 1991). In so far as these perturbations are rare and probably highly diverse in nature, there is no opportunity to evolve, by means of natural selection, organisms specifically resistant to them: thus 'bad luck' to happen to be in the wrong place at the wrong time. But in so far as the perturbations *are* environmental events, presumably some organisms happen to have a biology that is resistant to them, such as the power to disperse elsewhere, to remain dormant, or whatever: thus selection of those with 'good genes'. How it could be possible to tell in a given case which of these alternative concepts applies is not clear. Gould is clearly right to stress how mass extinctions disrupt the normal flow of evolution and restart it along unpredictable lines. But it is hard to imagine a latterday Darwin having much difficulty with the idea.

## The global Phanerozoic pattern

When Raup and Sepkoski (1982) first began to plot the percentage of marine families and later genera that went extinct in each successive segment of geological time, the supposed mass-extinction events immediately appeared as spikes standing out from a general level of extinction (Fig. 9.1b). However crude the plots were at that time, there were certainly grounds for distinguishing two modes of extinction. Background extinction occurs at a low, roughly constant rate and is the pattern associated with the taxonomic turnover discussed in Chapter 8. Mass extinctions, however, are brief periods of greatly enhanced rates of extinction of up to 30 per cent of the families and 60 per cent of the genera. As noted in the previous section, Jablonski (1986, 1989) even described certain differences in the kinds of groups most affected by the two respective modes. This difference in susceptibility implies that the underlying mechanism is different in the two respective modes. But Raup (1995) cast doubt on this simple division by showing that there is a continuum in amplitude and frequency, from the largest mass extinctions to small events not significantly different from the background. His kill curve (Fig. 9.4a), which illustrates the frequency of occurrence (as mean waiting time) against the intensity of extinction plotted for all extinction events, is a smooth curve with no evident discontinuity between qualitatively distinguishable kinds of event.

Raup and Sepkoski (1984) used the plots of extinction rates against geological time from the Permian to the Holocene to argue that the major extinction peaks occur precisely every 26 million years (Fig. 9.4b). What they actually claimed in effect was that there is a highly significant goodness of fit between the timing of the extinction phases as seen in the fossil record, and a model based on a periodicity of 26 million years. The fit is not exact, but the mismatch can be attributed to inaccuracies in the absolute time-scale used, and in the level of resolution of the dating of the last occurrences of the individual fossil taxa. Necessarily these dates come from the literature and are generally no more accurate than to the geological Stage in which the fossil was found. Yet Stages themselves are not very precisely dated, and vary in length from as low as 2 million years to over 20 million years, with an average of 6–7 million years.

Indeed, the imprecision of the data is such that other palaeobiologists have been able to show that alternative models can fit it equally well. Kitchell and Pena (1984) found that a stochastic model with an autoregressive component gives a better fit, and Hoffman and Ghiold (1985) argued that a null model in which the respective rates of extinction and origination of families vary randomly and independently cannot be rejected by the data. For all the implications that would follow acceptance of exact periodicity of occurrence of mass extinctions, it must be concluded that at present the data are simply not good enough either to corroborate or to refute the hypothesis. This is a pity because it means that the several ingenious proposals for possible extraterrestrial triggers to mass extinction are at best on hold. The Nemesis hypothesis (Whitmire and Jackson 1984), for example, holds that an as-yet undiscovered Black Dwarf companion star of the Sun triggers cometary bombardment of the Earth from the Oort cloud of comets that lies on the outer margins of the Solar System.

## The Big Seven bio-events

Although there is no longer thought to be a clear distinction between mass extinctions and smaller biotic crises, it is convenient to follow Raup

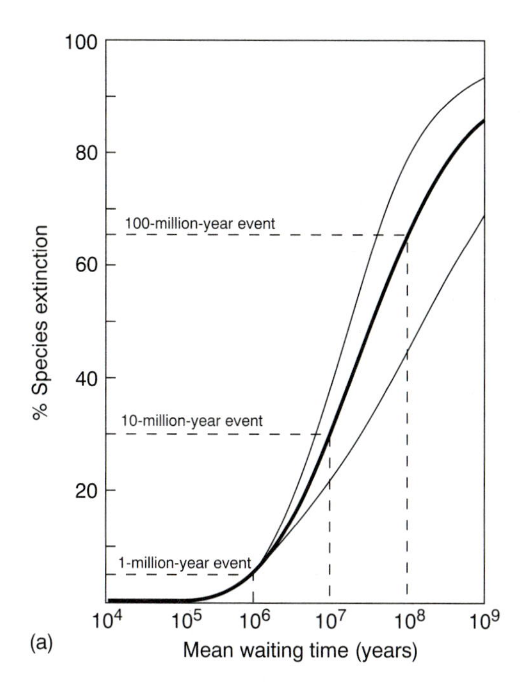


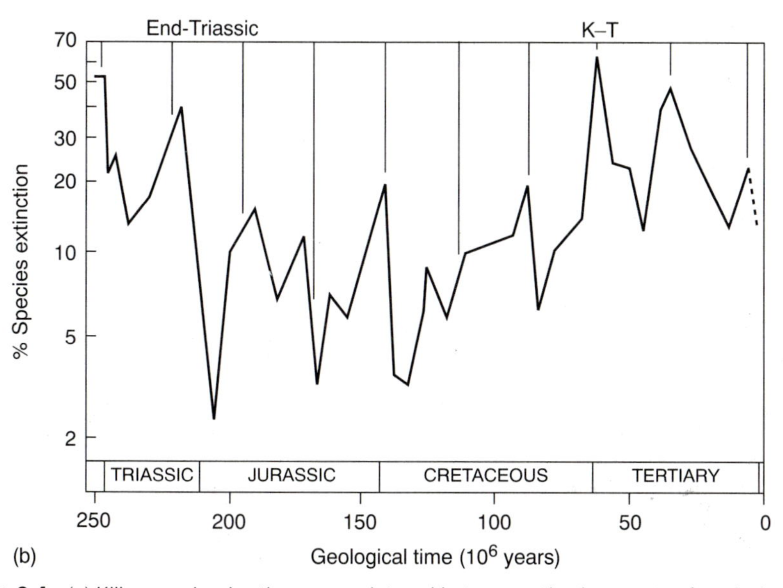


**Fig. 9.4** (a) Kill curve showing the average interval between extinction events of particular intensity. (b) The pattern of mass extinctions compared with a model of 26-million-year periodicity. The vertical bars represent the 26 million year intervals. ((a) After Raup 1995; (b) after Raup and Sepkoski 1984.)

(1995) and arbitrarily select those events in which the extinction is calculated to have been 75 per cent or more of the species (see Fig. 9.1). This gives the so-called 'Big Five' events. There is also the possibility, very hard to substantiate at present, of up to two earlier but equally severe events. These occur at the end of the Precambrian, and in the Lower Cambrian.

## Precambrian–Cambrian event

There was certainly a major change in the biota 555–545 million years ago, at or shortly before the start of the Phanerozoic. The hitherto world-wide Ediacaran fauna (Fig. 9.5a) with its peculiar frond-like, ribbon-like, and

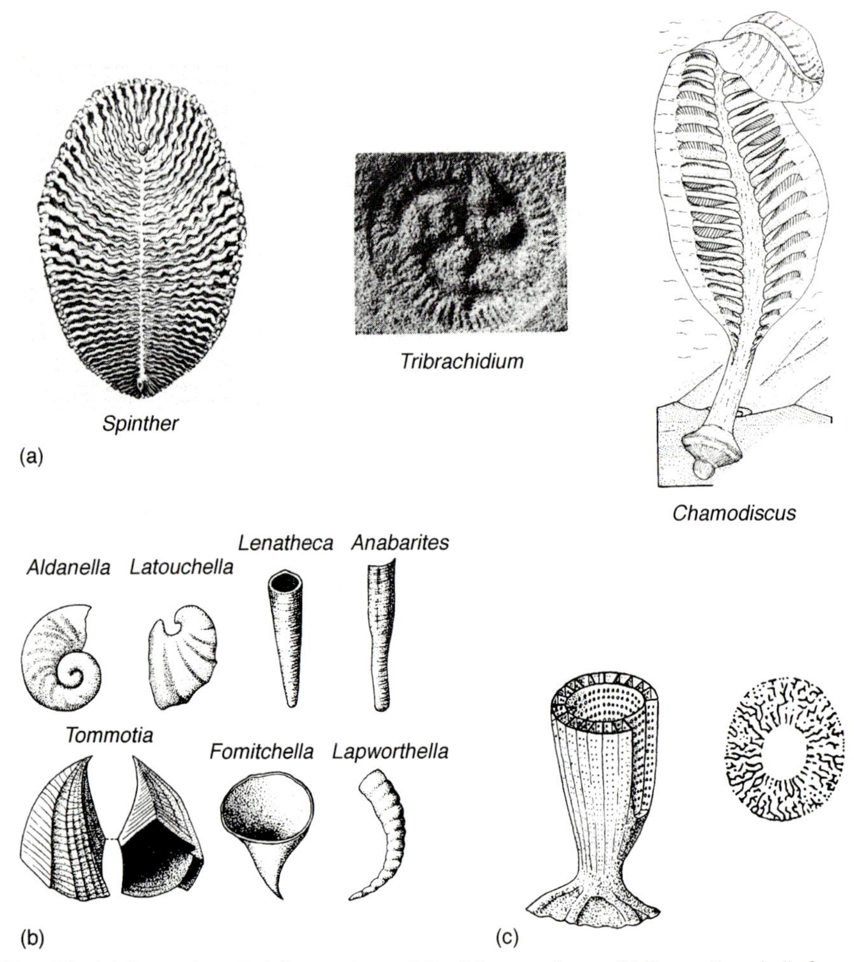


**Fig. 9.5** (a) Some characteristic members of the Ediacaran fauna. (b) Tommatian shelly fauna. (c) Archaeocyathan sponge, whole organism and transverse section. ((a) After Glaessner 1984; (b) after Clarkson 1993; (c) after Kuhn *et al.* 1986.)

jellyfish-like forms disappeared, apart from a few apparently related taxa surviving into the middle Cambrian (Conway Morris 1993). Most of the larger-sized, spiny acritarch algal cells (see Fig. 8.1c) that had formed a major part of the plankton also disappeared, and there was a reduction in the Precambrian kinds of trace fossils (McMenamin 1990). Unfortunately, temporal resolution of, and correlation between end-Cambrian sections is very poor and the time-course of this extinction phase is unclear. It may well have been a stepwise extinction event over a period of several million years.

Whatever the exact pattern of loss of species, there are also indications of major abiotic changes to the environment associated with the extinction, as reviewed by Brasier (1989). These include a global regression of the sea, and changes in stable isotope rations. The $\delta^{13}C$ rises, and there is a high level of $\delta^{34}S$, which could be related to a fall in oxygen levels. There is also some evidence of cooling in the form of glaciations at this time, at least in some parts of the world. On the negative side there is as yet no evidence for an impact nor for high levels of volcanic activity.

At present, therefore, any perception of a mass extinction and its causes at this time must remain vague and imprecise (see, for example, McMenamin 1990).

## Lower Cambrian (Botomian–Toyonian) event

The next phase of large-scale extinction occurred at about 523–520 million years ago (Brasier 1996) and affected all taxa, particularly the archaeocyathan sponges (Fig. 9.5c) and the characteristic species forming the small-shelled Tommotian fauna (Fig. 9.5b) of the early part of the Cambrian. Several families of trilobites also succumbed. Like the end-Precambrian extinction, resolution and correlation are poor, and good sections spanning the boundary are uncommon. So, once again, it is unclear how extensive, rapid, or stepwise the extinction actually was. Signor (1992) regarded it as certainly one, possibly the most severe, of the Phanerozoic. Brasier (1996) tentatively described it as a stepped event, with two distinct phases—one mid-Botomian and the other at the top of the Toyonian stages, respectively.

As to causes, there are a few indications of environmental changes. A rise in sea level is associated with black shales, and this is apparently succeeded by a massive regression. The $\delta^{13}C$ falls, which may indicate Strangelove Ocean conditions with a collapse of phytoplanktonic productivity.

## End-Ordovician (Ashgillian) event

The mass extinction in the latest Ordovician, 440 million years ago, was the last of several significant extinction events during the Ordovician, and was second only to the end-Permian event in severity, with the loss of about 28 per cent of families and an estimated 85 per cent of species (Barnes *et al.* 1996; Raup 1995). The graptoloids were reduced to three or four species, the trilobites (Fig. 9.6) suffered the greatest decline of their history, and the conodonts and corals were also severely depleted.

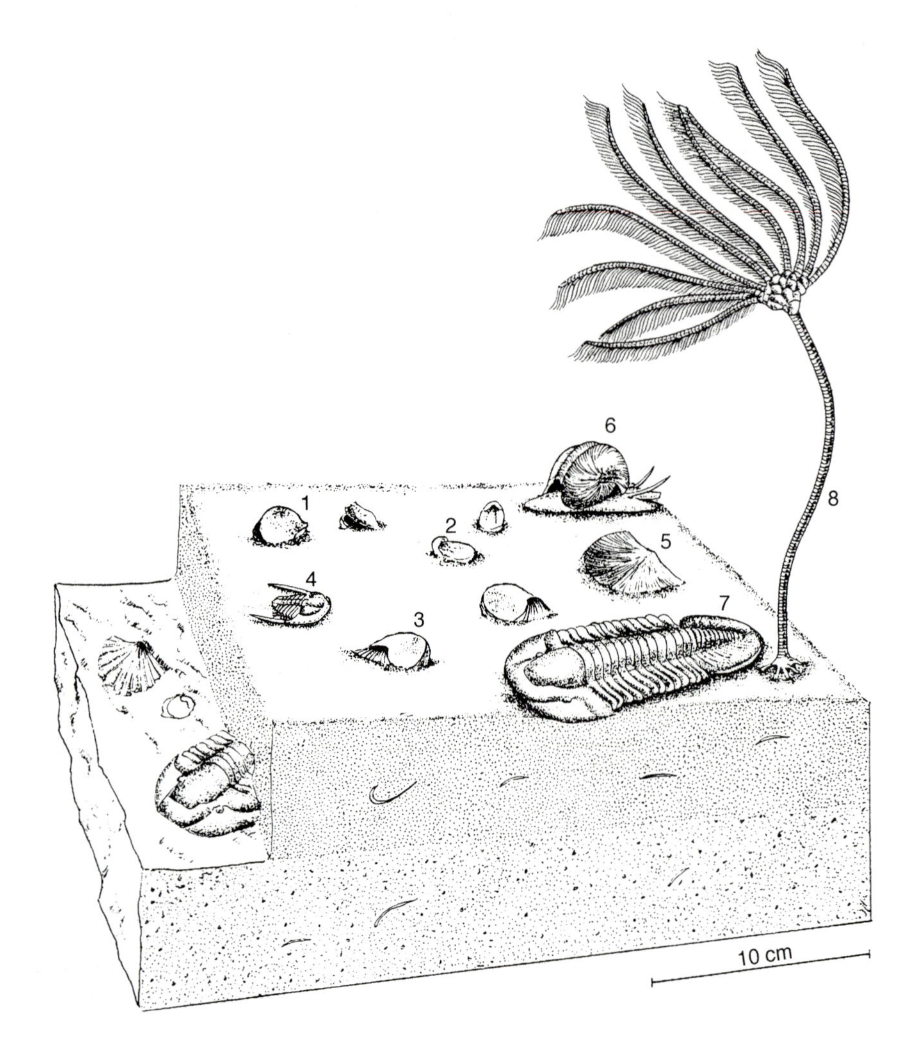


**Fig. 9.6** Ashgillian (latest Ordovician) *Christiania–Sampo* brachiopod community: 1, *Sampo* (strophomenid brachiopod); 2, *Christiania* (strophomenid brachiopod); 3, *Dinorthis* (orthid brachiopod); 4, *Tretaspis* (trilobite); 5, *Rafinesquina* (strophomenid brachiopod); 6, bellerephontid (monoplacophoran mollusc); 7, *Brongniartella* (trilobite); 8, crinoid (crinozoan). (After Cocks and McKerrow 1978.)

Hitherto, the environment of the Ordovician had consisted of generally greenhouse conditions, but the last phase, during the Ashgillian, was marked by the onset of severe glaciation. This was accompanied by a large fall in sea level, followed by a rise at the very end of the period that is associated with deposits of black shale. It is not certain, but likely, that the extinction was to some degree stepwise, and Brenchley (1989) recognized two main phases—one associated with the regression and the other with the subsequent transgression—separated by some hundreds of thousands of years. As far as negative evidence is concerned, there is no indication at all of bolide impact or of extensive volcanic activity, and it was a time at which the continents were dispersed, neither congealing nor fragmenting to any marked degree.

## Late Devonian (Frasnian–Famennian) event

The third largest mass extinction occurred during the Late Devonian, culminating at the boundary between the Frasnian and Famennian stages about 365 million years ago. Detailed studies quoted by Walliser (1996*a*) and McGhee (1996) indicate that there had already been a prolonged biotic crisis, consisting of a more or less stepwise reduction in diversity of many groups over the preceding few million years. This, the late Frasnian crisis, coincided with a series of marine transgressions during the early Frasnian and resulted in the virtual cessation of the growth of coral reefs by the late Frasnian. Many other groups, including pelagic forms such as ammonoids, suffered major losses.

The actual Frasnian–Famennian boundary has been well defined on the basis of conodont species, and it was then that the largest extinction pulse occurred, over the course of at the most 300 000 years and possibly much less. In Europe it is known as the Upper Kellwasser event, a particularly well-preserved section that shows the final extinction of the dominant reef-forming species (Fig. 9.7). Some estimates for other major taxa are that 88 per cent of the remaining ammonoid species, 86 per cent of brachiopod species, and similar proportions of trilobite species went extinct at that time, in what may itself have been a stepped pattern spanning up to 0.5 million years (McGhee 1996).

The most prominent environmental feature of the Upper Kellwasser is a series of black shale deposits indicating widespread anoxic conditions in the oceans. There are also large stable isotope excursions, which Wang *et al.* (1996) analysed across particularly good sections in Canada that encompass the Frasnian–Famennian boundary. $\delta^{13}C$ shows a gradual rise from 1. to 4 per cent, beginning 10 metres below and ending 23 metres above the boundary, which represents a period of about 1–2 million years. After that, it gradually returns to its original value during the next 10 million years. At the boundary itself there is a brief, transitory fall in $\delta^{13}C$, and

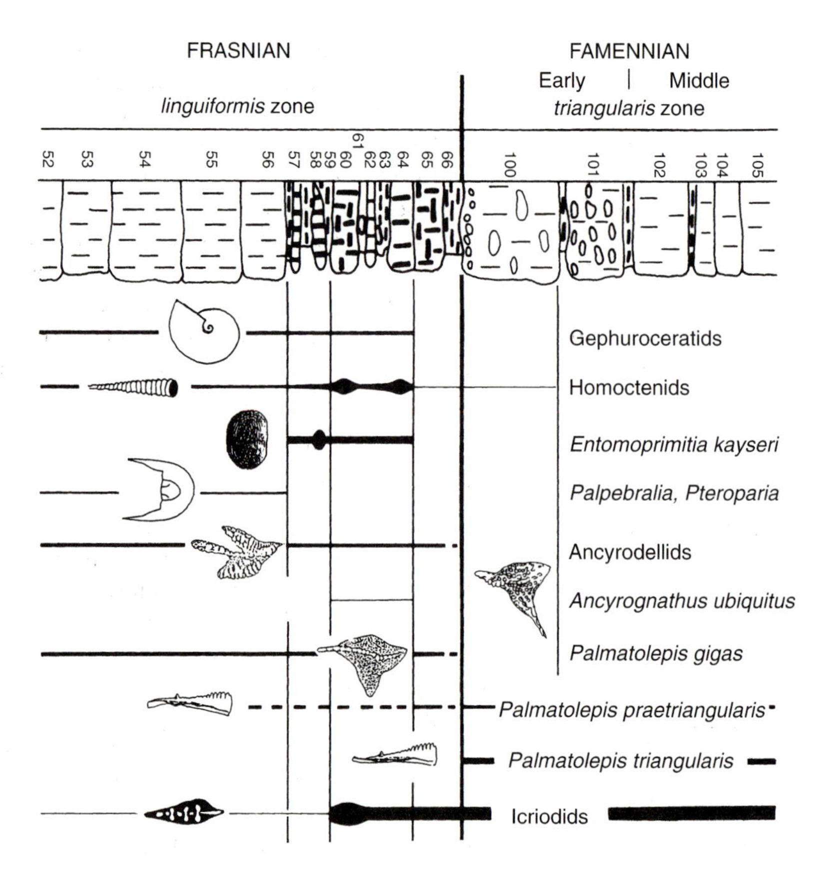


**Fig. 9.7** The Upper Kellwasser event of the Late Devonian of Europe. Different species disappeared at at least three slightly different levels in the sequence that spans the Frasnian and Famennian stages. The zones indicated are based on conodont species. (After Walliser 1996*a*)

also an increase in $\delta^{34}S$. Wang *et al.* interpreted these figures as supporting the evidence of the organic-rich black shales of the Upper Kellwaser pointing to a high level of anoxia in the oceans.

There is also some evidence for a bolide impact at the Frasnian–Famennian boundary (McGhee 1996) in the form of microtektites found in Belgium, and possible craters of the same date in Sweden and Quebec. Very small iridium anomalies have been described from China and Australia, but these are of doubtful significance and may be a result of biotic concentration of the element. Other environmental indicators are that the sea levels were probably high and temperatures, as indicated by $\delta^{18}O$ levels, were also high, indicating possible greenhouse conditions.

Wang *et al*. (1996) concluded that the overall explanation for the final Frasnian–Famennian extinction was a period of environmental stress lasting for perhaps half a million years and characterized by high temperatures and widespread anoxic conditions. They suggested that bolide impacts at the final boundary might have exacerbated the rate of extinction. In contrast, McGhee (1996) suggested global cooling as the likely immediate cause, resulting perhaps from a series of bolide impacts. Others find evidence for a marine regression, and some do not find the evidence for any impact at all overwhelming (Walliser 1996*a*).

## End-Permian event

With an estimated extinction of 96 per cent of species (Raup 1996)—though Erwin (1994) believes it to be up to 10 per cent less than this figure —the end-Permian event is the largest mass extinction of all, and therefore the nearest point to total obliteration of metazoan life in the whole of the Phanerozoic (Fig. 9.8). It is no coincidence that it marks the boundary

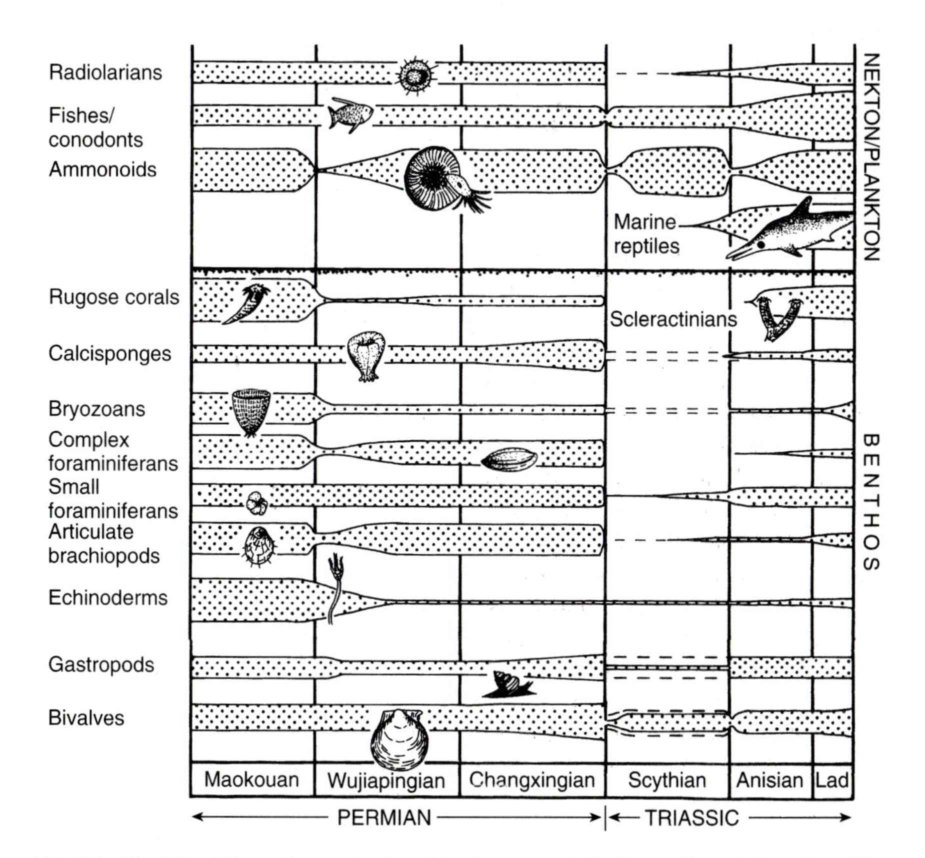


**Fig. 9.8** The fate of the major marine invertebrate groups at the Permo-Triassic boundary. (After Hallam and Wignall 1997.)

between the Palaeozoic and the Mesozoic eras, around 250 million years ago. The detailed picture, however, is obscured by the difficulties of accurate correlations on a global basis. Even the more recent suggestions about the time-course vary from several million years (Teichert 1990) to virtually an instantaneous catastrophe (Xu 1991).

Erwin (1994,1996) extensively reviewed this mass extinction, and concluded that it was a phased event lasting a total of 1–2 million years. Both marine and terrestrial faunas were severely affected. In the sea, inshore taxa including the major reef builders were the most heavily affected, and amongst the eminent victims were almost all foraminiferan, trilobite, and ammonoid species. On land, the insects suffered the greatest decline in diversity of their history: out of a total of 27 Permian orders, 8 became extinct and another 10 were severely reduced. Also 75 per cent of tetrapod families were lost at or around this time. The flora, on the other hand, had been changing more gradually throughout the Permian, from the palaeophytic flora of ferns, etc. to the mesophytic flora dominated by plants more adapted to drier conditions, such as conifers, ginkoes, and cycads. At the actual boundary between the Permian and the Triassic there is a large but brief increase in fungal spores.

There was a general geographic stability at this time. The supercontinent of Pangaea had formed and most of the large land masses were still connected. Also, there are no reliable indications at all of bolide impacts coinciding with the mass extinction. On the other hand, there are several, indeed embarrassingly many indications of changes to the abiotic environment around the end of the Permian. They include a pronounced sea-level regression at about the actual boundary, followed by a transgression in the earliest Triassic. It was also the time of one of the greatest episodes of volcanic activity. The Siberian flood basalts cover some 2.5 million square kilometres to a depth of 3000 metres; and at about the same time, though precise correlation is impossible, there was extensive pyroclastic volcanic activity in what is now South China. The occurrence of evaporite rocks seems to indicate a general climatic warming rather then a cooling and this is supported by estimates of increased $CO_2$ levels (Berner 1994), suggesting greenhouse conditions. Shifts in stable isotope ratios point towards a rise in anoxic conditions. $\delta^{13}C$ falls from + 3 per cent to − 1 per cent, and $\delta^{34}S$ shows a gradual fall throughout the Late Permian. Finally, there was a shift in the strontium isotope ratio.

Putative explanations for the end-Permian mass extinction have been numerous, and have invoked all the observed abiotic features, and many assorted combinations of them. What certainly is clear is that no single, simple cause is likely to explain the entire pattern of extinctions. Erwin (1994) proposed that there was a multitude of interacting causes, which resulted in a triple-phased event. The first phase was associated with the marine regression, which removed a large fraction of the habitat area and caused increasing climatic instability. The second phase was associated with the volcanic activity, causing a rise in $CO_2$, global warming and therefore further climatic deterioration and ecological collapse. The third and final phase was due to the marine transgression at the start of the Triassic, causing the spread of anoxic water and removing large, land areas, thereby affecting the terrestrial fauna.

## The end-Triassic event

According to Raup's (1995) calculation the mass extinction that occurred around 205 million years ago, at the end of the Triassic period, caused the loss of about 76 per cent of species. The effect was most felt amongst marine groups; for example, Hallam (1996) noted that six superfamilies of ammonoids became extinct, and only a single genus apparently survived into the succeeding Jurassic period. Almost all the European species of bivalves disappeared and half the genera world-wide. In the Alps, the widespread fossilized reef ecosystems disappear from the record. It is not yet clear whether this event was catastrophic or stepped, and nor is it certain whether there was an equally dramatic extinction on land at the same time. Certainly many families of tetrapods became extinct in the late part of the Triassic, but Benton (1991) believes that the larger part of the extinction occurred somewhat earlier than the end of the Triassic. The terrestrial plants were generally much less affected, although some sections spanning the boundary do show a significant change, including a fall in the diversity of the seed ferns.

Compared with some of the other major events, the end-Triassic shows a somewhat smaller range of indications of biotic change (Hallam 1996). The most prominent is a large regression right at the boundary, followed by a rapid transgression in the earliest part of the Jurassic associated with anaerobic conditions. For the rest, the evidence is largely negative. There are no signs of global cooling, such as glaciers. It was once believed that massive volcanic activity in southern Africa and North America coincided with the extinction, but both are now known to have occurred later in time. There is also a complete absence of evidence for a major bolide impact, either iridium anomaly, microtektites, or shocked quartz. The Manicouagan Crater in Quebec has been associated with the end-Triassic, but it is now believed to be a few million years too old to have been involved in the event. No strong shifts in stable isotope ratios have yet been discovered. There is a fall in $\delta^{34}S$, but this is only the culmination of a gradual fall that had been occurring throughout the Triassic, and also a small negative shift in $^{87}Sr/^{86}Sr$ (Morrow *et al.* 1996). No $\delta^{13}C$ change at all has yet been recorded.

Given the lack of clear correlation between the terrestrial extinctions and those in the seas, the end-Triassic event is perhaps the one most easily accounted for by the environmental effects of a regression reducing the area of marine basins, followed by a transgression introducing anoxic waters over wide areas of the seabed.

## The end-Cretaceous (K–T) event

The combination of the extinction of the dinosaurs and the discovery of the most extensive and unambiguous signs of a massive bolide impact conspire to make this by far the best known and most studied of the Big Seven events, even though it is only fourth or fifth in magnitude. Raup (1995) calculated a 76 per cent loss of species, although others suspect it was somewhat less severe, with perhaps 65 per cent loss. As well as being the time of final extinction of some of the most prominent groups of both land and sea—notably dinosaurs and ammonoids (Fig. 9.9), respectively—many very different kinds of taxa were more or less equally badly affected. Of these, the planktonic and benthic foraminiferans, most of the then dominant mollusc groups such as the bivalve rudists, the echinoids, and the corals were all severely depleted. On land, however, apart from dinosaurs few reptile or mammal families were lost. The terrestrial plant record indicates that about 60 per cent of angiosperm species became extinct, and there is a prominent spike of fern spores at the boundary. The detailed investigations of the last decade or so have increasingly suggested that there was a sequence of stepped extinction events prior to the actual K–T boundary (Kauffman and Hart 1996).

Evidence of abiotic changes at this time is abundant, including, in fact, almost every signature possible. Most prominently, and uniquely amongst the mass extinctions, there is virtually incontrovertible evidence of a major bolide impact, probably a comet. Building on the discovery of Alvarez *et al.* (1980), enhanced levels of iridium and other elements associated with impacts are now known to occur world-wide and almost certainly synchronously, and both shocked quartz grains and microtektites are found. Furthermore, there is an excellent candidate for the crater caused by the collision, in the Chixulbub structure that has been discovered on the Yucatan Peninsula of Mexico (Hildebrand *et al.* 1991), and there are large tsunami deposits in the Caribbean, suggesting a massive tidal wave following an impact. The end-Cretaceous is also one of the mass extinctions associated with huge volcanic activity, in this case responsible for the formation of the Deccan Traps of much of central India. There is a soot layer suggestive of great forest fires, which could have been caused by either an impact or volcanoes. As for eustatic changes, this period is marked by a major regression. Finally, there are significant shifts in the stable isotope ratios. $\delta^{13}C$ suffered a fall, indicating the onset of a Strangelove Ocean (page 195), and $\delta^{18}O$ rose substantially, probably indicating a period of cold conditions.

**Fig. 9.9** Maastrichtian (latest Cretaceous) chalk community: 1, *Belemnella lanceolata* (belemnite); 2, belemnite guard; 3, *Hercoglossa* (nautiloid); 4, *Bostrychoceras* (ammonoid); 5, *Enoploclytia* (decapod); 6, *Seliscothon* (demosponge); 7, hexactinellid sponge; 8, '*Ostrea*' *lunata* (oyster); 9, *Inoceramus* fragments (pterioid bivalve); 10, *Echinocorys ciplyensis* (echinoid echinoderm); 11, *Galerites* (echinoid); 12, *Cretirhynchia* (rhynchonellid brachiopod); 13, *Pycnodonte vesicularis* (oyster); 14, *Ostrea* (oyster). (After Kennedy 1978.)

Given this complex picture of biotic and abiotic changes, there is unlikely to be a single, simple explanation for the end-Cretaceous mass extinction. Most palaeobiolgists accept that an impact occurred, and that it probably played a significant part in causing the final mass extinction. Nevertheless, it is generally thought to have been only one factor in a complex series of environmental changes, causing a sequence of extinction pulses over the final 2–3 million years of the Cretaceous (Kauffman and Hart 1996; MacLeod 1996).

## Overview of the causation of mass extinctions

Nothing is universally accepted about the cause of mass extinction, although two things are fairly widely agreed. The first is that they are not usually single, instantaneous events but occur over a period of time that is very long on the ecological time-scale, perhaps of the order of 10 000–100 000 years, and often consist of more than one pulse over a period of one to a few million years. The second thing is that they were not all due to the same, single cause. The differences in the abiotic signals between the different events are such as to suggest that different combinations of environmental perturbations can trigger a mass extinction. Indeed, about the only unified explanation that might apply is that a certain level of disturbance of the environment leads to a collapse of the Earth's ecosystem; the disturbance itself can be any of a great variety of kinds.

The most cogent attempt to discover a unified cause of mass extinctions along these lines is the concept of long-term climatic cycling developed by Fischer and Arthur (1977). They proposed that the global climate alternates between greenhouse phases when high levels of $CO_2$ in the atmosphere increase global temperatures, and icehouse phases when $CO_2$ is low, and temperatures fall (Fig. 9.10). The engine driving the cycle is regarded as entirely Earth-based and is due to convection events in the mantle affecting both the volumes of the ocean basins and therefore sea levels, and also the levels of volcanic activity. The mass extinctions occur at times when one phase is giving way to the other, and so the climate is relatively rapidly changing, either to warmer or to cooler conditions as the case may be. While there is a reasonable-looking fit between this model and the timing of some of the big mass extinctions, it is far from exact. Furthermore, even if such a mechanism does fundamentally underlie the environmental perturbations, it could hardly fail to have many other processes superimposed on it. There are the well-known, short-term Milankovitch cycles of environmental change resulting from minor alterations in the Earth's axis, which have periodicities in the order of 10 000–100 000 years. There are extraterrestrial impacts. And presumably there are various unpredictable, chaotic outcomes from the complexities of continental movements, mountain-building and erosion, biotic activity itself, and so on.

Potentially the existence of mass extinctions throughout the Phanerozoic says much about the nature of the global ecosystem and its level of instability. It also says much about its ultimate resistance to complete collapse: as yet no mass extinction has achieved 100 per cent species extinction, despite some extraordinarily near misses.

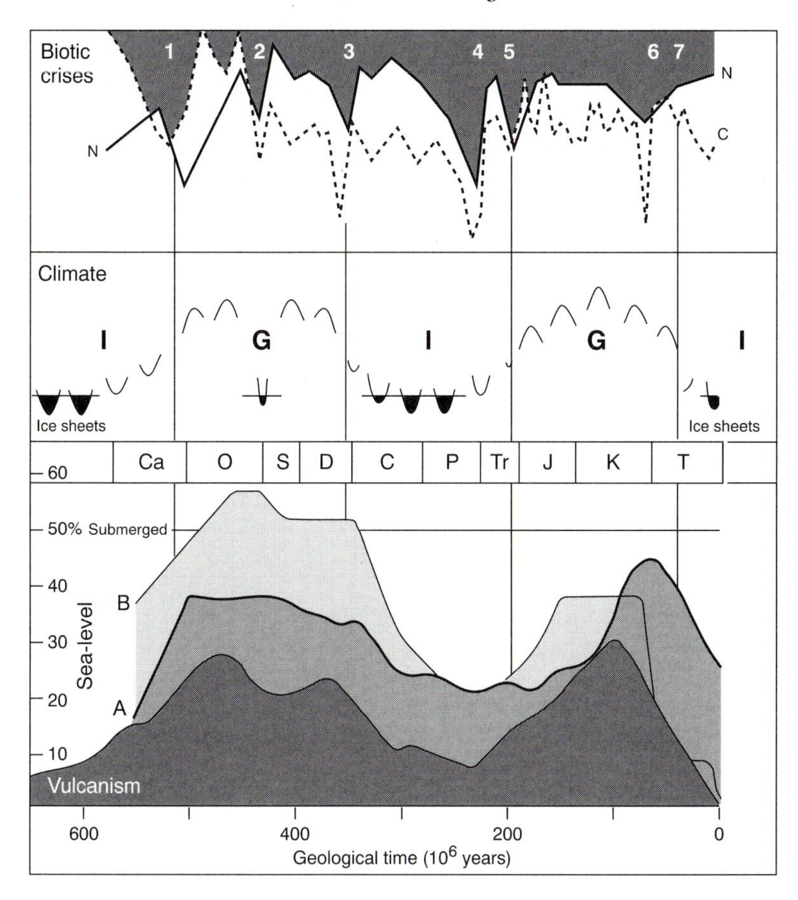


**Fig. 9.10** Fischer and Arthur's theory of Phanerozoic megacycles of global climate change. Note the succession of ice-house (I) and greenhouse (G) conditions, with a tendency for mass extinctions to occur at the points of transition from one to another. (After Fischer 1984.)

# Further reading

The literature on mass extinctions, both the direct and the indirect, is huge and multidisciplinary. Fortunately it is a popular-enough topic to have attracted an on-going sequence of both edited volumes and review texts. Chaloner and Hallam's (1989) edited volume arose from a Royal Society symposium, and contains discussions of several general topics such as Hoffman (1989*b*) on the definition, Jablonski (1989) on differential extinction susceptibility, and Raup (1989) on the Signor–Lipps and hiatus effects. Donovan's (1989) edited volume has a useful palaeobiological overview by himself, but is mostly a series of reviews of the major events. Walliser (1996*b*) has a set of very up-to-date review papers covering all the extinction events, not just the larger ones, and thereby keeping the conventional Big Seven mass extinctions more in context.

Erwin's (1993) monograph is devoted to the end-Permian, and that by McGhee (1996) to the Late Devonian mass extinctions: the latter is particularly readable and contains useful discussion of more-general issues, such as the problems of dating. Hallam and Wignall (1997) is the latest text, and it comprehensively pulls together into one place a huge amount of information about all the major events.

# 10
# The origin of new higher taxa: the ultimate question

Any definition of the category 'higher taxon' is bound to be rather subjective, and will be along the lines of: a monophyletic group of organisms that is sufficiently distinctive and morphologically discontinuous from other contemporaneous taxa. As with all concepts in systematic biology that involve degrees of difference, or disparity (page 67), it is difficult to see how to get away from some such expression as 'sufficiently distinctive'. Nevertheless, it is uncontroversial that at any particular instant in geological time, such as the present, some taxa can sensibly be described as 'higher', because there certainly are large morphological gaps between such respective clusters of species in morphospace. Amongst animals, the phyla and classes are such, as are the roughly corresponding ranks for plants. The essential point of interest is that the morphological distance between these groups is not spanned by *contemporary* intermediate forms; it is this very absence of intermediates that puts palaeontology at the forefront of any contemplation of the processes by which higher taxa and the gaps between them arise over evolutionary time.

Another area of biology that also comes into prominence when discussing the differences between higher taxa is developmental biology. Of course, all heritable evolutionary change involves a modification of the developmental programme of the organism, simply because this is how a genetic mutation expresses itself in the phenotype. At lower taxonomic levels such as species and genus, the differences in the development between related taxa are small and often impossible even to identify. At the level of living higher taxa, however, there are often considerable, easily observable differences in the way the respective organisms are assembled during development. Understanding these differences and the inferred modifications to the ancestral developmental programmes that caused them adds another potential level of understanding of the complex of events that is responsible for the diversification of higher taxa from one another (see, for example, Thomson 1988; Raff 1996; Carroll 1997).

No part of evolutionary theory has suffered more from lists of speculative concepts being passed off as explanations than the study of the origin of higher taxa: concepts that tended to be of more philosophical interest than empirical utility. However, there are grounds for believing that empirical study is catching up with conjecture, as the molecular genetic revolution increasingly embraces developmental biology. At any event, an appropriate framework for study is needed and this can be developed by temporarily disentangling three closely interconnected conceptual questions. After that, some real examples in the fossil record will be used to illustrate the way in which the questions can be addressed and the extent to which they can be answered.

## Questions and speculations

### Does the evolution of a higher taxon differ from 'normal' evolution?

In principle, the morphological distinctiveness of a higher taxon, and therefore the taxon itself, could arise by a succession of perfectly normal speciations coupled with normal extinctions of the intermediate species. After enough time had elapsed, the new taxon would be sufficiently distinguishable from any other to be regarded as 'higher'. This interpretation implies that there is no difference between the respective processes by which lower and higher taxa evolve. On the other hand, it is also conceivable that the distinction between higher taxa comes about because of some difference in the process of evolution from that which causes microevolutionary change and speciation. Numerous authors have proposed that either special environmental circumstances or special kinds of genetic changes are associated with the emergence of the radically new kinds of organisms represented by higher taxa.

It may simply be that at certain times the sequence of speciation and extinction events was particularly rapid, so that the emergence of the new higher taxon itself was exceptionally fast, even to the extent that it appears in the fossil record to have been instantaneous. One imaginable cause of an accelerated rate of change is the invasion of a new adaptive zone, for which many new adaptations are required (Fig. 10.1). The concept of a key innovation is sometimes invoked as part of this explanation. It is supposed that some particular modification to a character acts as the initial adaptation to the new habitat, opening the door as it were to the evolution of many subsequent adaptations within the habitat. The result is the new kind of organisms constituting the new higher taxon, which then rapidly diversifies to fill the many niches available within the new habitat. Long ago, Simpson (1944, 1953) termed this process quantum evolution. Another possible cause of the rapid origin of a new higher taxon might be the circumstances prevailing after a mass-extinction event. Again it may be imagined in simple terms that many vacated niches became available to a surviving taxon whose member species had acquired suitable adaptations for existence in the new circumstances.

As an alternative to this idea of the origin of higher taxa being due to an acceleration in the rate of occurrence of a series of small evolutionary steps,

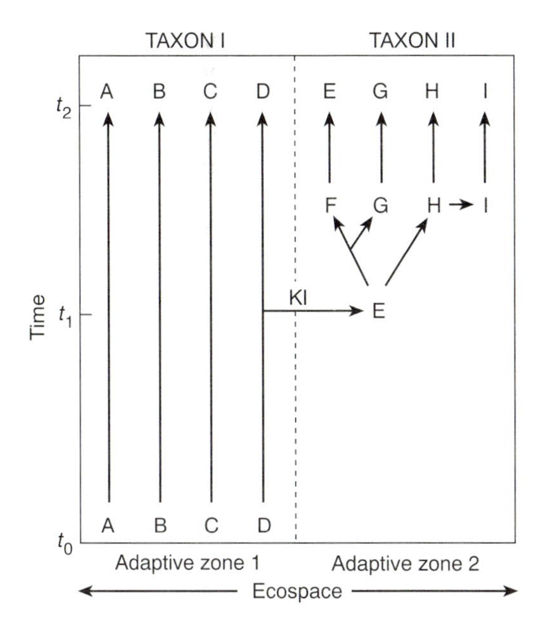


**Fig. 10.1** Model of quantum evolution, in which a key adaptation (KI) occurs, permitting the invasion of, and rapid expansion within a new adaptive zone, leading to a new higher taxon.

it could be due to a small number of relatively large individual transformations. In its extreme form, the instant origin of the new kind of organism in a single macromutational jump is regarded as at best heterodox, and at worst unscientific. Schindewolf's typostrophism (page 16) is usually considered to be barely more worthwhile than, say, Hoyle and Wickramasinge's (1986) panspermia theory that the major evolutionary transformations were caused by virus-like particles from Outer Space. Over the last decade or so, however, molecular genetics and developmental biology have begun to reveal some distinctly possible mechanisms for the origin of relatively large, instantaneous morphological transitions, and these now have to be considered rather more seriously.

One of them is heterochrony (Fig. 10.2), in which the rate and timing of development of different parts of the organism can vary as a result of mutations in control genes. This can cause the organism to achieve sexual maturity and therefore adulthood at a juvenile stage of development, referred to generally as paedomorphosis. Alternatively, the process of peramorphosis occurs when sexual maturity is delayed while morphological development continues beyond what was the adult stage of the ancestor. Heterochrony can also have the effect of shuffling combinations of juvenile and adult characters in the same adult body, with effectively increased and decreased expression of different attributes. A surprising number of evolutionary transitions seen in the fossil record can be explained by heterochrony (McKinney and McNamara

1991; McNamara 1995; Raff 1996). Indeed it should be recalled that the most widely respected proposal of all for the macromutational origin of a new higher taxon is Garstang's neoteny theory of the origin of the vertebrates. His hypothesized evolution of precocious sexual maturity in the motile tadpole larva of a sessile ascidiacean sea squirt is a case of heterochrony.

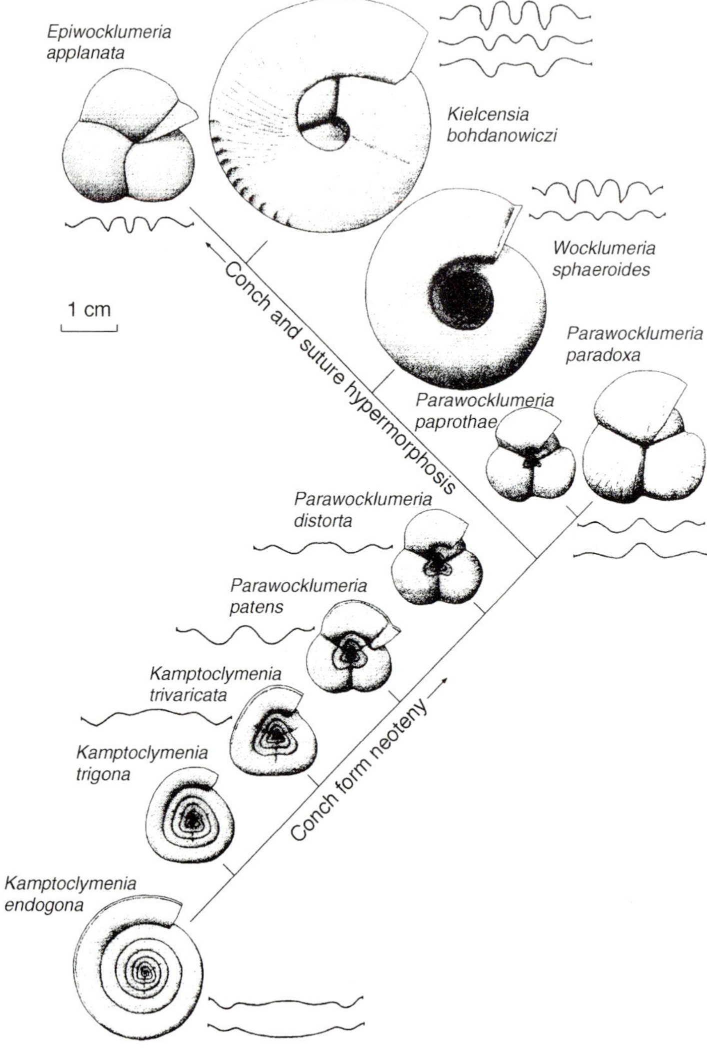


**Fig. 10.2** Evolution of certain late Devonian ammonites showing apparent heterochrony. The transition from *Kamtoclymenia endogona* to the genus *Parawocklumeria* involves retention of the juvenile triangular shell form in the adult of the descendant, which is neoteny. Evolution of the several genera from *Parawocklumeria* involves extension of the growth period of *Parawocklumeria* as indicated by the increasing complexity of the suture lines, a process of hypermorphosis. (After Korn 1995.)

The homeotic (Hox) genes control positional information in developing animals, and homologous versions of these genes are now known in virtually all major animal taxa (see Fig. 3.4). Comparative studies of the role of the Hox genes in different taxa, and particularly the way in which they control the development of patterns of segmentation, is beginning to hint strongly that significant changes in the body plans of various organisms could have arisen by relatively simple patterns of mutations in these genes (S. Carroll 1995; Holland and Garcia-Fernandez 1996). It is premature to assert that this is the case, but in the near future it may well become clear from hard experimental evidence that such macromutational modification of development by single mutations in regulator genes is not as far-fetched as had been supposed. Indeed, it may eventually prove that the apparent suddenness of the appearance of new taxa in the fossil record is a close reflection of reality after all, based on mechanisms such as these.

At the level of the whole genome as well, there is some evidence that at least twice in the evolution of vertebrates a polyploidy event causing multiplication of the entire genome was involved at critical stages. Comparative evidence suggests that the first occurred at the time of origin of the group as a whole, and the second at the time of origin of the jawed fishes (Ohno 1970, Arthur 1997). If substantiated, these would also surely count as macromutational events.

All in all, there are no empirical grounds for dismissing the possibility of a macromutational origin of new groups by such genetic processes. The argument that in principle the probability of a macromutational shift being successful is too low, loses ground to these discoveries at the gene level. It loses even more ground in the light of growing evidence about the ways in which organisms can maintain their structural and functional integrity in the face of major modifications, as discussed shortly.

A third way in which the evolution of a new higher taxon might imaginably differ from normal evolution is by an imposed directionality on the course of evolution, from ancestral towards derived morphology. This possibility is discussed in the form of the second of the three questions.

## What causes the evolutionary trend that results in a higher taxon?

Leaving aside the possibility of an instantaneous macromutational event, the origin of a higher taxon must be assumed to consist of a sequence of intermediate forms, between the ancestral and the descendant morphologies. Therefore the evolutionary pattern can be described as a trend, which is to say a period during which the organisms evolved in an approximately constant morphological direction for long enough to accumulate the differences they have from the related contemporary taxa. It must be stressed that in the present context, the idea of the evolutionary trend does not necessarily imply constancy of rate, or even absolute constancy of direction. It merely indicates that evolutionary change over a relatively long span of morphospace occurred via a series of intermediate points. Generally speaking, there have been four kinds of explanations offered for long-term trends, as so defined (Kemp 1988*b*)—chance, orthogenesis, natural selection, and species selection.

### *Chance*

Care must be taken that what is described as a trend actually is a significantly different pattern from the overall evolutionary pattern within the rest of the phylogeny. In the case of a complex phylogenetic tree that consists of many diverging, splitting, and disappearing lineages, it is always possible to pick out a wandering, arbitrary route through it, remove that particular sequence and, as it were, straighten it out. What is now seen in isolation is an apparently special evolutionary trend compared with the rest of tree. Viewed in the context of the whole tree, however, there may be no reason to suppose that this selected lineage evolved in any significantly different way from any other lineage that could have been extracted and looked at in isolation, and therefore no reason to believe that any special process was at work generating it (McNamara 1990).

The risk of this fallacy is probably greatest when considering the origin of vertebrate groups, such as tetrapods, mammals, and, probably worst of all, humans. A combination of the special interest shown to taxa of which humans are members, with a subconscious vestige of the *scala natura* of pre-evolutionary times, may lead to a very distorted impression. Cladograms, by their very form, may also unintentionally convey this false impression that there is something special about the taxa arranged directly off the main stem compared with those occupying various lateral branches.

### *Orthogenesis*

In the days when palaeontologists investigated evolution in the fossil record with little reference to contemporary ideas being developed by neontologists (page 14), a frequent explanation of a trend was that there are internal 'forces' driving evolutionary change in a particular direction. From Eimer's original concept of orthogenesis, through such proposals as the aristogenesis of Osborn and the nomogenesis of Berg, it was believed by many that vitalistic forces were at work, and no more explanation was possible or required. As has been discussed elsewhere (page 31), more important, strictly mechanistic versions of orthogenetic-like explanations have been proposed, not based on unknown laws verging on the non-material but on the idea of the existence of constraints to evolutionary change. If an organism is so constructed that there are narrow limits to the ways in which it could change without losing viability, then evolutionary change would in practice only be possible in those permitted directions.

At present, however, there are no serious cases of the evolution of higher taxa where there is any evidence at all for such extreme constraints, and therefore of a role for even the most respectable versions of 'neo-orthogenesis'.

### *Natural selection*

The standard neo-Darwinian explanation for evolutionary trends is that they are the result of natural selection acting on organisms and generating continually improving adaptation to some particular environment, a process that has been referred to as amelioration. An alternative kind of circumstance that would have a comparable effect is adaptive tracking of a changing environment. If the environment in question changed slowly enough and extensively enough in a certain consistent direction through space or time, the evolving lineage could be imagined as evolving to maintain an adequate level of adaptation in a trend-like fashion, until sufficiently divergent from other taxa to be recognizable as a new higher taxon.

For most evolutionary biologists, this category of explanation is automatically accepted by default. As is clear from the case studies below, however, the necessary empirical evidence that a trend leading to some particular higher taxon was indeed driven by natural selection and was therefore a response to specified environmental features is virtually impossible to acquire because it is not preserved in the fossil or stratigraphic record. Accepting natural selection as the explanation for a particular trend is based on an assumption about biology in general, not on adequate observations of fossils and their palaeoenvironments in particular (see Hoffman 1989*a* and Carroll 1997 for extended essays based on the adaptive point of view of the origin of new higher taxa).

### *Species selection*

The most recent kind of explanation offered for trends lies in the possibility that certain kinds of species may have higher probabilities of speciating than others, leading to a continuing process of species selection (page 41). For this category of explanation to apply, the reason why one species has a greater probability of speciating than another has to be due to a difference between the species in some unambiguously species-level character. The obvious species-level characters that could affect the probability of speciation are population size and structure. For the process of species selection to create a continuous morphological trend, however, there would also have to be some recognizable morphological characteristic of the organisms that in turn was the cause of the different population characteristics (Gould 1990).

Suppose, for example, that organisms with larger body size tended to live in the sort of population whose population structure made it more probable that speciation would occur. At any instant, the species consisting of the larger organisms would be the most likely to speciate. Any resulting daughter species with yet larger body-sized members would in turn be the one with the highest probability of speciation. As this process was repeated over evolutionary time, the effect would be a tendency to evolve a succession of species of increasingly large organisms. An explanation such as this is reasonably acceptable, although not necessarily correct, of course, for a simple morphological trait such as body size. Even a more-complex character such as digit reduction in the lineage within the horse clade from the Eocene *Hyracotherium* to the Holocene *Equus* could conceivably be explained by species selection. If it were the case that reducing numbers of toes correlates with an increasing tendency to form a certain population pattern, and that that pattern itself tended to increase the probability of speciation occurring, then a trend in toe reduction, driven by species selection, would occur over time.

It appears at first sight rather far-fetched that the process of species selection should play any part in the explanation of trends leading to higher taxa. But at least one of the case histories to be discussed shortly, that of the origin of mammals, indicates that species selection must certainly be taken seriously as a possible mechanism.

## How is morphological integration maintained?

What distinguishes a higher taxon in the sense used here is the relatively large number and extent of character differences that its members show compared with those of its contemporary sister group, and therefore the extent of character evolution that has occurred in at least one of them. In any individual organism, the characters are highly integrated with one another to produce the complex, multifunctional entity itself, and this integration is presumably as true of any intermediate stages as it is of the ancestral and its descendant organism. Therefore the question arises of how numerous separate characters can change, and yet at every stage the respective intermediate organism was still adequately integrated. One response to this problem has been the claim that intermediates could not be viable, integrated organisms, and therefore could not actually have existed. New higher taxa must necessarily arise spontaneously from their ancestors by a macromutational jump through morphospace. The idea is that the ancestral species, and the derived species initiating the new higher taxon, represent two respective forms of stable organization, and that transition from the one to the other must be by a revolutionary reorganization of the whole organism in one go. It is actually impossible to refute this explanation in most cases beyond an assertion about the biological improbability of such events (but, of course, how this probability is to be measured remains vague). At any event, no such extreme solution is necessary, for there are at least two moderate, and mutually compatible, mechanisms available for maintaining integration in a series of intermediate forms—correlated progression and developmental feedback.

## *Correlated progression*

This concept (Fig. 10.3a) explains the maintenance of integration between many characters during extensive evolutionary change in all of them by supposing that there is some degree of looseness in the functional interrelationships between the various structures and processes of an organism (Thomson 1966, 1988; Kemp 1982*a*, 1985*b*). Therefore, despite the need to remain integrated with all the others, any particular character is capable of some degree of variation within the organism, and can change to a small extent, in response, for example, to a selection pressure. Only a relatively small degree of change will be possible, however, beyond which the character would cease to be well integrated with the rest of the organism. At this point, further evolution of that character would be impossible unless and until other characters, each behaving in the same way, caught up with it, as it were. The first character would now be able to evolve a further incremental step.

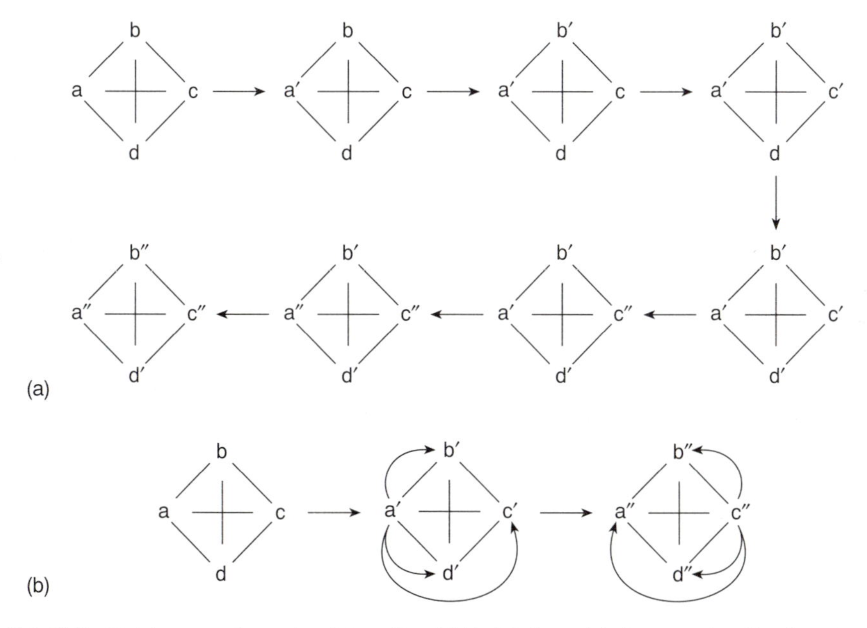


**Fig. 10.3** Maintenance of organism integration. (a) Model of correlated progression: the characters a, b, c, and d are loosely integrated with each other. No more than a small evolutionary change in any one may occur unless and until similarly small changes in the others have occurred and accumulated, maintaining integration during phylogeny. (b) Model of developmental feedback: again, the characters a, b, c, and d are integrated. A change in one character such as 'a' induces appropriate compensatory changes in other characters during the process of development, so the adult is still an integrated organism. A further change such as in c′ has the same effect.

If all the characters behave in this way, then they can all be imagined as evolving in partial but not completely tight correlation with one another, rather like a line of people walking hand in hand can progress forwards as a whole without ever losing go of one another, but with some lagging slightly behind at times and others a little further ahead. Throughout the evolution of the trend, adequate functional and structural integration is maintained to ensure the viability of all the intermediate-stage organisms.

### *Developmental feedback*

A century of experimental embryology has taught that the development of an organism involves a great deal in the way of feedback systems to maintain integration of the resultant adult (Fig. 10.3b). Mechanical stresses acting on a bone directly cause it to adopt a shape suitable for resisting those very stresses; spinal nerves do not grow in predetermined directions, but are caused to go to wherever the developing muscles lie; tissues and organs adopt forms appropriate to their respective positions in the embryo; and so on.

The detailed ways in which these relationships arise and are maintained remain obscure although molecular genetics is starting to elucidate the mechanisms involved. For example, the Hox genes that occur in most animal groups are examples of high-level regulators affecting the position and form of whole segments. Mutations in these kinds of genes can generate large morphological changes, but without losing functional integration between the different tissues and organs within the segment. At the lower levels in the hierarchy of developmental processes, the role of surface interactions between cells and molecular controls of induction of one tissue by another are beginning to be understood as mechanisms involved in feedback control (see, for example, Lawrence 1992; De Robertis 1996).

Applied to the origin of new kinds of organisms, the control and feedback mechanisms could certainly in principle ensure that organism-level integration is maintained, even in organisms in which one or more of the characters has been affected to a large degree by a mutation. As an explanation for how higher taxa can evolve, there is nothing new about invoking developmental feedbacks, or about expressing confidence that a great deal about them is shortly to be learned. With the advent of the field of molecular embryology, this time it may be true (Depew and Weber 1996).

## Case studies

Armed with these various concepts bearing on possible mechanisms of the origin of higher taxa, some real cases may now be considered. The vertebrate fossil record offers the most detailed examples. In part, this is because there are so many characters preservable in the vertebrates, and because they are largely internal skeletal characters that are relatively informative about biological aspects of the organism, such as modes of feeding, locomotion, brain size, etc. It is also in part due to the fact that the main vertebrate groups generally evolved much later in time than the invertebrate higher taxa, and so a greater volume of relevant sediments containing fossils from the critical periods may be available. It is also possible that the environmental circumstances and evolutionary processes involved in the origin of vertebrate taxa tend to be different from those in the origin of invertebrate groups and plants, perhaps because of the higher levels of motility and metabolic rate of the former. Examples of all kinds of need to be considered before attempting to generalize about major evolutionary transitions.

## The origin of mammals

There can be little argument that the class Mammalia is a higher taxon in the sense in which this term is being used, given the number of differences in characters between its members and those of its living sister group the Sauropsida (extant 'reptiles' plus birds). The origin of the mammals is also by far the best-represented transition of all in the fossil record at this general taxonomic level. Moreover, quite apart from any conclusions to be drawn about processes of evolutionary change that may have been involved, it offers an extraordinarily good illustration of the way in which hypotheses of process at this scale of events can be inferred from the taxonomic pattern of distribution of fossils and their characters over time. The whole field was reviewed by Kemp (1982*a*, 1985*b*) and this case has been used in several textbooks as a paradigm for major evolutionary change (see, for example, Levinton 1988; Ridley 1996). All these authors interpreted the fossil record along straightforward neo-Darwinian lines, attributing the inferred evolutionary changes in anatomy to simple adaptation by natural selection. Alternative interpretations cannot be ruled out, however, and the story is certainly a lot less simple than hitherto supposed.

The first step in the analysis is to recognize all the fossils constituting the stem group Mammalia—namely those that have some but not all the characters defining modern mammals (page 127). The fossils in question are widely, a little misleadingly, but affectionately called the mammal-like reptiles (synapsids), and occur from Late Carboniferous (Pennsylvanian) times onwards (Fig. 10.4). By the latest Triassic and earliest Jurassic, skeletons of small animals such as *Megazostrodon* (Fig. 10.5) are found with nearly all, if not completely all, the skeletal characters of the crown group Mammalia. Throughout this considerable period of time of about 100 million years, a complex pattern of originations and extinctions of synapsid taxa occurred, with an overall but erratic trend through time of organisms with increasing numbers of mammalian characters (see Sidor and Hopson (1998) for a more detailed analysis). A cladogram (Fig. 10.5) based strictly

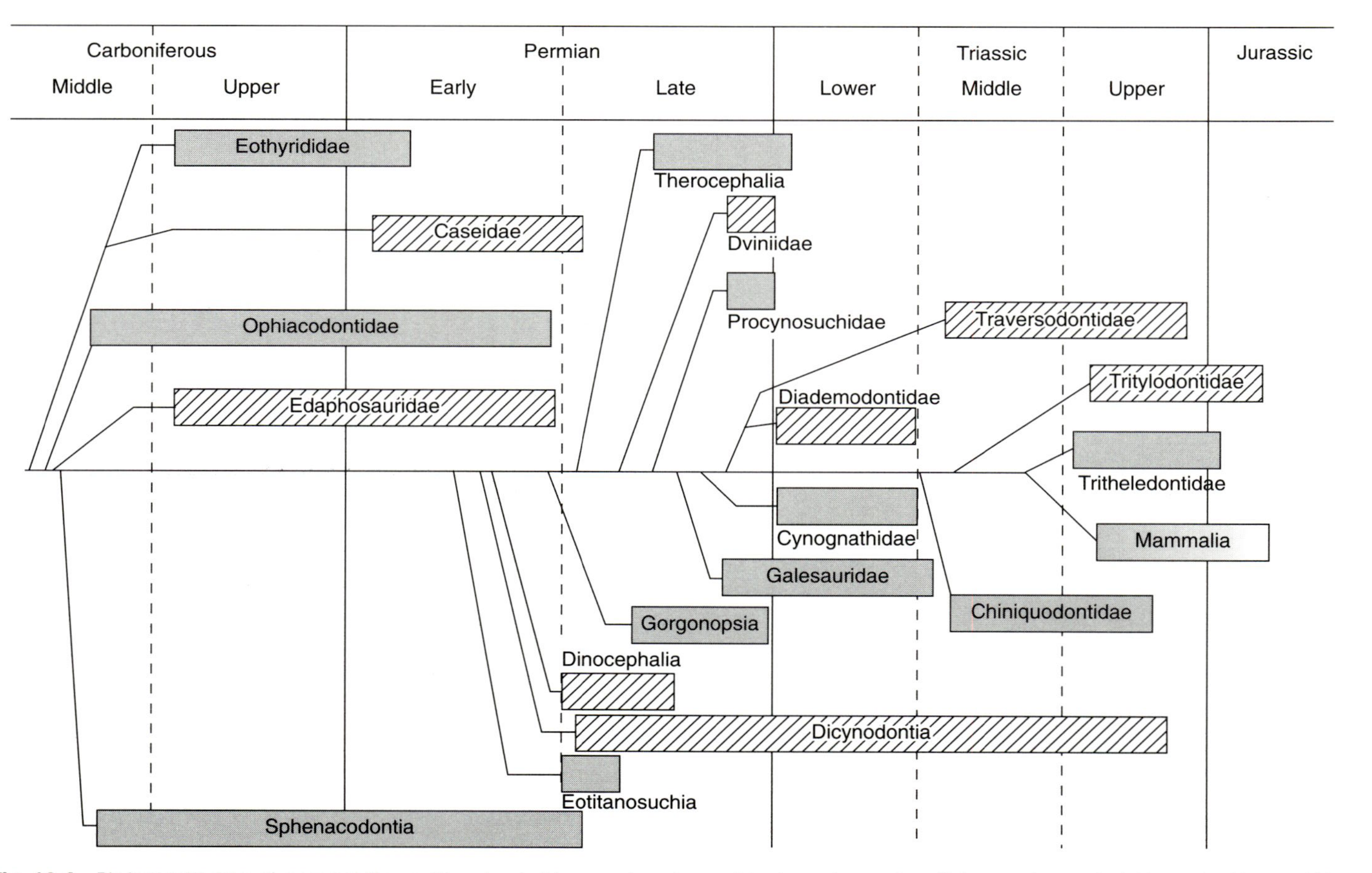


**Fig. 10.4** Phylogenetic tree of mammal-like reptiles showing temporal ranges and basic ecotypes: Grey tint = carnivores; hatching = herbivores. (After Kemp 1982*a*.)

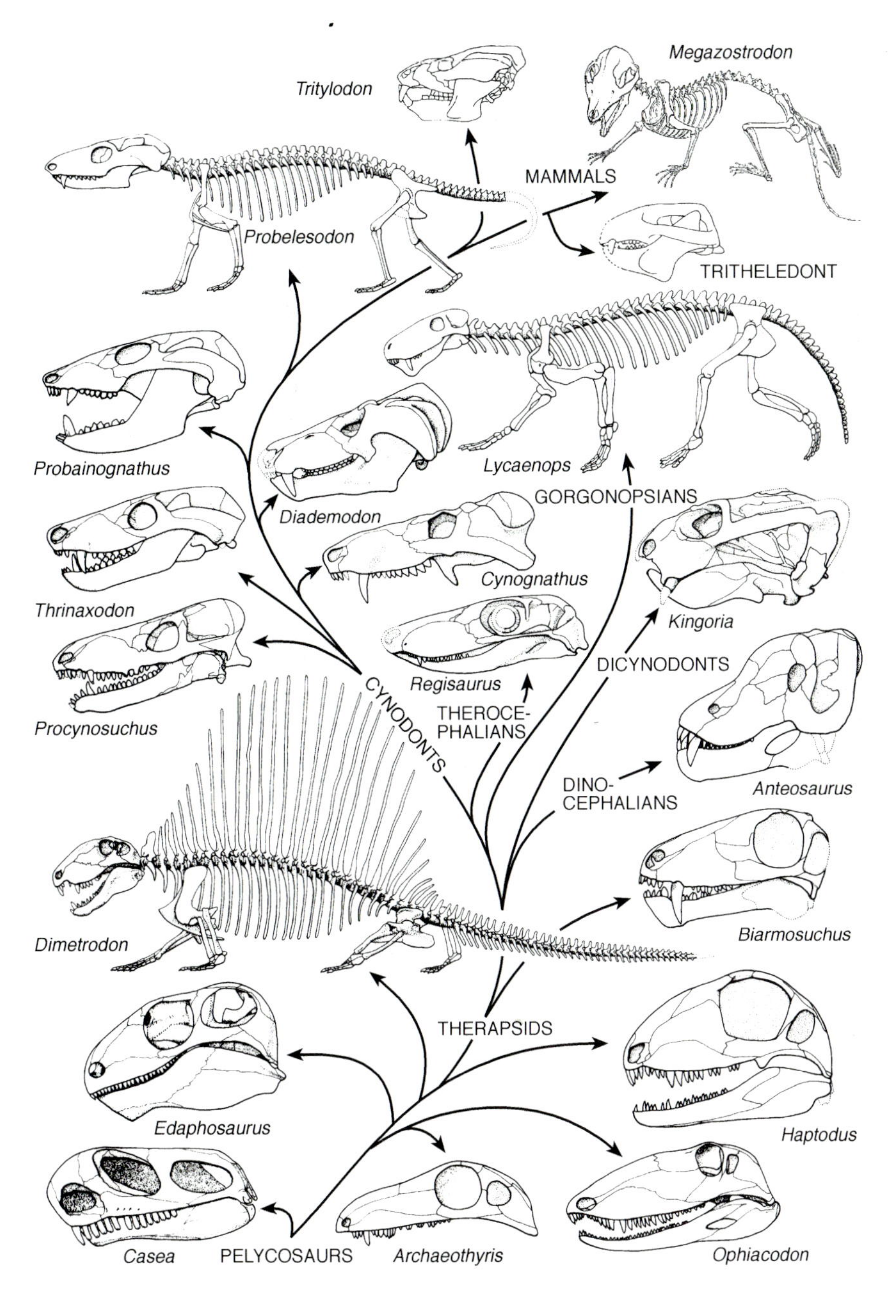


**Fig. 10.5** Phylogenetic relationships of the stem-group mammals. Although unconventionally attractive, this diagram is rigorously cladistic in content! (After Kemp 1982*b*.)

on morphological characters, and ignoring the possibly misleading relative stratigraphic positions and geographical localities, can be constructed. This constitutes the basis for all subsequent inference and speculation about the origin of mammals.

First, by assuming that the cladogram represents the best hypothesis of evolutionary relationships of the included organisms, it can be read as a phylogenetic tree with hypothetical ancestral taxa occupying the respective nodes. From this point in the analysis, there are two ways of looking at the particular sequence of hypothetical ancestors and descendants that constitute the lineage from the most primitive (plesiomorphic) mammal-like reptile to the first mammal. One way is to see it in the context of the overall evolutionary tree of the mammal-like reptiles, set against the stratigraphic column (Fig. 10.4). The lineage has no strikingly orthogenetic appearance, but is apparently picked out from a highly complex tree with major radiations arising, flourishing, and declining around it. While the absolute times of the branchings within the tree cannot be known, the latest dates for these points are given by the ages of the various derived fossils, and there is no more an impression of constancy of rate of evolution than there is of constancy of direction.

If, however, the hypothetical lineage is looked at in isolation of the rest of the tree, extracted and straightened up as it were (Fig. 10.6), a more coherent pattern emerges. The sequence of hypothetical ancestral forms as reconstructed from the actual fossils gives the sequence of acquisition of mammalian characters along the lineage, or to be exact, gives all the information available from the fossil record about this sequence. Sidor and Hopson (1998) compared the number of character changes with the estimated length of time between successive nodes, and showed a reasonably good correlation. Overall, this implies a fairly constant rate of morphological change. As far as the particular character changes are concerned, the first regularity that emerges is that one of the character states at each of the nodes (in other words the plesiomorphic character state for that part of the cladogram) seems always to emerge as small body size relative to the range of sizes of the various immediately descendant fossil species. A second feature is that the dentition and the indications of jaw musculature are those of a carnivorous rather than a herbivorous animal. A third feature of the pattern is that each node represents the simultaneous addition of an incremental step towards the definitive mammalian state in characters of more than one morphological system. The morphological systems that impress themselves on the skeleton of the mammal-like reptiles, and whose evolution can therefore be followed, are primarily the feeding and locomotory systems. But quite a lot can be inferred more indirectly from skeletal features about the ventilation system, from the ribcage, the olfactory and hearing sensory equipment from the skull, and the brain size from its cranial cavity. None of these features undergoes major transitions towards

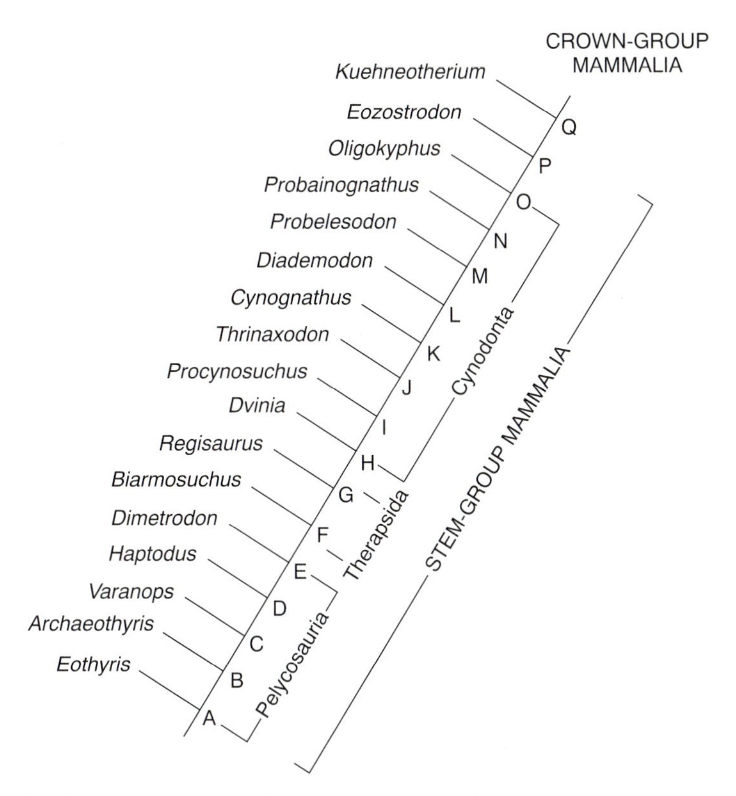


**Fig. 10.6** The lineage of hypothetical ancestors for inferring the sequence of acquisition of mammalian characters. (After Kemp 1982*a*.)

the mammalian condition in isolation, but each gives the appearance of evolving in correlation with the others.

The overall sequence of acquisition of characters, therefore, corresponds to the pattern described earlier as a correlated progression (page 225). The cladogram as a hypothesis of relationships of the stem-group mammals therefore implies that the evolutionary trend culminating in the origin of mammals consisted of a sequence of relatively small carnivores that evolved increasingly mammalian characteristics of different structures and systems of the body more or less simultaneously. There is no indication of a single key adaptation leading to the rapid evolution of the mammalian condition, nor of a macromutational jump from non-mammalian to mammalian form in a single event. It is apparent, however, that, so far as can be extrapolated from the dates of the actual fossils, the rate of evolution of the morphology was not at all constant. At certain times—for instance, during the Late Permian—rather greater morphological transitions are found than during other times, such as the Late Carboniferous.

The next step towards understanding the origin of mammals is to interpret the functional and biological significance of the evolving mammalian characters. Mammals are essentially organisms that have evolved a very high ability to regulate their body temperature, internal chemical composition, and position in space in the face of the heterogeneous and fluctuating terrestrial environment in which they occur. Temperature is maintained constant with the help of the high metabolic rate coupled with a series of devices for controlling the conductivity of the body surface, such as variable insulation and rate of blood flow through the skin. The elevated metabolic rate is achieved by, among other features, a high rate of food collection made possible by the sophisticated locomotory, sensory, and central nervous systems. A high rate of food ingestion is achieved by the complex dentition and jaw musculature. Elaboration of the rate of gas exchange is achieved by the enlarged lung, the diaphragm, the modified locomotory gait that allows higher levels of ventilation whilst active, and the rapid rate of blood flow and gas exchange at the tissues that is possible with complete double circulation. The high blood pressure is also important for a high ultrafiltration rate of the kidney and consequently for the rapid and precise regulation of the body fluids.

All these characteristics of mammals are interrelated with one another and all contribute to the very high level of overall homeostatic (regulatory) ability typical of the group (Fig. 10.7). Many of them are also discernible from the skeletal characters preserved in the fossil mammal-like reptiles. Many others are not preserved either directly or indirectly in the fossilized skeleton but, given the essential integration of the organism, it can be assumed that the soft characters underwent the same pattern of evolution as the associated hard characters. From this functional point of view, the lineage of hypothetical ancestors culminating in the mammals is inferred to have consisted of a sequence of forms with ever-increasing homeostatic abilities.

To summarize, the origin of the mammals is inferred to have involved a trend of relatively small carnivores that gradually increased their level of homeostasis by a correlated evolutionary progression of all the characters observable in the fossils and, by implication, of the non-preserved soft characters as well. This translates into a trend of ever-increasing ability to regulate body temperature accurately, to maintain a constant internal chemical environment, and to live an active life in a spatially heterogeneous physical habitat. It also embraces, by implication, the evolution of the characteristic mammalian method of provision of a high level of nutrition and a controlled physical environment for the vulnerable juvenile stages. Although viviparity evolved later, within the crown group Mammalia itself, lactation and use of a nest or burrow must surely have appeared within the stem group. The whole process took some 100 million years from the origin of the stem group to the achievement of virtually complete mammalian status. The evolving lineage is part of an overall phylogenetic

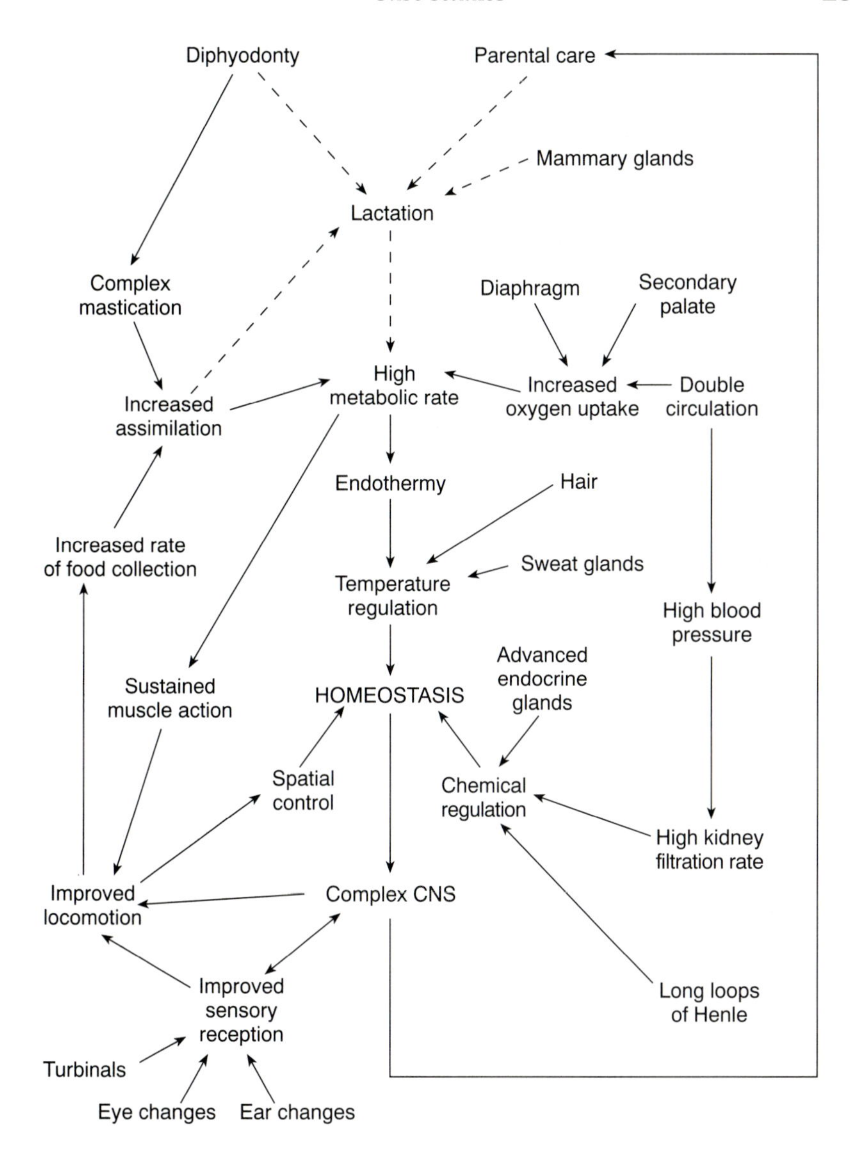


**Fig. 10.7** Integration of the structures and functions of mammals. The dashed arrows relate to the juvenile state. After Kemp 1982*a*.

pattern in which other kinds of mammal-like reptiles, particularly herbivores and large carnivores, occur, and which often show their own high levels of diversity and abundance. These latter ecotypes, however, did not themselves undergo long-term evolutionary trends towards increasingly mammalian biology. Instead, they appeared and flourished at several respective levels of 'mammalness', only to become extinct and to be replaced by new taxa occupying the same roles that had evolved afresh from the persisting lineage of small carnivores (Fig. 10.4).

The cause of the main trend has usually been taken to be a simple neo-Darwinian process, whereby the more homeostatic the organism, the better adapted it is to the particular environment occupied by the species forming the lineage. There are certain features of the pattern that do not support this explanation. One is the variable rate and general slowness of the trend to reach maturity, although this could always be accounted for by an *ad hoc* assumption that the rate of appearance of suitable mutations limited the rate of evolution. A much more serious problem is why only relatively small carnivores are involved in the inferred sequence of ancestral and descendant stages. If natural selection favours the more mammalian forms, why do the herbivore and the large carnivore taxa not also undergo the same selection process, rather than these ecotypes being successively replaced by new, more mammal-like versions that had evolved directly from the main lineage of small carnivores?

The simplest explanation for the anomaly is species selection, such that relatively small, relatively more homeostatic carnivores have a greater probability of speciating. This, and this alone, would be enough to cause a significant trend over a long period of time, whereby small carnivores constituted the lineage leading to ever-increasing mammalness. This is not to say that the other ecotypes were less well adapted as organisms, only that they had a lower probability of splitting into new species. The slow and irregular time-course of the trend may also be accountable for by the species-selection hypothesis. First, species selection is a much longer term process than organism selection, being based on a speciation rate that could feasibly average one every couple of million years. Second, the variance in speciation rate is likely to be high enough to prevent anything like a constant rate of change.

Here is an inferred pattern of evolution of characters that is most easily accounted for by a true species-selection process (page 223) superimposed upon organism-level selection, and explaining otherwise inexplicable aspects of the pattern. Whether there is any reason to expect relatively small, relatively homeostatic terrestrial tetrapods to speciate at a higher rate must be speculative, but potentially susceptible to testing from modern organsims. Such organisms do tend to form populations with low home ranges and low levels of vagility. They may also be expected to have lower population densities than either equivalent ectothermic species with lower food requirements, or similar-sized herbivores that have more abundant food resources. Whatever the detailed mechanism may be, species-level factors of this nature could reasonably be expected to have an influence on probability of speciation.

## The origin of other vertebrate groups

In the last few years, the fossil record bearing on other major transitions in the vertebrates has been improving, although none yet approaches the quality or quantity of the stem-group mammal record.

It is now overwhelmingly accepted that birds are related to theropod dinosaurs (Fig. 10.8a), indeed that dinosaurs themselves are strictly speaking stem-group birds because they all had some modern avian characters, such as digitigrade hind legs. Until relatively recently, the only significant fossil form between the dinosaur grade and the modern birds was *Archaeopteryx*, with its well-known combination of primitive amniote and avian features. Its anatomy has been long-debated in the context of the origin of flight, but on its own it cannot throw much light at all on the evolutionary processes involved in the origin of modern birds beyond illustrating a single hypothetical stage in a presumed lineage consisting of many stages.

An ever-increasing number of new stem-group birds showing other combinations of ancestral and derived avian characters have been found in recent years, so a clearer picture is beginning to emerge (Carroll 1997; Padian and Chiappe (1998 *a,b*)). At levels below *Archaeopteryx*, a sequence of dinosaurs with increasingly avian characters can be identified. They are all members of the Theropoda, the bipedal carnivorous group. The closest related group to *Archaeopteryx* (Fig. 10.8a) are the dromaeosaurs such as *Deinonychus*, which have three well-developed fingers, reduced caudal vertebrae and stiffened tail, and some avian features of the hindlimb and pelvis. The very lightly built and superficially bird-like *Compsognathus*, which is found in the same Solnhofen lithographic limestone as *Archaeopteryx*, is a possible sister group of dromaeosaurs plus birds, while genera such as *Tyrannosaurus* are more plesiomorphic still. There are several forms of bird-like fossils now known that lie between *Archaeopteryx* and modern birds (Hou *et al.* 1996). *Confuciusornis* is a Chinese genus. It is about the same age as *Archaeoptreyx* and is similar in most respects except for the replacement of the toothed jaws by a horny beak. A slightly later Lower Cretaceous Chinese form is *Sinornis*, which also has many of the primitive features found in *Archaeopteryx*, but in combination with the modern features of a reduced number of vertebrae, loss of the elongated tail, a wing-folding mechanism, and an advanced perching type of foot. *Iberomesornis* from Spain is at a similar level.

Cladistic analysis of these new forms is beginning to make possible a hypothesis for the sequence of acquisition of avian characters. The first question to ask is whether the origin of birds was a single lineage embedded within a large and complex overall phylogeny and as part of the Mesozoic dinosaur radiation; the answer is clearly that it was. Amongst the stem group were dinosaurian taxa that themselves underwent major radiations. Even above the level of *Archaeopteryx*, there is growing evidence that the phylogeny

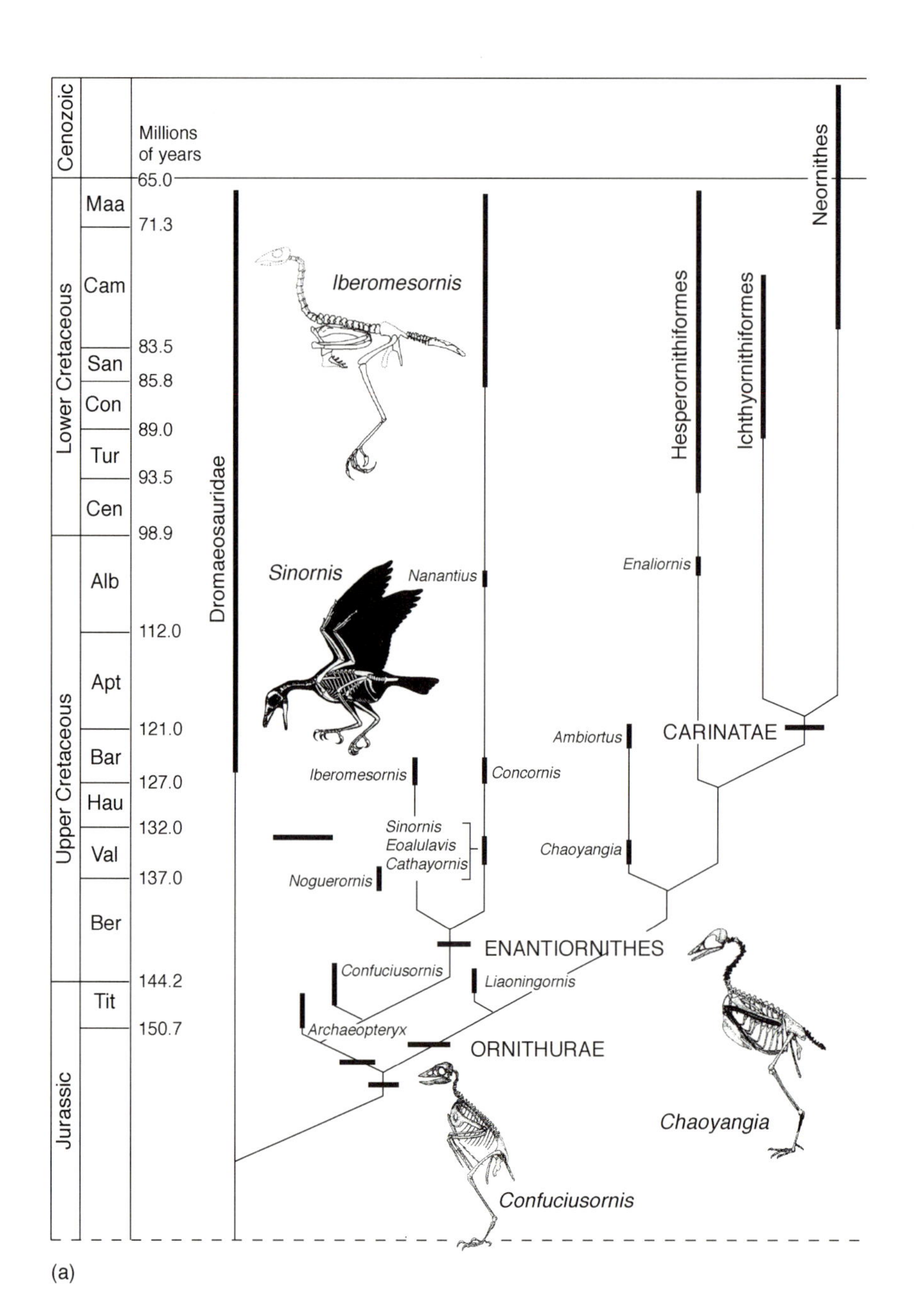
Cenozoic
Millions of years
65.0
Maa
71.3
Cam
Lower Cretaceous
83.5
San
85.8
Con
89.0
Tur
93.5
Cen
98.9
Alb
112.0
Apt
Upper Cretaceous
121.0
Bar
127.0
Hau
132.0
Val
137.0
Ber
144.2
Tit
150.7
Jurassic
Dromaeosauridae
Iberomesornis
Sinornis
Nanantius
Iberomesornis
Concornis
Sinornis
Eoalulavis
Cathayornis
Noguerornis
Confuciusornis
Liaoningornis
Archaeopteryx
ENANTIORNITHES
ORNITHURAE
Confuciusornis
Hesperornithiformes
Ichthyornithiformes
Neornithes
Enaliornis
Ambiortus
CARINATAE
Chaoyangia
Chaoyangia

(a)

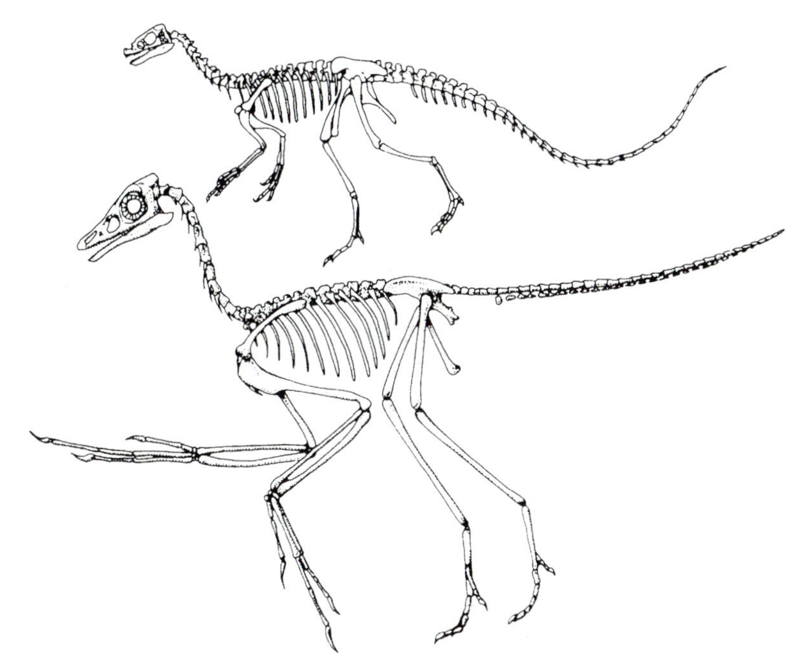

**Fig 10.8** (a) Cladogram of stem-group birds. (b) Comparison of the skeletons of a juvenile coelurosaur dinosaur *Coelophysis* (above), and *Archaeopteryx* (below). ((a) After Carroll 1997; (b) after Long and McNamara 1995.)

was complex with at least one major division into two subgroups of 'birds' (Fig. 10.8a). The Enantiornithes include *Sinornis* and many Cretaceous birds, while the Ornithurae independently evolved advanced avian features and includes other Cretaceous forms and also all the modern birds. Even at the current modest level of knowledge of the phylogenetic tree, there is no evidence for any uniquely orthogenetic-like trend to modern birds. As in the case of the mammals, the transition had a somewhat erratic aspect about it.

Whether the transition was a correlated progression of acquisition of avian features is even less clear. The only functional system that is adequately demonstrated by the fossils themselves is locomotion. There were evidently correlated changes between different parts of the locomotory equipment, forelimb and flight, and hindlimb and running, respectively. In this case, however, there is very little indeed that can be inferred about other biological systems such as feeding, ventilation, temperature physiology, and sense organs. Whether these, too, were undergoing correlated changes towards the modern avian state is unknown, however probable it might seem that they were. One particularly intriguing inference from the morphology is that heterochrony (page 219) may have played an important part in the transition from dinosaur-grade to *Archaeopteryx*-

grade (Fig. 10.8b). Long and McNamara (1995) pointed out that *Archaeopteryx* shows a greater resemblance to the juvenile of the coelurosaur *Coelophysis* than to the adult in such features as skull shape and tooth morphology, suggesting paedomorphosis was the cause of the miniaturization of birds. On the other hand, the longer forearms coupled with the reduced body size could be due to a peramorphic increase in forearm growth.

There are two points about the sequence of hypothetical ancestral forms that are particularly reminiscent of the mammal case: one is that all the stem-group birds appear to have been carnivorous; the second is that an important stage involved miniaturization.

Another important vertebrate transition under intensive study at present is the origin of the tetrapods (Ahlberg 1995; Carroll 1997). For many years now, the Upper Devonian stem-group tetrapod *Ichthyostega* has played a conceptually similar part for tetrapods to that of *Archaeopteryx* for birds; it has most of the tetrapod characters, but retains some more primitive features found in a 'rhipidistian' fish such as *Eusthenopteron*. Several other stem-group tetrapods (see Fig. 6.9) have been described recently, notably *Acanthostega*, which is the same age and from the same East Greenland locality as *Ichthyostega*. There are considerable and unexpected differences between these two contemporary animals, which indicates that the transition from fish to tetrapod was part of a radiation and not a single, simple lineage. Other stem-group tetrapods, mostly represented by very incomplete remains so far, are being added to the cladogram and indicate a little more about the sequence of acquisition of tetrapod characters. The picture is still extremely unclear and the pattern too hazy for confident inferences to be made. Thomson (1966, 1993) believes that the acquisition of tetrapod characters from the fish grade shows a correlated progression comparable to that of the origin of mammals. One interesting suggestion about this transition is that it may have occurred over a much shorter time span than other vertebrate transitions of comparable degree (Ahlberg *et al.* 1996). Although the amount of morphological and biological change is unquantifiable, and the precise dates of origin of particular character states cannot be known, there is nevertheless an impression that from fully aquatic fish-like grade to an animal fully competent to move, feed, and generally survive on land may have been as little as about 10 million years during the Late Devonian. If so, it would be tempting to speculate that this relates to the unusually extreme nature of the ecological transition from water to land.

The concept of preadaptation has been discussed widely in the context of the origin of tetrapods. It presumes that an adaptation for some environmental feature happened by chance to be able to act as an adaptation for a different feature in a different environment, and therefore to behave as a version of a key adaptation (page 218). In the case of the lobed-finned fishes that preceded the land-living tetrapods, the ability to breath atmospheric air, to locomote by pushing on the substrate, and perhaps to deal with terrestrial food are putative preadaptations. The difficulty with the idea of preadaptation is that, as with key adaptations generally, it is impossible to test whether any given character is any more important than any other in creating the opportunity for subsequent evolutionary radiation. It is also doubtful if the concept actually has any meaning beyond 'adaptation' anyway. The three functions mentioned are all functions that have to be possessed by animals living in very shallow parts of the aquatic habitat, in which dissolved oxygen levels will be low at times, and which will contain food items derived from the adjacent bank or shore. These are the same functions, related to the same environmental requirements as found in primitive terrestrial animals. What these characters indicate is that the transition from water to land was possible because there is enough ecological overlap in parameters between the two for an ecological gradient to exist between them. An organism adapted to shallow, muddy water is automatically largely adapted to the damp, muddy bank.

## The origin of arthropods

Amongst the invertebrate animals, arthropods are the closest rivals to vertebrates in the amount of biological information that potentially can be preserved directly and indirectly in the fossilized skeleton. They have the additional advantage of having been common and highly diverse in marine environments throughout the Phanerozoic, and are well represented as complete body fossils in some extremely important *Lagerstätten* formed during Cambrian times. The best known of these is the Burgess Shale of Middle Cambrian times in Canada, but others have been discovered more recently (see Conway Morris 1992, 1998), including the Lower Cambrian Chengjiang fauna of China (Jiang 1992), which contains soft-bodied arthropods of various sorts, medusa-like forms, and several 'worms'. For various reasons, non-mineralized structures such as chitinous exoskeletons and appendages are preserved in these deposits (page 72), producing very detailed and often beautiful fossils.

Phylogenetic analysis of the interrelationships between the major arthropodan groups has undergone a series of conceptual changes. Traditionally, a monophyletic phylum Arthropoda was recognized for all the segmented, jointed-limbed, chitinous exoskeletal organisms. But in a classic series of papers, Sidnie Manton and others (see, for example, Manton and Anderson 1979) argued for a polyphyletic origin of arthropods, with the Chelicerata (spiders, mites, king crabs, etc.), Crustacea, Uniramia (insects, centipedes, and millipedes), and the entirely Palaeozoic Trilobita all having evolved separately from some pre-arthropodan, annelid-like ancestor. This conclusion was based on the considerable differences between the patterns of tagmatization (the grouping of segments into distinct regions along the body), and on the large differences in the nature of the head appendages and the trunk limbs in the different groups. There are also notable differences in details in the embryology of the respective groups.

Although all these differences are indeed sufficiently great to imply that the 'arthropodan' characters are not homologous, but were evolved independently, it was soon pointed out that in cladistic terms this need not imply that the organisms themselves must constitute a polyphyletic group. There could have been a single ancestor unique to all the arthropodan groups that was not itself arthropodized. Each descendant lineage would then have evolved its own particular version of exoskeleton, tagmatization, and appendages. This is the view that largely (for example, Waggoner 1998, Wills *et al.* 1994) although not exclusively (for example, Willmer 1990) holds sway at present, based on the possession by all the arthropodan groups of several uniquely shared characters, such as the chitinous skeleton, preoral segments, compound eyes, and absence of cilia. Early indications from molecular sequence data also strongly supports monophyly. Wheeler *et al.* (1993) combined morphological characters with 18S rDNA and the protein ubiquitin and found a good consensus tree with the arthropodan groups clustering as monopyletic compared with onychophorans and annelids.

Turning to the bearing of the fossil record on the origin of arthropods, the scene is dominated by the great Cambrian 'explosion' and what this might imply about evolutionary processes. As has been known and puzzled about for well over a century, virtually all the animal phyla and most of the contained classes first appear as fossils in the Cambrian in an extraordinarily small window of time of only a few million years (Bowring *et al.* 1993). At the time of first appearance the members of each such group had practically the full complement of the definitive characters of that group. Furthermore, the *Lagerstätten* preservation of arthropodan animals indicates that there was an astonishing array of different groups of these organisms at that time, most of which left no post-Cambrian representatives. Thus the disparity (page 66) was high and there is an as yet unresolved argument about whether it was in fact greater at this early stage in the history of the arthropods than at any time since. Taken together, both the sudden appearance in the fossil record and the early high level of disparity of the group are the subject of an intense contemporary debate about whether the actual origin of the arthropods, and by implication of other groups, involved special evolutionary mechanisms that ceased to operate subsequent to the full evolutionary expression of the group.

Gould (1989, 1991) reconsidered the Middle Cambrian Burgess Shale, with its extraordinarily rich fauna of some 120 genera, many of them non-calcified organisms that would not normally have been preserved as fossils. This list includes members of the four major arthropodan groups, but also, he argued, representatives of as many as 20 additional arthropodan groups without any subsequent members known, fossil or living. Whilst accepting that there is no objective method of measuring disparity (and incidentally pleading that one should be derived as soon as possible), it is clear in Gould's mind that the range of arthropodan body plans or, as it can be expressed, the volume of morphospace occupied by arthropodan animals at that time, really did far exceed any subsequent disparity, including that found within the group today. The analogy he proposed (Gould 1991) was that of an inverted cone of disparity (Fig. 10.9a), in which the phenetic range of the group was initially high, and subsequently declined as extinction removed subgroups with particular body plans, without any accompanying origination of radically new kinds. Instead, the diversity, measured as number of species of arthropods, was maintained by radiation of species in the surviving subgroups.

To the extent that this perception is true, the question arises of why at the very dawn of the origin of arthropods, and in a geologically brief length of time, a wide variety of arthropods with many different patterns of segmentation, tagmatization, and form of appendages evolved (Fig. 10.9c). Furthermore, why was this evolutionary lability subsequently replaced by a stability in which no new major taxa arose, so that all subsequent evolution concerned only a small sample of the initial set of body plans? Gould's own answer was that this evolutionary pattern is underlain by a changing degree of developmental stability, a process he called 'congealing', and which involves as yet unexplicated genetic changes causing a reduction in the ability to generate radical new morphological forms. The implication is that the rapid origin of the initial range of major arthropod groups involved an evolutionary process that ceased to operate once the essentially new body plans had been completed.

This challenging interpretation, based largely on the Burgess Shale fauna but extended to the process of origination of new higher taxa generally, is itself challenged on two predictable fronts: is the pattern as perceived by Gould the true pattern and, if so, is the inferred process the correct inference? Wills *et al.* (1994) applied various methods to measure disparity of both Cambrian and living arthropods. Some of these were standard phenetic methods for discovering how much morphospace the group occupied, and included principal component analysis techniques. Another method was cladistic in form and consisted essentially of creating a cladogram for all the relevant genera and noting how many of the exclusively fossil forms appear on branches sufficiently remote from the four standard arthropodan groups to represent uniquely evolved body plans (Fig. 10.9b). They concluded that none of these methods indicates markedly greater arthropodan disparity in the Cambrian than the present. To stress once more, however, there is at present no objective means of measuring disparity available that can be justified by reference to any real, numerically definable property of groups of organisms, and so it is as easy to criticize Wills

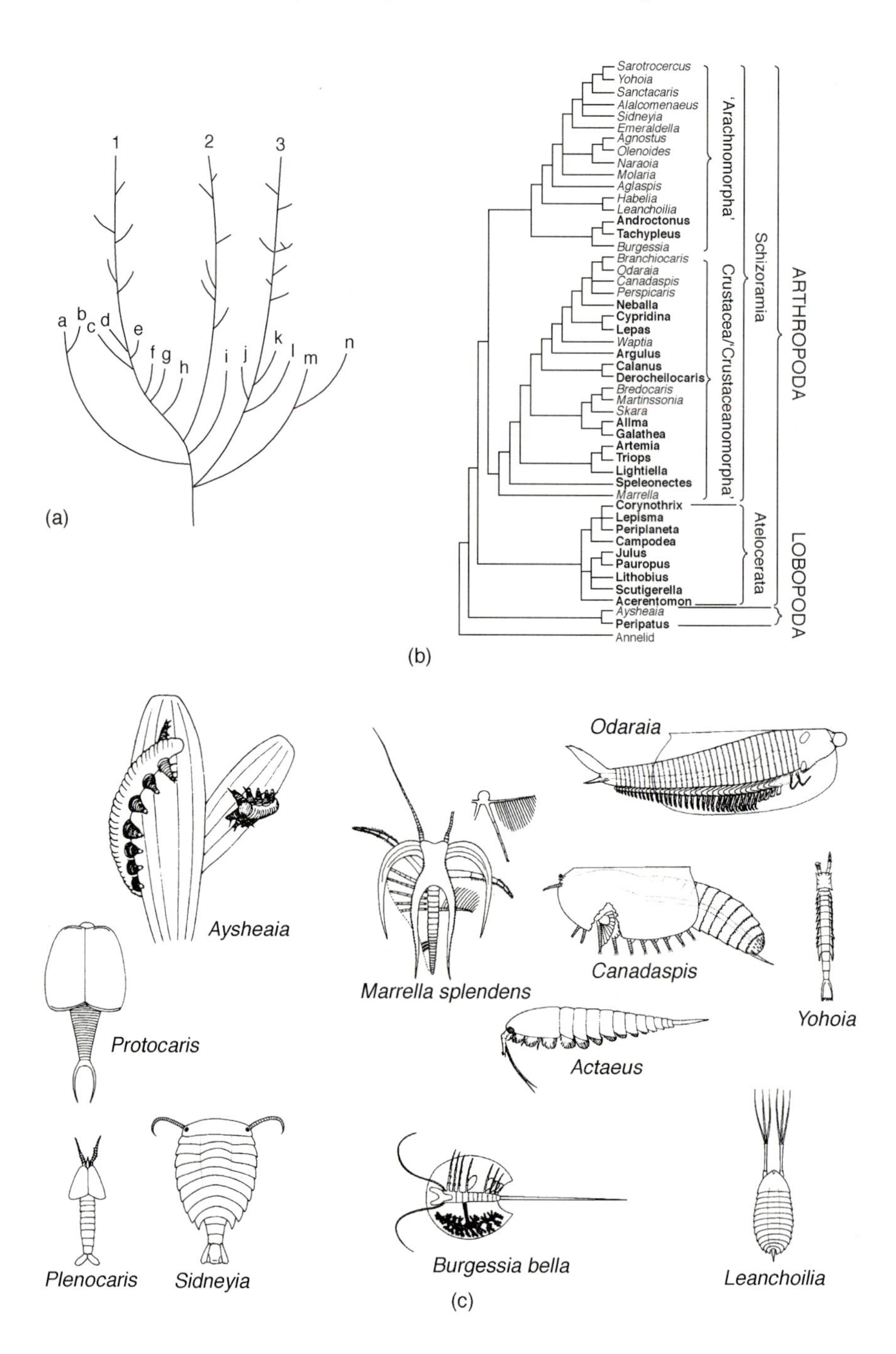
1
2
3
a
b
c
d
e
f
g
h
i
j
k
l
m
n
(a)
Sarotrocercus
Yohoia
Sanctacaris
Alalcomenaeus
Sidneyia
Emeraldella
Agnostus
Olenoides
Naraoia
Molaria
Aglaspis
Habelia
Leanchoilia
Androctonus
Tachypleus
Burgessia
Branchiocaris
Odaraia
Canadaspis
Perspicaris
Neballa
Cypridina
Lepas
Waptia
Argulus
Calanus
Derocheilocaris
Bredocaris
Martinssonia
Skara
Allma
Galathea
Artemia
Triops
Lightiella
Speleonectes
Marrella
Corynothrix
Lepisma
Periplaneta
Campodea
Julus
Pauropus
Lithobius
Scutigerella
Acerentomon
Aysheaia
Peripatus
Annelid
'Arachnomorpha'
Crustacea/'Crustaceanomorpha'
Schizoramia
Atelocerata
ARTHROPODA
LOBOPODA
(b)
Aysheaia
Protocaris
Plenocaris
Sidneyia
Marrella splendens
Odaraia
Canadaspis
Yohoia
Actaeus
Burgessia bella
Leanchoilia
(c)

*et al.*'s methods as it is for them to criticize Gould's impressions. In fact, the impression, however superficial, remains that there was a disconcertingly large variety of arthropods at this extremely early stage in their known history.

Fortey *et al.* (1996) went on look in as much detail as they could at the Cambrian 'explosion', primarily to see if there really was an extremely rapid origin of the higher taxa that appeared at that time, or whether there had been a longer period of largely unrecorded history extending back into the Precambrian. The evidence for such an earlier phase is to say the least sparse. For example, there are apparently arthropodan trackways preserved in the Ediacaran fauna of the late Precambrian. According to Fortey *et al.* the Precambrian organism *Parvancorina* (Fig. 10.10a) may have been derived from a trilobite larval form by heterochrony, suggesting that there must have been a Precambrian existence of trilobites themselves. Cladistic analysis of the major arthropodan groups points to a relatively advanced position for the trilobites, so that if trilobites did indeed occur in the Precambrian, it follows that virtually all the other arthropod groups also had their origin before the Cambrian 'explosion'. Their absence as fossils from earlier rocks must be due to non-preservation, and they are therefore examples of 'ghost' lineages.

Fortey *et al.* (1996) also appealed to molecular evidence bearing on the relationships and times of divergence of the modern groups. Wray *et al.* (1996) used seven different molecular sequences in the hope that between them the average rate of evolution is constant enough to indicate acceptable divergence times for animal phyla. Their analysis suggests that chordates diverged from echinoderms about 1000 million years ago, which is far back into the Precambrian (Fig. 10.12a). The protostome group, consisting mainly of annelids, arthropods, and molluscs, diverged still earlier. At present, there are no estimates for the times of separation of the individual protostome phyla, but, even as it is, this astonishing result implies that the history of these groups began up to twice as far back in time than widely supposed, long before the appearance of any trace of their fossils in the Cambrian. On the other hand, Ayala *et al.* (1998) have calculated from other sequences that the protostome divergence was only around 670 million years ago, and therefore that there is a much briefer period of missing fossils.

**Fig. 10.9** (a) The 'inverted cone' model of Gould: the disparity or range of morphological forms of a taxon such as the Arthropoda declines through time even though the diversity or number of species may not. Only three of the major subtaxa of arthropods, 1–3, are extant, compared with the much larger number, a–n, now extinct. (b) Wills *et al.*'s cladogram of arthropods. This supposedly indicates that the disparity of arthropods has not decreased since the Cambrian because the extant forms (in bold) are more or less as widely distributed in the cladogram as the Cambrian forms (in italics). (c) A selection of Burgess Shale, Middle Cambrian arthropods. ((a) After Gould 1991; (b) after Wills *et al.* 1994; (c) after Clarkson 1993.)

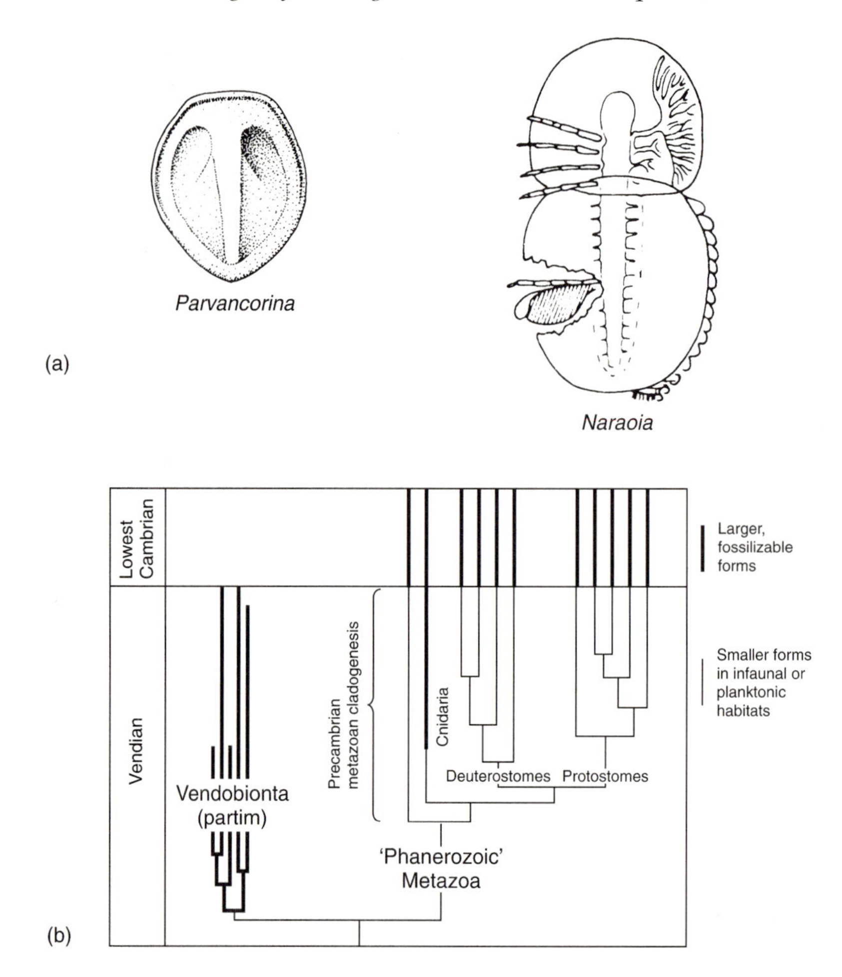


**Fig. 10.10** (a) *Parvancorina*, a Precambrian genus possibly related to juvenile trilobites such as the soft-bodied *Naraoia*. (b) Fortey *et al.*'s interpretation of the Cambrian 'explosion', as due to a long history of small, unskeletonized and therefore unfossilized members during the Precambrian. ((a) After Clarkson 1993; (b) after Fortey *et al.* 1996.)

The main difficulty to accepting this kind of evidence at face value is the assumption that the molecular clock is reliable when extrapolated so far back in time from modern taxa. Even if it is accepted that the various major arthropodan groups originated during a much earlier Precambrian radiation, the subsequent Cambrian 'explosion' was a remarkable event in their subsequent histories. It could nevertheless be explained by conventional evolutionary processes, and to complete their analysis of the situation, Fortey *et al.* (1996) proceeded to suggest a scenario in which the initial members of the lineages were very small organisms (Fig. 10.10b). They were envisaged as meiofaunal organisms, living in the interstitial or planktonic habitat (Boaden 1989), and therefore it is not surprising that they should be unknown as fossils. They assume that the start of the Cambrian period was associated with an environmental change that encouraged increase in body size, causing the simultaneous, rapid evolution of larger, well-skeletonized, and therefore fossilizable members of all the arthropodan groups and many non-arthropod groups as well.

In the context of growing knowledge about the developmental mechanisms of arthropods, palaeontologists propose that the apparent ease with which size increased phylogenetically, as witnessed by the number of groups assumed to have undergone the process, was due to heterochrony under the influence of mutations in the Hox genes (Minelli and Fusco 1995). As far as environmental change is concerned, there is evidence for a large increase in oxygen levels at that time. The increased metabolic rates that this permitted could well account for the evolution of large body size independently in many lineages.

Any unbiased overview of the origin of the arthropods has to admit that the questions posed earlier about the general process of the origin of new higher taxa cannot be answered adequately by the fossil record of these particular organisms. Taken at face value, the pattern of evolution shown is one of geologically instantaneous appearance of all the main kinds of arthropods, only a small sample of which survived beyond Cambrian times. The inference that the evolutionary processes involved were at the very least unusual if not radically different from 'normal' times has to be considered entirely consistent with the fossil evidence. On the other hand, if it is insisted that natural selection acting in a 'normal' fashion must have been the driving force, then it has to be assumed that the fossil record for the Precambrian is not giving a true picture. The actual evidence that this is the case is far from adequate at present. More optimistically, proponents of each of these respective views can point to potential tests. Gould (1993) believes that the key to understanding the origin of the new groups lies in learning more about the role of genes in development, and therefore of ways in which flexible, easily modified developmental programmes in early organisms could congeal into more inflexible, well-proven ones as taxa settle down into genetically more stable configurations. Conversely, Fortey *et al.* (1996) expect that a search for very small Precambrian fossils along with improving molecular analysis of modern groups will elucidate the early phylogenetic history of arthropods. No doubt they would also expect geochemical evidence to provide more details about the environmental conditions that ushered in the Cambrian, and provoked the rapid evolution of large body size that is, for them, what the Cambrian 'explosion' actually consists of.

## The origin of other higher taxa

The fossil record bearing on the origin of other higher taxa (invertebrate, plant, or lower vertebrate) adds little in principle to the examples discussed. In virtually all cases a new taxon appears for the first time in the fossil record with most definitive features already present, and practically no known stem-group forms. For example, the echinoderms (Fig. 10.11) fossilize well, and up to about 20 classes are recognized. Echinoderms make their initial appearance in the Lower Cambrian, and by Middle Cambrian times there are already about half a dozen of the classes present (Paul and Smith 1984; Sprinkle 1992). At best these can be classified cladistically into five subphyla, with nothing at all inferable about the pattern of acquisition of characters of the phylum as a whole and precious little regarding the emergence of the individual subphyla and classes.

Wills (1998) has considered the priapulid worms and shown that there has been a distinct increase in morphological disparity between the Cambrian forms and the modern ones. Brachiopods and molluscs are about the most readily fossilizable of all phyla, but because they possess so few preserved characters, no useful assessment of the sequence of acquisition of their respective characters is possible. Neither is it likely that any biologically meaningful measurements of the disparity of the whole organisms can be made from these shells alone. As far as brachiopods are concerned, the paucity of information from fossils about the nature of their origin is highlighted by the failure even to agree on whether the group is monophyletic, diphyletic, or polyphyletically derived from several separate soft-bodied phoronid worms (Popov 1992).

The origin of the higher taxa of plants, notably the Angiosperma, has become a great deal clearer in recent years as a result of careful cladistic analysis (e.g. Kendrick and Crane 1997; Niklas 1997). However, most of the useful information about the sequence of acquisition of angiosperm characters is derived from the living representatives of the more primitive groups, rather than from fossils. The latter are important in giving estimates of dates of divergences, and certainly a few of the relevant groups are exclusively fossil. Otherwise, they add little. Nor is there yet a study of possible underlying molecular genetic mechanisms causing the major transitions. No doubt it will not be long before one is undertaken.

# Generalizations

Returning to the questions posed at the start of this chapter, the most fundamental one is whether the origin of a new higher taxon involves processes different to the microevolutionary and speciation mechanisms of neo-Darwinism. Obviously, if it is first *assumed* that natural selection alone was

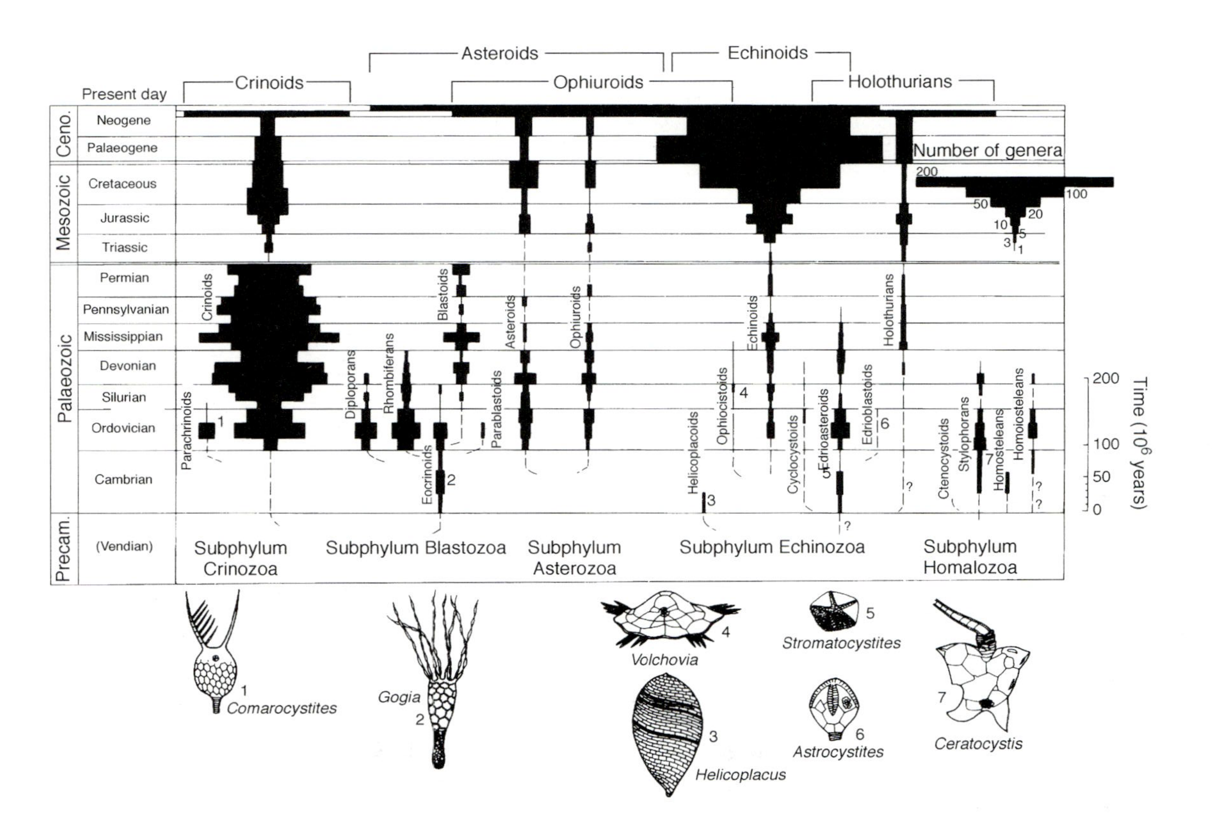


**Fig. 10.11** Echinoderm phylogeny. (After Clarkson 1993.)

the cause, then the fossil record of the transitions discussed can be so explained, with appropriate *ad hoc* assumptions about what environmental features might have been present to drive natural selection in the direction it supposedly went. Despite its very limited testability, this mode of argument has been used on countless occasions and continues to be so used, at least by someone to explain the origin of every major taxon considered. If, however, the fossils are to be taken seriously as a source of evidence about the origin of higher taxa, then they do indicate at the very least that a simple neo-Darwinian process alone cannot always provide a complete and acceptable explanation, because certain aspects of the pattern of fossils observed by the palaeobiologist are not as would be predicted.

In both the examples dealt with at length here, there are unexpected observations to be explained. In the case of mammals, an important aspect of the taxonomic pattern is most simply explained by a higher probability of speciation in small carnivores than in other kinds of organisms. This is the process of species selection, and not organism-level natural selection. In the case of arthropods, it may be simpler to accept early evolutionary lability allowing rapid diversification, followed by a process of genetic congealing, rather than invoking unaccountable and, on the face of it, rather improbable non-fossilization circumstances in the late Precambrian. This is a process involving a level of genetic constraint, rather than pure natural selection.

Cooper and Fortey (1998) have considered the Cambrian 'explosion', and also the divergence of the modern orders of placental mammals and birds. They note that in all three cases the fossil record of the first appearances of the members of the radiation post-date the currently available molecular estimates of divergence dates by a considerable period of time (Fig. 10.12). The expression 'phylogenetic fuse' refers to this interpretation, and they claim that there are enough hints in other groups of organisms to suggest that a long, hidden period during which major evolutionary change and diversification was occurring is a widespread phenomenon. This concept may be the most important result to date of the combined use of molecular and palaeobiological evidence. There are several conceivable causes of a phylogenetic fuse. These include the obvious possibility that the species involved in the early part of a radiation were extremely rare, so that their fossils are not found. Alternatively, the fuse could be an artefact due to changing rates of molecular evolution (Ayala *et al.* 1998). If the rate was high at some stage during the radiation—for example, at the beginning or perhaps after completion of the morphological changes—then the apparent molecular divergent date would be earlier than the palaeontological date. At least these two possibilities are potentially testable, and the outcome will have considerable impact on ideas about the origin of new higher taxa.

It should be clear that two cases, mammals and arthropods, and maybe others too, such as tetrapods and land angiosperms, are very different

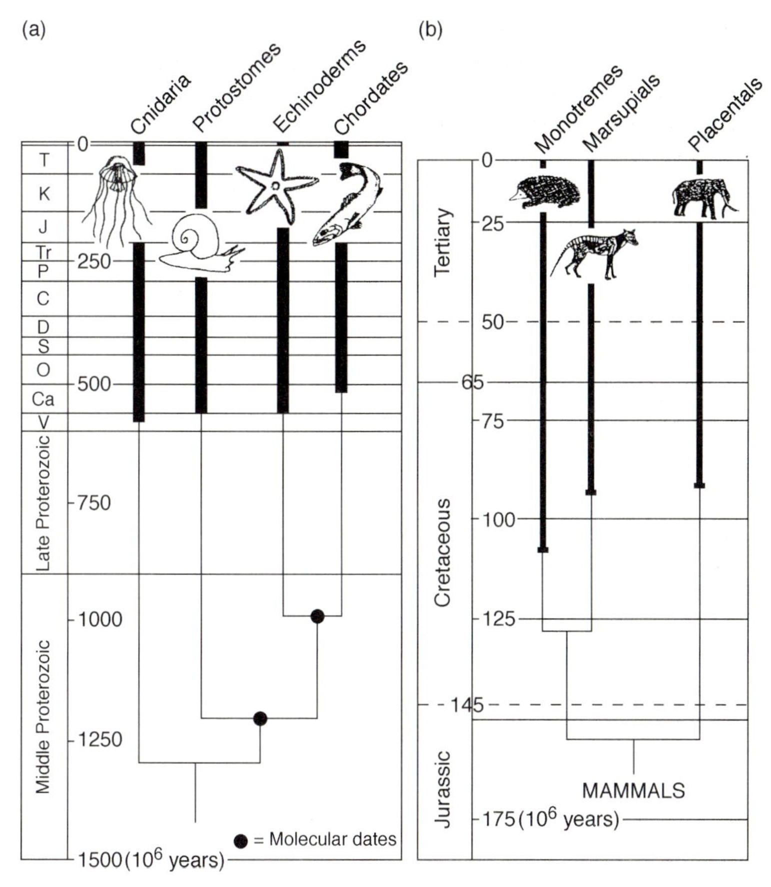


**Fig. 10.12** The phylogenetic fuse. (a) Fossil ranges of major invertebrate taxa in thick lines, and extensions to ranges as inferred from the two molecular dates shown in thin lines. (b) The same for monotreme, marsupial, and placental mammals. (After Cooper and Fortey 1998.)

from one another. As with other palaeobiological questions discussed in other chapters, one of the principal messages from the fossil record is that a single, simple process is unlikely to be the cause of the evolution of new kinds of organisms. The origin of a new higher taxon undoubtedly results from a complex of various different, interlinked processes; each case must have its own unique blend of ingredients.

# Further reading

Wake and Roth (1989) edited a Dahlem Conference volume called *Complex organismal functions: integration and evolution in vertebrates*, which contains several papers struggling with the concepts, but very little of substance emerges, beyond the obvious need for more genetic and developmental information about how organisms are actually assembled and therefore modified by evolution. Raff and Kaufman (1983) and Thomson (1988) were in the same position: writing directly about development and evolution, they had a good idea what the questions were but not how to find the answers. A decade on, and Marshall and Schopf (1996) briefly and easily, and Raff (1996) at considerable length review what has since been discovered about how genes work in development. Arthur (1997) applies much of this modern genetic knowledge for an overall review of the origin of animal body plans, a book which is pointing very much in the right direction.

Carroll (1997), particularly his chapter 7, is a palaeontologist beginning to look again at the fossil record armed with ideas from molecular genetics, although still seeing life from a rather conventional neo-Darwinian view.

# 11
# Epilogue: where next?

It is not difficult for palaeobiologists to become rather depressed about how little is known and how much seems unknowable about life in the past, and its long-term patterns and processes of evolutionary change. Indeed, much of the foregoing text has been about coming up against the severe limitations imposed by the various categories of incompleteness and lack of resolution of the fossil record. But things are much more cheering when attention is focused on what actually is known thanks to that same fossil record, things that could not be discovered and would not even have been predicted from the nature of the living biota alone. And one may verge on positively optimistic in the light of three new techniques that have been introduced quite recently, which are still in the early stages of development but which promise quantum leaps forward in understanding past evolutionary changes, and the palaeoenvironmental settings in which they occurred.

## The known

The single most-important aspect of evolution that can be appreciated only with the help of palaeobiology is undoubtedly the epistemological gap—that chasm in the continuity of knowledge between the high-resolution but extremely brief time span of events revealed by modern organisms on the one hand, and the low-resolution but unimaginably long time span of other events visible only in the fossil record. For some reason this seems to be the hardest message to get across to most evolutionary biologists. Indeed, many of the bitterest contemporary arguments in evolutionary biology hinge on the mutual misunderstanding between those wedded to the neontological viewpoint and others seeing evolution from the vantage point of palaeobiology. To account for evolutionary changes that take millions to hundreds of millions years to completion solely by reference to processes that can be studied only over tens of years requires an extraordinary faith. It needs the assumption that no processes that are either very slow or highly improbable exist that could be the cause of any evolutionary pattern.

Certainly it is possible to offer an explanation of the whole of the fossil record by processes occurring on the ecological time-scale. Unfortunately, any such explanation is completely untestable because the detailed information necessary to do so is not discoverable. Furthermore, the very assumption itself that all is caused by short-term processes, excludes the possibility of ever discovering whether any long-term processes exist, which is a pity. Causes of evolution in addition to neo-Darwinian selection acting at the level of the organism would be rather interesting phenomena, well worth knowing about. Simple extrapolation from short-term processes to explain long-term patterns is as absurd as attempting to understand the history of the international economic system by referring to the events that occurred on some particular day on the London Stock Exchange. One would hardly see much evidence of the effect of the wars, the colonial expansions, or the conflicting ideological dogmas that have played such a part. This is not to say that the observations made on that day are untrue or irrelevant, but that they are not enough.

It is equally unacceptable, of course, to offer an explanation of evolution processes on the assumption that long-term evolutionary patterns found in the fossil record illustrate the only processes at work. All those old, naive variations on the themes of vitalism, orthogenetic forces, typostrophism and so on were of this form, and have largely and rightly been abandoned. No explanation of the fossil record advanced by palaeobiologists can ever legitimately ignore the nature of living organisms as such; for example, the likelihood that natural selection affecting interbreeding populations is a feature of at least most of the biota for at least most of the time.

Given the reasonableness of treating the fossil record as a source of information bearing on the discovery of long-term evolutionary processes, some particular aspects of the pattern that are not predictable from living organisms alone exist, and must therefore be incorporated into any overall theory of evolution.

1. *The longevity of species.* The length of time over which a species, or at least a population of fossils of constant morphology, exists is hugely variable, but in nearly all cases is from 1 to 25 million years. This is a great deal longer than might have otherwise been supposed. Furthermore, the species of different taxonomic groups often have quite different mean longevities; for example, the characteristic values of around 2 million years for mammals and 25 million years for foraminiferans.
2. *The extent of taxonomic turnover.* An extraordinary and what would have been an entirely unpredictable number of taxa have disappeared over the hundreds of millions of years of evolutionary history recorded in the fossil record. These include taxa of every rank from a few phyla that did not even survive the Cambrian, through increasing percentages of classes, orders, families, and genera. Many of the modern groups are highly impoverished remnants of what were once much more diverse taxa. Conversely, and accounting for an apparently ever-

increasing number of species through time to the present, there are groups that today are far more diverse than they have ever been.

3. *The phenomenon of mass extinction.* Possibly the most surprising and least predictable events of all those that are revealed by the fossil record is the sequence of mass extinctions occurring every few tens of millions of years. What exactly caused them, and whether they were catastrophic, stepped or gradual declines in species numbers remain to be resolved. What is virtually certain is that from time to time a reduction in diversity has occurred of such magnitude that it can only be described as a pretty narrow squeak for the continuance of multicellular life on Earth. Loss of possibly as many as 96 per cent of species at the end of the Permian is quite astounding.
4. *The abruptness of the appearance of new higher taxa.* With few exceptions, radically new kinds of organisms appear for the first time in the fossil record already fully evolved, with most of their characteristic features present. Whether this is because evolution of such new forms is extremely rapid or whether because new higher taxa remain as very rare groups for a long initial period of their evolution is not clear. Whichever category of explanation is true, however, it is remarkably interesting, and not at all what might have been expected.

## The knowable

Progress in a scientific discipline can often be seen to consist of alternating phases. In one period the exciting thrust of research is dominated by the development of new techniques for acquiring hitherto unavailable knowledge. In the next it is the development of ideas and concepts that is seen to be leading the way to new understanding. Since about 1972, palaeobiology has been in a phase of conceptual development. The introduction of ideas such as punctuated equilibria, species selection, causes and correlations of diversity changes over time, and so on led directly to debates over the interpretation of existing data, and the search for more detailed information of much the same kind as that which existed. Simultaneously, the introduction of cladistic methodology has resulted in far more objective, testable hypotheses about evolutionary patterns and their possible interpretation. Even more recently, the application of sophisticated statistical methods such as bootstrapping and maximum likelihood to a range of questions, from estimating sedimentary completeness to the combination of different data sets in phylogenetic inference, has refined hypothesis testing even further.

Things changed in the 1990s, and palaeobiology is undergoing a distinct and exciting shift towards a techniques-led phase of progress. The application of three particular new techniques is leading to the accumulation of new kinds of knowledge, knowledge that is going to permit assessment of a range of evolutionary hypotheses considered virtually immune to testing until now. Within 10–20 years, palaeobiology is going to be a much rejuvenated subject.

## Stable isotope ratios

The ratio of pairs of non-radioactive isotopes of carbon, oxygen, sulphur, and strontium are already producing previously undreamed of information about past environments. Sea temperature and salinity, atmospheric oxygen levels, and photosynthetic productivity are some of the estimable parameters and there is certain to be an ever-increasing range of environmental indicators developed. These and other geochemical signals may yet permit serious—that is to say testable—hypotheses about possible selection forces and adaptive responses of species throughout the geological record.

Furthermore, they are already leading to vastly improved stratigraphy and temporal correlation as they provide more and more unique markers of contemporaneous and successive fossiliferous strata.

## Molecular taxonomy

If anything is capable of spanning the epistemological gap it is DNA sequences. Within the same genome there are molecules whose evolutionary rate of change is measured in terms of generations, and others where significant changes are detectable only between organisms that diverged from millions to thousands of millions of years ago. It is perfectly likely that with enough sequence data, and suitable statistical treatment, macromolecular evolution really can be used as a clock for estimating actual times of divergence of taxa with living representatives, even where these times date far back into the Precambrian. With an independently derived temporal framework such as this, it will become possible to interpret the fossil groups in terms of their actual times of origin and rates of morphological evolution.

## Molecular developmental genetics

Ultimately palaeobiology is about phenotypes because these are what fossil specimens actually represent. Therefore, the more that is known about the nature of organisms the more the subject benefits. Evolution of organisms consists initially of genetic modifications that affect the way the organism develops; evolutionary forces such as natural selection constitute the next stage—namely the ecological consequence of the phenotypic change.

Contemporary study of the molecular basis of development is beginning to throw light on the processes acting at the initial stage, that of the production of viable variation on which the evolutionary forces can act. It is revealing the presence of homologous genes controlling aspects of the development of whole regions of organisms from very different taxa, such as different phyla, suggesting that the different classical body plans may actually have a greater underlying genetic unity that hitherto supposed. It is becoming apparent that radically new morphotypes could appear as a consequence of effectively re-ordering the basic units of a body plan, under the influence of very simple patterns of mutations in appropriate genes. This, to the extent that it proves true, will expand considerably the menu of possible processes underlying the origin of new organisms, processes that may prove to be compatible with some of the more unexpected quirks in the fossil record of what really happened.

## A conclusion

If the fossil record reveals one single thing about evolution, it is that no single thing exists! Typical of the debates between holders of opposing, respectively rather simple views is that the resolution always starts to look as if both parties are right in particular cases and wrong in others. Alternative explanations seem to apply to different examples of what had been taken to be cases of the same phenomenon. Speciation: punctuated or gradual? Diversity changes over geological time: random or determinate? Mass extinctions: catastrophic, gradual, or stepped? Macroevolution: the result of the same processes as microevolution or of higher-level processes? Origin of higher taxa: rapid evolution or unpreserved initial phase?

For the time being, it seems that the search for unified causes is itself a lost cause and should be abandoned as a research programme. Instead, the circumstances associated with one kind of explanation or another should be sought. Not: is speciation punctuated or gradual? Rather: under what environmental and biotic circumstances is speciation one or the other, or some third mode? Universal evolutionary laws and the like probably apply to some different realm in the hierarchy of being than evolutionary biologists are yet accustomed to thinking about.

# *References*


Alvarez, L. W., Alvarez, W., and Michel, H. V. (1980). Extraterrestrial cause for the Cretaceous-Tertiary extinction. *Science*, **208**, 1095–108.

Ackery, P. R. and Vane-Wright, R. I. (1984). *Milkweed butterflies: their cladistics and biology*. British Museum (Natural History), London.

Ager, D. V. (1993). *The nature of the stratigraphical record*, (3rd edn). Wiley, Chichester.

Ahlberg, P. E. (1995). *Elginerpeton pancheni* and the earliest tetrapod clade. *Nature*, **373**, 420–5.

Ahlberg, P. E. and Milner, A. R. (1994). The origin and early diversification of tetrapods. *Nature*, **368**, 507–14.

Ahlberg, P. E., Clack, J. A. and Luksevics, E. (1996). Rapid braincase evolution between *Panderichthys* and the earliest tetrapods. *Nature*, **381**, 61–4.

Algeo, T. J. (1993). Quantifying stratigraphic completeness: a probabalistic approach using paleomagnetic data. *Journal of Geology*, **101**, 421–33.

Allison, P. A. and Briggs, D. E. G. (ed.) (1991). *Taphonomy: releasing the data locked in the fossil record.* Plenum, New York.

Allmon, W. D. (1989). Paleontological completeness of the record of Lower Tertiary molluscs, U.S. Gulf and Atlantic coastal plains: implications for phylogenetic studies. *Historical Biology*, **3**, 141–58.

Anders, M. H., Kreuger, S. W., and Sadler, P. M. (1987). A new look at sedimentation rates and the completeness of the stratigraphic record. *Journal of Geology*, **95**, 1–14.

Anstey, R. L. and Pachut, J. F. (1995). Phylogeny, diversity history, and speciation in Paleozoic bryozoans. In *New approaches to speciation in the fossil record*, (ed. D. H. Erwin and R. L.Anstey), pp. 239–84. Columbia University Press, New York.

Arthur, W. (1997). *The origin of animal body plans.* Cambridge University Press, Cambridge.

Ausich, W. I. and Bottjer, D. J. (1982). Tiering in suspension feeding communities on soft substrata throughout the Phanerozoic. *Science*, **216**, 173–4.

Ausich, W. I. and Bottjer, D. J. (1985). Phanerozoic tiering in suspension-feeding communities on soft substrata: implications for diversity. In *Phanerozoic diversity patterns: profiles in macroevolution*, (ed. J. W. Valentine), pp. 255–74. Princeton University Press, Princeton, N.J.

Avise, J. C. (1994). *Molecular markers, natural history and evolution.* Chapman and Hall, London.

Ax, P. (1987). *The phylogenetic system: the systematization of organisms on the basis of their phylogenies*, (transl. R. P. S. Jefferies). Wiley, Chichester.

Ayala, F. J., Rzhetsky, A. and Ayala, F. J. (1998). Origin of the metazoan phyla: molecular clocks confirm paleontological estimates. *Proc. Natl. Acad. Sci. USA*, **95**, 606–11.

Bambach, R. K. (1983). Ecospace utilisation and guilds in marine communities through the Phanerozoic. In *Biotic interactions in Recent and fossil benthic communities*, (ed. M. J. S. Tevesz and P. L. McCall), pp. 719–46. Plenum, New York.

Bambach, R. K. (1985). Classes and adaptive variety: the ecology of diversification in marine faunas through the Phanerozoic. In *Phanerozoic diversity patterns: profiles in macroevolution*, (ed. J. W.Valentine), pp. 191–253. Princeton University Press, Princeton, NJ.

Bambach, R. K. and Bennington, J. B. (1996). Do communities evolve? A major question in evolutionary biology. In *Evolutionary paleobiology*, (ed. D. Jablonski, D. H. Erwin, and J. H. Lipps). Chicago University Press, Chicago.

Barnes, C. R., Fortey, R. A., and Williams, S. H. (1996). The pattern of global bioevents during the Ordovician Period. In *Global events and global stratigraphy in the Phanerozoic*, (ed. O. H. Walliser), pp. 139–72. Springer-Verlag, Berlin.

Barton, N. H. (1988). Speciation. In *Analytical biogeography*, (ed. A. A. Myers and P. S. Giller), pp. 185–218. Chapman and Hall, New York.

Barton, N. H. (1989). Founder effect speciation. In *Speciation and its consequences*, (ed. D. Otte and J. A. Endler), pp. 229–56. Sinauer, Sunderland, Mass.

Benton, M. J. (1983). Dinosaur success in the Triassic: a non-competitive ecological model. *Quarterly Review of Biology*, **58**, 29–55.

Benton, M. J. (1985). Mass extinctions among non-marine tetrapods. *Nature*, **316**, 811–14.

Benton, M. J. (1987). Progress and competition in macroevolution. *Biological Reviews*, **62**, 305–38.

Benton, M. J. (1989). Mass extinctions among tetrapods and the quality of the fossil record. *Philosophical Transactions of the Royal Society*, **B325**, 369–86.

Benton, M. J. (1990). Reptiles. In *Evolutionary trends*, (ed. K. J. McNamara), pp. 279–300.

Benton, M. J. (1991). End-Triassic. In *Palaeobiology: a synthesis*, (ed. D. E. G. Briggs and P. R. Crowther), pp. 194–8. Blackwell Scientific, Oxford.

Benton, M. J. (1996). On the nonprevalence of competitive replacement in the evolution of tetrapods. In *Evolutionary paleobiology: in honour of James W. Valentine*, (ed. D. Jablonski, D. H. Erwin, and J. H. Lipps), pp. 185–210. Chicago University Press, Chicago.

Berger, W. H. *et al.* (1984). Short-term changes affecting atmosphere, oceans, and sediments during the Phanerozoic. In *Patterns of change in Earth evolution*, (ed. H. D. Holland and A. F. Trendall), pp. 171–205. Springer-Verlag, Berlin.

Berggren, W. A. *et al.* (1979). *Treatise on invertebrate paleontology. Part A. Introduction: fossilisation (taphonomy), biogeography and biostratigraphy.* The Geological Society of America and University of Kansas Press, Boulder, Colo. and Kansas.

Beringer, D. J. B. A. (1726). *Lithographiae Wirceburgensis*. University of Wurtzburg.

Berner, R. A. (1994). 3 Geocarb II: a revised model of atmospheric CO over Phanerozoic time. *American Journal of Science*, **294**, 56–91.

Boaden, P. J. S. (1989). Meiofauna and the origin of the Metazoa. *Zoological Journal of the Linnean Society of London*, **96**, 217–27.

Bowring, S. A., Grotzinger, J. P., Isachsen, C. E., Knoll, A. K., Pelechaty, S. M., and Kolosov, P. (1993). Calibrating rates of early Cambrian evolution. *Science*, **261**, 1293–8.

Brandon, R. N. and Burian, R. M. (1984). *Genes, organisms, populations.* MIT Press, Cambridge, Mass.

Brasier, M. D. (1989). On mass extinctions and faunal turnover near the end of the Precambrian. In *Mass extinctions: processes and evidence*, (ed. S. K. Donovan), pp. 73–88. Belhaven Press, London.

Brasier, M. D. (1996). The basal Cambrian transition and Cambrian bio-events (from terminal Proterozoic extinctions to Cambrian biomeres). In *Global events and event stratigraphy in the Phanerozoic*, (ed. O. H. Walliser), pp. 113–38. Springer-Verlag, Berlin.

Brenchley, P. J. (1989). The Late Ordovician extinction. In *Mass extinctions: processes and evidence*, (ed. O. H. Walliser), pp. 104–32. Belhaven Press, London.

Brett, C. E. (1990). Destructive taphonomic processes and skeletal durability. In *Palaeobiology: a synthesis*, (ed. D. E. G. Briggs and P. R. Crowther), pp. 223–6. Blackwell Scientific, Oxford.

Brett, C. E. and Baird, G. C. (1995). Coordinated stasis and evolutionary ecology of Silurian to Middle Devonian faunas in the Appalachian Basin. In *New approaches to speciation in the fossil record*, (ed. D. H. Erwin and R. L. Anstey), pp. 283–315. Columbia University Press, New York.

Briggs, D. E. G. and Crowther, P. R. (ed.) (1990). *Palaeobiology: a synthesis.* Blackwell Scientific, Oxford.

Bromley, R. G. (1996). *Trace fossils: biology and taphonomy and applications*, (2nd edn). Chapman and Hall, London.

Bryant, H. N. and Russell, A. P. (1992). The role of phylogenetic analysis in the inference of unpreserved attributes of extinct taxa. *Philosophical Transactions of the Royal Society*, **B337**, 405–18.

Bull, J. J., Huelsenbeck, J. P., Cunningham, C. W., Swofford, D. L., and Waddell, R. J. (1993). Partitioning and combining data in phylogenetic analysis. *Systematic Biology*, **42**, 384–397.

Carroll, R. L. (1988). *Vertebrate paleontology and evolution.* W. H. Freeman, New York.

Carroll, R. L. (1997). *Patterns and processes of vertebrate evolution.* Cambridge University Press, Cambridge.

Carroll, S. (1995). Homeotic genes and the evolution of arthropods and vertebrates. *Nature*, **376**, 479–85.

Carson, H. L. and Templeton, A. R. (1984). Genetic revolutions in relation to speciation phenomena: the founding of new populations. *Annual Review of Ecology and Systematics*, **15**, 97–131.

Chalonor, W. G. and Hallam, A. (ed.) (1989). Evolution and extinction. *Philosophical Transactions of the Royal Society*, **B325**, 239–488.

Charlesworth, B. and Lande, R. (1982). Punctuationism and Darwinism reconciled? The Lake Turkana mollusc sequence. Morphological stasis and developmental constraint: no problem for neo-Darwinism. *Nature*, **296**, 610.

Cheetham, A. H. (1986). Tempo of evolution in a Neogene bryozoan: rates of morphologic change within and across species boundaries. *Paleobiology*, **12**, 190–202.

Cheetham, A. H. (1987). Tempo of evolution in a Neogene bryozoan: are trends in single morphological characters misleading? *Paleobiology*, **13**, 286–96.

Clarkson, E. N. K. (1993). *Invertebrate palaeontology and evolution*, (3rd edn). Chapman and Hall, London.

Clyde, W. C. and Fisher, D. C. (1997). Comparing the fit of stratigraphic and morphologic data in phylogenetic analysis. *Paleobiology*, **23**, 1–19.

Cocks, L. R. M. and McKerrow, W. S. (1978). Ordovician. In *The ecology of fossils*, (ed. W. S. McKerrow), pp. 62–92. Duckworth, London.

Coddington, J. A. (1988). Cladistic tests of adaptational hypotheses. *Cladistics*, **4**, 3–22.

Cohen, A. S. and Schwartz, H. L. (1983). Speciation in molluscs from Turkana Basin. *Nature*, **304**, 659–60.

Conway Morris, S. (1992). Burgess Shale-type faunas in the context of the 'Cambrian explosion': a review. *Journal of the Geological Society of London*, **149**, 631–6.

Conway Morris, S. (1993). Ediacaran-like fossils in Cambrian Burgess Shale-type faunas of North America. *Paleontology*, **36**, 593–635.

Conway Morris (1998). *The crucible of creation: the Burgess Shale and the rise of animals*. Oxford University Press, Oxford.

Coope, G. R. (1970). Interpretations of Quaternary insect fossils. *Annual Review of Entomology*, **15**, 97–120.

Coope, G. R. (1979). Late Cenozoic fossil Coleoptera: evolution, biogeography, and ecology. *Annual Review of Ecology and Systematics*, **10**, 247–67.

Cooper, A. and Fortey, R. A. (1988). The phylogenetic fuse. *Trends in Ecology and Evolution*, **13**, 151–6.

Cracraft, J. and Eldredge, N. (1979). *Phylogenetic analysis and paleontology*. Columbia University Press, New York.

Crowson, R. A. (1970). *Classification and biology*. Heinemann, London.

Dawkins, R. (1986). *The blind watchmaker*. Longman, Harlow, Essex.

Dawkins, R. (1996). *Climbing Mount Improbable*. Penguin Books, London.

Depew, D. J. and Weber, B. H. (1995). *Darwinism evolving*. MIT Press, Cambridge, Mass.

De Robertis, E. M. (1996). Homeotic genes and the evolution of body plans. In *Evolution and the molecular revolution*, (ed. C. R. Marshall and J. W. Schopf), pp. 109–24. Jones and Bartlett, Boston.

Dingus, L. (1984). Effects of stratigraphic completeness on interpretations of extinction rates across the Cretaceous-Tertiary boundary. *Paleobiology*, **10**, 420–38.

Dingus, L. and Sadler, P. M. (1982). The effects of stratigraphic completeness on estimates of evolutionary rates. *Systematic Zoology*, **31**, 400–12.

Dodd, J. R. and Stanton, R. J. (1990). *Paleoecology: concepts and applications*, (2nd edn). Wiley, New York.

Donoghue, M. J., Doyle, J. A., and Gauthier, J. (1989). The importance of fossils in phylogeny reconstruction. *Annual Review of Ecology and Systematics*, **20**, 431–60.

Donovan, S. K. (ed.) (1989). *Mass extinctions: processes and evidence*. Belhaven Press, London.

Donovan, S. K. (ed.) (1994). *The palaeobiology of trace fossils*. Wiley, Chichester.

Donovan, S. K. and Paul, C. R. C. (eds) (1998). *The adequacy of the fossil record*. Wiley, Chichester.

Doyle, P. and Bennett, M. R. (ed) (1998*a*). *Unlocking the stratigraphical record: advances in modern stratigraphy*. Wiley, Chichester.

Doyle, P. and Bennett, M. R. (1998*b*). Interpreting palaeoenvironments from fossils. In *Unlocking the stratigraphical record: advances in modern stratigraphy*, (ed. P. Boyle and M. R. Bennett). Wiley, Chichester.

Doyle, P., Bennett, M. R., and Baxter, A. N. (1994). *The key to earth history: an introduction to stratigraphy*. Wiley, Chichester.

Eldredge, N. (1979). Cladism and common sense. In *Phylogenetic analysis and paleontology*, (ed. J. Cracraft and N. Eldredge), pp. 165–98. Columbia University Press, New York.

Eldredge, N. (1995*a*). Species, speciation, and the context of adaptive change in evolution. In *New approaches to speciation in the fossil record*, (ed. D. H. Erwin and R. L. Anstey), pp. 39–63. Columbia University Press, New York.

Eldredge, N. (1995*b*). *Reinventing Darwin*. Wiley, New York.

Eldredge, N. (1996). Hierarchies and macroevolution. In *Evolutionary palaeontology*, (ed. D. Jablonski, D. H. Erwin, and J. H. Lipps), pp. 42–61. Chicago University Press, Chicago.

Eldredge, N. and Cracraft, J. (1980). *Phylogenetic patterns and the evolutionary process*. Columbia University Press, New York.

Eldredge, N. and Gould, S. J. (1972). Punctuated equilibria: an alternative to phyletic gradualism. In *Models in paleobiology*, (ed. T. J. M. Schopf), pp. 82–115. Freeman, Cooper, San Francisco.

Endler, J. A. (1986). *Natural selection in the wild*. Princeton University Press, Princeton, N. J.

Ereshefsky, M. (ed.) (1992). *The units of evolution*. MIT Press, Cambridge Mass.

Erwin, D. H. (1993). The great Paleozoic crisis: life and death in the Permian. Columbia University Press, New York.

Erwin, D. H. 1994. The Permo-Triassic extinction. *Nature*, **367**, 231–6.

Erwin, D. H. (1996). Permian global bio-events. In *Global events and event stratigraphy in the Phanerozoic*, (ed. O. H. Walliser), pp. 251–64. Springer-Verlag, Berlin.

Erwin, D. H. and Anstey, R. L. (1995*a*). Speciation in the fossil record. In *New approaches to speciation in the fossil record*, (ed. D. H. Erwin and R. L. Anstey), pp. 11–38. Columbia University Press, New York.

Erwin, D. H. and Anstey, R. L. (ed.) (1995*b*). *New approaches to speciation in the fossil record*. Columbia University Press, New York.

Farris, J. S. (1983). The logical basis of phylogenetic analysis. In *Advances in cladistics II*, (ed. V. Funk and N. Platnick), pp. 7–36. Columbia University Press, New York.

Fischer, A. G. (1984). The two Phanerozoic supercycles. In *Catastrophes and earth histroy* (ed. W. A. Berggren and J. A. van Couvering) pp. 129–50. Princeton University Press, Princeton.

Fischer, A. G. and Arthur, M. A. (1977). Secular variation in the pelagic realm. *Society of Economic Paleontologists and Mineralogists Special Publication*, **25**, 19–50.

Fisher, D. C. (1994). Stratocladistics: morphological and temporal patterns and their relationship to phylogenetic process. In *Interpreting the hierarchy of nature: from systematic patterns to evolutionary process theories*, (ed. L. Grande and O. Rieppel), pp. 133–71. Academic Press, San Diego.

Flessa, K. W. and Levinton, J. S. (1975). Phanerozoic diversity patterns: tests for randomness. *Journal of Geology*, **83**, 239–48.

Foote, M. (1997). The evolution of morphological diversity. *Annu. Rev. Ecol. Syst.*, **28**, 129–52.

Foote, M. (1993*a*). Contribution of individual taxa to overall morphological disparity. *Paleobiology*, **19**, 403–19.

Foote, M. (1993*b*). Discordance and concordance between morphological and taxonomic diversity. *Paleobiology*, **19**, 185–204.

Forey, P. L. (1992). Fossils and cladistic analysis. In *Cladistic analysis: a practical course in systematics*, (ed. P. L. Forey *et al.*), pp. 124–36. Clarendon Press, Oxford.

Forey, P. L., Gardiner, B. G., and Patterson, C. (1991). The lungfish, the coelacanth, and the cow revisited. In *Origins of the higher groups of tetrapods: controversy and consensus*, (ed. H.-P. Schultze and L. Trueb), pp. 145–72. Comstock, Ithaca, N.Y.

Forey, P. L., Humphries, C. J., Kitching, I. J., Scotland, R. W., Seibert, D. J., and Williams, D. M. (1992). *Cladistics: a practical course in systematics*. Oxford University Press, Oxford.

Fortey, R. A. (1985). Gradualism and punctuated equilibria as competing and complementary theories. *Special papers in palaeontology*, No. 33, 17–28.

Fortey, R. A. and Owens, R. M. (1990). Evolutionary radiations in the Trilobita. In *Major evolutionary radiations*, (ed. P. D. Taylor and G. P. Larwood), pp. 139–64. Oxford University Press, Oxford.

Fortey, R. A., Briggs, D. E. G., and Wills, M. A. (1996). The Cambrian evolutionary 'explosion': decoupling cladogenesis from morphological disparity. *Biological Journal of the Linnean Society*, **57**, 13–33.

Fryer, G., Greenwood, P. H., and Peake, J. F. (1983). Punctuated equilibria, morphological stasis and the palaeontological documentation of speciation: a biological appraisal of a case history in an African lake. *Biological Journal of the Linnean Society*, **20**, 195–205.

Gardiner, B. G. (1982). Tetrapod classification. *Zoological Journal of the Linnean Society*, **74**, 207–32.

Gauthier, J., Kluge, A. G., and Rowe, T. (1988). Amniote phylogeny and the importance of fossils. *Cladistics*, **4**, 105–209.

Gavrilets, S. (1997). Evolution and speciation on holey adaptive landscapes. *Trends in Ecology and Evolution*, **12**, 307–12.

Geary, D. H. (1990*a*). Evaluating intrinsic and extrinsic factors in the evolution of *Melanopsis* in the Pannonian Basin. In *Causes of evolution: a paleontological perspective*, (ed. R. M. Ross and W. D. Allmon), pp. 305–21. Chicago University Press, Chicago.

Geary, D. H. (1990*b*). Patterns of evolutionary tempo and mode in the radiation of *Melanopsis* (Gastropoda; Melanopsidae). *Paleobiology*, **16**, 492–511.

Gilinsky, N. L. and Bambach, R. K. (1986). The evolutionary bootstrap: a new approach to the study of taxonomic diversity. *Paleobiology*, **12**, 251–68.

Gingerich, P. D. (1979). The stratophenetic approach to phylogeny reconstruction in vertebrate paleontology. In *Phylogenetic analysis and paleontology*, (ed. J. Cracraft and N. Eldredge), pp. 41–77. Columbia University Pres, New York.

Gingerich, P. D. (1985). Species in the fossil record: concepts, trends, and transitions. *Paleobiology*, **11**, 27–41.

Givnish, T. J. and Sytsma, K. J. (ed) (1997). *Molecular evolution and adaptive radiation.* Cambridge University Press, Cambridge.

Glaessner, M. F. (1984). *The dawn of animal life.* Cambridge University Press, Cambridge.

Gleik, J. (1988). *Chaos: making a new science.* Heinemann, London.

Golubic, S. and Knoll, A. M. (1993). Prokaryotes. In *Fossil prokaryotes and protists*, (ed. J. H. Lipps), pp. 51–76. Blackwell Scientific, Oxford.

Gould, S. J. (1983). Irrelevance, submission and partnership: the changing role of paleontology in Darwin's three centennials, and a modest proposal for macroevolution. In *Evolution from molecules to man*, (ed. D. S. Bendall), pp. 347–66. Cambridge University Press, Cambridge.

Gould, S. J. (1985). The paradox of the first tier: an agenda for paleobiology. *Paleobiology*, **11**, 2–12.

Gould, S. J. (1989). *Wonderful life.* Hutchinson Radius, London.

Gould, S. J. (1990). Speciating and sorting as the source of evolutionary trends, or 'things are seldom what they seem'. In *Evolutionary trends*, (ed. K. J. McNamara), pp. 3–27. Belhaven Press, London.

Gould, S. J. (1991). The disparity of the Burgess Shale arthropod fauna and the limits of cladistic analysis: why we must strive to quantify morphospace. *Paleobiology*, **17**, 411–23.

Gould, S. J. (1993). How to analyse Burgess Shale disparity—a reply to Ridley. *Paleobiology*, **19**, 522–3.

Gould, S. J. and Calloway, C. B. (1980). Clams and brachiopods—ships that pass in the night. *Paleobiology*, **6**, 383–96.

Gould, S. J. and Eldredge, N. (1977). Punctuated equilibria: the tempo and mode of evolution reconsidered. *Paleobiology*, **3**, 115–51.

Gould, S. J. and Lewontin, R. C. (1979). The spandrels of San Marco and the Panglossian paradign: a critique of the adaptationist programme. *Proc. R. Soc. Lond.*, **B205**, 581–98.

Gould, S. J., Gilinsky, N. L., and German, R. Z. (1987). Asymmetry of lineages and the direction of evolutionary time. *Science*, **236**, 1437–41.

Grande, L. and Reippel, O. (1994). Summary and comments on systematic pattern and evolutionary process. In *Interpreting the hierarchy of nature: from systematic patterns to evolutionary process theories*, (ed. L. Grande and O. Reippel), pp. 227–55. Academic Press, San Diego.

Greive, R. A. F. (1995). The record of terrestrial impact cratering. *GSA Today*, **189**, 194–6.

Greive, R. A. F. and Robertson, P. B. (1987). Terrestrial impact structures. *Geological Survey of Canada Map 1658A.*

Grene, M. (1958). Two evolutionary theories. *British Journal of the Philosophy of Science*, **9**, 110–27 and 185–93.

Griffiths, P. (1993). The question of *Compsognathus*. *Revue de Paléobiologie*, **7**, 85–94.

Grimaldi, D. A., Shedrinsky, A., Ross, A. and Norbert, N. S. (1994). Forgeries of fossils in 'amber': history, identification and case studies. *Curator*, **37**, 251–74.

Hallam, A. (1982). Patterns of speciation in Jurassic *Gryphaea*. *Paleobiology*, **8**, 354–66.

Hallam, A. (1989). The case for sea-level as a dominant causal factor in mass extinction of marine invertebrates. *Philosophical Transactions of the Royal Society*, **B325**, 437–55.

Hallam, A. (1990). Biotic and abiotic factors in the evolution of early Mesozoic marine molluscs. In *Causes of evolution: a paleontological perspective*, (ed. R. M. Ross and W. D. Allmon), pp. 249–69. Chicago University Press, Chicago.

Hallam, A. (1996). Major events in the Triassic and Jurassic. In *Global events and event stratigraphy in the Phanerozoic*, (ed. O. H. Walliser), pp. 265–83. Springer-Verlag, Berlin.

Hallam, A. and Wignall, P. B. (1997). *Mass extinctions and their aftermath*. Oxford University Press, Oxford.

Harper, C. W. (1976). Phylogenetic inference in paleontology. *Journal of Paleontology*, **50**, 180–93.

Harvey, P. H. and Pagel, M. D. (1991). *The comparative method in evolutionary biology*. Oxford University Press, Oxford.

Hecht, M. K. and Hoffman, A. (1986). Why not neo-darwinism? A critique of paleobiological challenges. *Oxford Surveys in Evolutionary Biology*, **3**, 1–47

Hennig, W. (1966). *Phylogenetic systematics*. University of Illinois Press, Chicago.

Hewitt, G. M. (1989). The subdivision of species by hybrid zones. In *Speciation and its consequences*, (ed. D. Otte and J. A. Endler), pp. 85–110. Sinauer, Sunderland, Mass.

Hickman, C. S. (1988). Analysis of form and function in fossils. *American Zoologist*, **28**, 775–93.

Hildebrand, A. R., Penfield, G. T., Kring, D. A., Pilkington, M., Camargo, Z. A., Jacobsen, S. B., and Boynton, W. V. (1991). Chicxulbub Crater: a possible Cretaceous/Tertiary boundary impact crater on the Yucatan Peninsula, Mexico. *Geology*, **19**, 867–71.

Hillis, D. M., Moritz, C., and Mable, B. K. (1996). *Molecular systematics, 2nd edition*. Sinauer, Massachusetts.

Hitchin, R. and Benton, M. J. (1997). Congruence between parsimony and stratigraphy: comparisons of three indices. *Paleobiology*, **23**, 20–32.

Hoffman, A. (1989*a*). *Arguments on evolution: a paleontologist's perspective*. Oxford University Press, Oxford.

Hoffman, A. (1989*b*). In. 'Evolution and extinction' (ed. W. G. Chalonor and A. Hallam), *Philosophical Transactions of the Royal Society*, **B325**, 239–488.

Hoffman, A. and Ghiold, J. (1985). Randomness in the pattern of 'mass extinctions' and 'waves of originations'. *Geological Magazine*, **122**, 1–4.

Hoffman, A. and Kitchell, J. A. (1984). Evolution in a pelagic planktic system: a paleobiologic test of models of multispecies evolution. *Paleobiology*, **10**, 9–33.

Holder, N. (1983). The vertebrate limb: patterns and constraints in development and evolution. In *Development and evolution*, (ed. B. C. Goodwin, N. Holder, and C. C. Wylie), pp. 399–421. Cambridge University Press, Cambridge.

Holland, P. W. H. and Garcia-Fernandez, J. (1996). Hox genes and chordate evolution. *Developmental Biology*, **173**, 382–95.

Holser, W. T., Magaritz, M., and Ripperdam, R. L. (1996). Global isotopic events. In *Global events and event stratigraphy*, (ed. O. H. Walliser), pp. 63–88. Springer-Verlag, Berlin.

Hou, L.-H., Martin, L. D., Zhou, Z., and Feduccia, A. (1996). Earliest adaptive radiation of birds revealed by newly discovered Chinese fossils. *Science*, **274**, 1164–67.

Hoyle, F. and Wickramasinge, C. (1986). *Archaeopteryx, the primordial bird: a case of fossil forgery*. Christopher Davies, Swansea.

Hsu, K. J. and McKenzie, J. A. (1990). Carbon-isotope anomalies at era boundaries: global catastrophes and their ultimate causes. In *Global catastrophes in earth history: an interdisciplinary conference on impacts, volcanism, and mass mortality*, (ed. V. L. Sharpton and P. D. Ward). *Geological Society of America Special Paper*, No. 247, 61–70.

Huelsenbeck, J. P. and Rannala, B. (1997). Maximum likelihood estimation of phylogeny using stratigraphic data. *Paleobiology*, **23**, 174–80.

Huelsenbeck, J. P., Swofford, D. L., Cunningham, C. W., Bull, J. J., and Waddell, P. J. (1994). Is character weighting a panacea for the problem of data heterogeneity in phylogenetic analysis? *Systematic Biology*, **43**, 288–91.

Huff, W. D., Bergstrom, S. M., and Kolata, D. R. (1992). Gigantic Ordovician volcanic ash fall in North America and Europe: biological, tectomagnetic, and event-stratigraphy significance. *Geology*, **20**, 875–8.

Hull, D. (1988). Science as a process: an evolutionary account of the social and conceptual development of science. University of Chicago Press, Chicago.

Jablonski, D. (1986). Background and mass extinctions: the alternation of macroevolutionary regimes. *Science*, **231**, 129–33.

Jablonski, D. (1989). The biology of mass extinction: a palaeontological view. *Philosophical Transactions of the Royal Society*, **B325**, 357.

Jablonski, D. and Bottjer, D. J. (1990*a*). Onshore-offshore trends in marine invertebrate evolution. In *Causes of evolution: a paleontological perspective*, (ed. R. M. Ross and W. D. Allmon), pp. 21–75. University of Chicago Press, Chicago.

Jablonski, D. and Bottjer, D. J. (1990*b*). The origin and diversification of major groups: environmental patterns and macroevolutionary lags. In *Major evolutionary radiations*, (ed. P. D. Taylor and G. P. Larwood), pp. 17–57. Clarendon Press, Oxford.

Jablonski, D. and Lutz, R. A. (1983). Larval ecology of marine benthic invertebrates: paleobiological implications. *Biological Reviews*, **58**, 21–89.

Jablonski, D., Sepkowski, J. J., Bottjer, D. J., and Sheehan, P. M. (1983). Onshore-offshore patterns in the evolution of Phanerozoic shelf communities. *Science*, **222**, 1123–4.

Jablonski, D., Gould, S. J., and Raup, D. M. (1986). The nature of the fossil record: a biological perspective. In *Patterns and processes in the history of life*, (ed. D. M. Raup and D. Jablonski), pp. 7–22. Springer-Verlag, Berlin.

Jablonski, D., Erwin, D. H., and Lipps, J. H. (ed.) (1996). *Evolutionary paleobiology*. University of Chicago Press, Chicago.

Jefferies, R. P. S. (1986). *The ancestry of the vertebrates*. British Museum (Natural) History. London.

Jiang, Z.-W. (1992). The Lower Cambrian fossil record of China. In *Origin and early evolution of the Metazoa*, (ed. J. H. Lipps and P. W. Signor), pp. 311–33. Plenum: New York.

Jones, E. M. and Kodis, J. W. (1982). Atmospheric effects of large body impacts: the first few minutes. In *Geological implications of impacts of large asteroids and comets on the Earth*, (ed. L. T. Silver and P. H. Schultz). *Geological Society of America Special Paper*, No. 190, 175–86.

Kauffman, E. G. and Hart, M. B. (1996). Cretaceous bio-events. In *Global events and event stratigraphy in the Phanerozoic*, (ed. O. H. Walliser), pp. 285–312. Springer-Verlag, Berlin.

Kauffman, S. A. (1993). *The origins of order*. Oxford University Press, Oxford.

Kelley, J. (1992). Evolution of apes. In *The Cambridge encyclopedia of human evolution*, (ed. S. Jones, R. Martin, and D. Pilbeam), pp. 223–30. Cambridge University Press, Cambridge.

Kemp, T. S. (1982*a*). *Mammal-like reptiles and the origin of mammals*. Academic Press, London.

Kemp, T. S. (1982*b*). The reptiles that became mammals. In *Darwin up to date*, (ed. J. Cherfas), pp. 31–4. IPC Magazines, London.

Kemp, T. S. (1985*a*). Models of diversity and phylogenetic reconstruction. *Oxford Surveys in Evolutionary Biology*, **2**, 135–58.

Kemp, T. S. (1985*b*). Synapsid reptiles and the origin of higher taxa. *Special Papers in Palaeontology*, No. 33, 175–84.

Kemp, T. S. (1988*a*). Haemothermia or Archosauria? The interrelationships of mammals birds and crocodiles. *Zoological Journal of the Linnean Society*, **92**, 67–104.

Kemp, T. S. (1988*b*). The origin of mammals: observed pattern and inferred process. In *L'évolution dans sa realité et ses diverses modalities*, pp. 65–91. Fondacion Singer-Polignac and Masson, Paris.

Kemp, T. S. (1979). The primitive cynodont *Procynosuchus*: functional anatomy of the skull and relationships. *Phil. Trans. R. Soc.*, **285B**, 73–122.

Kemp, T. S. (1989). The problem of the palaeontological evidence. In *Evolutionary studies: a centenary celebration of the life of Julian Huxley*, (ed. M. Keynes and G. A. Harrison), pp. 80–95. Macmillan, Basingstoke.

Kennedy, W. J. (1978). Cretaceous. In *The ecology of fossils* (ed. W. S. McKerrow), pp. 280–322. Duckworth, London.

Kendrick, P. and Crane, P. R. (1997). *The origin and early diversitfication of land plants: a cladistic study*, Smithsonian Press, Washington.

Kidwell, S. M. and Flessa, K. W. (1995). The quality of the fossil record: population, species, and communities. *Annu. Rev. Ecol. Syst.*, **28**, 495–516.

Kimura, M. (1983). *The neutral theory of molecular evolution*. Cambridge University Press, Cambridge.

Kitchell, J. A. (1990). The reciprocal interaction of organism and effective environment: learning more about 'and'. In *Causes of evolution: a paleontological perspective*, (ed. R. M. Ross and W. D. Aamon), pp. 151–69. Chicago University Press, Chicago.

Kitchell, J. A. and Carr, T. R. (1985). Nonequilibrium model of diversification: faunal turnover dynamics. In *Phanerozoic diversity patterns: profiles in macroevolution*, (ed. J. W. Valentine), pp. 277–309. Princeton University Press, Princeton, N.J.

Kitchell, J. A. and Pena, D. (1984). Periodicity of extinctions in the geologic past: deterministic versus stochastic explanations. *Science*, **226**, 689–92.

Kitts, D. B. (1974). Paleontology and evolutionary theory. *Evolution*, **28**, 458–72.

Kluge, A. (1989). A concern for evidence and a phylogenetic hypothesis of relationships among *Epicrates* (Boidae, Serpentes). *Systematic Zoology*, **38**, 7–25.

Knoll, A. H. (1995). Proterozoic and Early Cambrian protists: evidence for accelerating evolutionary tempo. In *Tempo and mode in evolution: genetics and paleontology 50 years after Simpson*, (ed. W. M. Fitch and F. J. Ayala), pp. 63–68. National Academy Press, Washington, DC.

Koch, C. F. (1991). Species extinction across the Cretaceous-Tertiary boundary: observed patterns versus predicted sampling effects, stepwise or otherwise? *Historical Biology*, **5**, 355–61.

Korn, D. (1995). Impact of environmental perturbations on heterochronic developments in palaeozoic ammonoids. In *Evolutionary change and heterochrony*, (ed. K. J. McNamara), pp. 245–60. Wiley, London.

Kreitman, M. and Akashi, H. (1995). Molecular evidence for natural selection. *Annual Review of Ecology and Systematics*, **26**, 403–22.

Kuhn-Schneider, E. and Rieber, H. (1986). *Handbook of paleozoology*. John Hopkins, Baltimore.

Lakatos, I. (1970). Falsification and methodology of scientific research programmes. In *Criticism and the growth of knowledge*, (ed. I. Lakatos and A. Musgrove), pp. 91–196. Cambridge University Press, Cambridge.

Lauder, G. V. (1995). On the inference of function from structure. In *Functional morphology in vertebrate paleontology*, (ed. J. J. Thomason), pp. 1–18. Cambridge University Press, Cambridge.

Lawrence, P. E. (1992). *The making of a fly: the genetics of animal design*. Blackwell Scientific, Oxford.

Leigh, E. G. (1986). Ronald Fisher and the development of evolutionary theory. I. The role of selection. *Oxford Surveys in Evolutionary Biology*, **3**, 187–223.

Leigh, E. G. (1987). Ronald Fisher and the development of evolutionary theory. II. Influences of new variation on evolutionary process. *Oxford Surveys in Evolutionary Biology*, **4**, 212–63.

Levinton, J. (1988). *Genetics, paleontology, and evolution*. Cambridge University Press, Cambridge.

Lidgard, S., McKinney, F. K., and Taylor, P. D. (1993). Competition, clade replacement, and a history of cyclostome and cheilostome bryozoan diversity. *Paleobiology*, **19**, 352–71.

Lieberman, B. S. (1995). Phylogenetic trends and speciation: analyzing macroevolutionary processes and levels of selection. In *New approaches to speciation in the fossil record*, (ed. D. H. Erwin and R. L. Anstey), pp. 316–37.Columbia University Press, New York.

Lieberman, B. S. and Dudgeon, S. (1996). An evaluation of stabilising selection as a mechanism for stasis. *Palaeogeography, Palaeoclimatology, Palaeoecology*, **127**, 229–38.

Lieberman, B. S., Brett, C. E., and Eldredge, N. (1995). A study of stasis and change in two species lineages from the Middle Devonian of New York State. *Paleobiology*, **21**, 15–27.

Lister, A. M. (1993). Patterns of evolution in Quaternary mammal lineages. In *Evolutionary patterns and processes*, (ed. D. R. Lees and D. Edwards), pp. 71–93. Academic Press, London.

Long, J. A. and McNamara, K. J. (1995). Heterochrony in dinosaur evolution. In *Evolutionary change and heterochrony*, (ed. K. J. McNamara), pp. 151–68. Wiley, New York.

Lovtrup, S. (1977). *The phylogeny of the Vertebrata.* Wiley, Chichester.

Maas, M. C., Krause, D. W., and Strait, S. G. (1988). The decline and extinction of plesiadapiformes (Mammalia: ?Primates) in North America: displacement or replacement? *Paleobiology*, **14**, 410–31.

McCune, A. R. (1982). On the fallacy of constant extinction rates. *Evolution*, **36**, 610–14.

MacFadden, B. J. (1992). Fossil horses: systematics, paleobiology, and evolution of the family Equidae. Cambridge University Press, Cambridge.

McGhee, G. R. (1996). *The Late Devonian mass extinction: the Frasnian/Famennian crisis*. Columbia University Press, New York.

McKinney, M. L. (1990). Classifying and analysing evolutionary trends. In *Evolutionary trends*, (ed. K. J. McNamara), pp. 28–58. Belhaven Press, London.

McKinney, M. L. (1997). Extinction vulnerability and selectivity: combining ecological and paleontological views. *Annu. Rev. Ecol. Syst.*, 28, 495–516.

McKinney, M. L. and McNamara, K. J. (1991). *Heterochrony in evolution; a multidisciplinary approach.* Plenum, New York.

MacLeod, N. (1991). Punctuated anagenesis and the importance of stratigraphy to paleontology. *Paleobiology*, **17**, 167–88.

MacLeod, N. (1996). K/T redux. *Paleobiology*, **22**, 311–17.

McMenamin, M. A. S. (1990). Vendian. In *Palaeobiology: a synthesis*, (ed. D. E. G. Briggs and P. R. Crowther), pp. 179–81. Blackwell Scientific, Oxford.

McNamara, K. J. (ed) (1990). *Evolutionary trends*. Belhaven, London.

McNamara, K. J. (ed) (1995). *Evolutionary change and heterochrony*. Wiley, Chichester.

McShea, D. W. (1993). Arguments, tests, and the Burgess Shale—a commentary on the debate. *Paleobiology*, **19**, 399–402.

McShea, D. W. and Raup, D. M. (1986). Completeness of the geological record. *Journal of Geology*, **94**, 569–74.

Malmgren, B. A. and Kennet, J. P. (1981). Phyletic gradualism in a Late Cenozoic planktonic foraminiferal lineage; DSDP site 284, southwest Pacific. *Paleobiology*, **7**, 230–40.

Malmgren, B. A., Berggren, W. A., and Lohmann, G. P. (1983). Evidence for punctuated gradualism in the Late Neogene *Globorotalia tumida* lineage of planktonic foraminifera. *Paleobiolgy*, **9**, 377–89.

Manton, S. M. and Anderson, D. T. (1979). Polyphyly and the evolution of the arthropods. In *The origin of major invertebrate groups*, (ed. M. R. House), pp. 269–321. Academic Press, London.

Marshall, C. R. and Schopf, J. W. (1996). *Evolution and the molecular revolution.* Jones and Bartlett, Sudbury, Mass.

May, R. M. (1975). Patterns of species abundance and diversity. In *Ecology and evolution of communities*, (ed. M. L. Cody and J. M. Diamond), pp. 81–120. Belknap Press, Cambridge, Mass.

May, R. M. (1987). Chaos and the dynamics of biological populations. In *Dynamical chaos*, (ed. M. V. Berry, I. C. Percival, and N. O. Weiss), pp. 27–43. Princeton University Press, Princeton, N.J.

Maynard Smith, J. (1983). Current controversies in evolutionary biology. In *Dimensions of Darwinism: themes and counterthemes in twentieth century evolutionary theory*, (ed. M. Grene), pp. 273–86. Cambridge University Press, Cambridge.

Maynard Smith, J., Burian, J., Kauffman, S., Alberch, P., Campbell, J., Goodwin, B., *et al.* (1983) Developmental constraints and evolution. *Quarterly Review of Biology*, **60**, 265–87.

Mayr, E. (1954). Change of genetic environment and evolution. In *Evolution as a process*, (ed. J. S. Huxley, A. C. Hardy, and E. B. Ford), pp. 157–80. Allen and Unwin, London.

Mayr, E. (1963). *Animal species and evolution.* Harvard University Press, Cambridge, Mass.

Mayr, E. (1982*a*). Processes of speciation in animals. In *Mechanisms of speciation*, (ed. C. Barigozzi), pp. 1–19. Alan R. Liss, New York. (Reprinted 1987 in Mayr, E. (ed.), *Towards a new philosophy of biology*, Harvard University Press, Cambridge, Mass.)

Mayr, E. (1982*b*). *The growth of biological thought: diversity, evolution, and inheritance.* Belknap Press, Cambridge, Mass.

Mayr, E. and Ashlock, P. D. (1991). *Principles of systematic zoology*, (2nd edn). McGraw-Hill, New York.

Mayr, E. and Provine, W. B. (ed.) (1980). *The evolutionary synthesis.* Harvard University Press, Cambridge, Mass.

Miller, A. I. (1997). Coordinated stasis or coincident relative stability? *Paleobiology*, **23**, 155–64.

Minelli, A. and Fusco, G. (1995). Body segmentation and segment differentiation: the scope for heterochronic change. In *Evolutionary change and heterochrony*, (ed. K. J. McNamara), pp. 49–63. Wiley, Chichester.

Morrow, J. R., Schindler, E., and Walliser, O. H. (1996). Phanerozoic development of selected global environmental features. In *Global events and event stratigraphy in the Phanerozoic*, (ed. O. H. Walliser), pp. 53–61. Springer-Verlag, Berlin.

Nelson, G. and Platnick, N. (1981). *Systematics and biogeography.* Columbia University Press, New York.

Newell, N. D. (1967). Revolutions in the history of life. *Geological Society of America Special Paper*, No. 89, 63–91.

Niklas, K. J. (1986). Large-scale changes in animal and plant terrestrial communities. In *Patterns and processes in the history of life*, (ed. D. M. Raup and D. Jablonski), pp. 383–405. Springer-Verlag, Berlin.

Niklas, K. J. (1995). Morphological evolution through complex domains of fitness. In *Tempo and mode in evolution: genetics and paleontology 50 years after Simpson*, (ed. W. M. Fitch and F. J. Ayala), pp. 145–68. National Academy Press, Washington, DC.

Niklas, K. J. (1997). *The evolutionary biology of plants.* Chicago University Press, Chicago.

Niklas, K. J., Tiffney, B. H., and Knoll, A. H. (1985). Patterns in vascular land plant diversification: an analysis at the species level. In *Phanerozoic diversity patterns: profiles in macroevolution*, (ed. J. W. Valentine), pp. 97–128. Princeton University Press, Princeton, NJ.

Norell, M. A. (1996). Ghost taxa, ancestors, and assumptions: a comment on Wagner. *Paleobiology*, **22**, 453–5.

Norell, M. A. and Novacek, M. J. (1992). The fossil record and evolution: comparing cladistic and paleontologic evidence for vertebrate history. *Science*, **255**, 1690–3.

Novacek, M. J. (1992). Fossils, topologies, missing data, and the higher level phylogeny of eutherian mammals. *Systematic Biology*, **41**, 58–73.

Novacek, M. J., Wyss, A. R., and McKenna, M. C. (1988). The major groups of eutherian mammals. In *The phylogeny and classification of the tetrapods. Volume 2. Mammals*, (ed. M. J. Benton), pp. 31–71. Oxford University Press, Oxford.

Officer, C. B. and Drake, C. L. (1985). Terminal Cretaceous environmental effects. *Science*, **227**, 1161–7.

Ohno, S. (1970). *Evolution by gene duplication*. Springer-Verlag, Berlin.

Orth, C. J. (1989). Geochemistry of the bio-event horizons. In *Mass extinctions: processes and evidence*, (ed. S. K. Donovan), pp. 37–72. Belhaven Press, London.

Ostrom, J. H. (1978). The osteology of *Compsognathus longiceps* Wagner. *Zitteliana und Abhandlungen der Bayerische Staatssammlung für Paläontologie*, **4**, 73–118.

Otte, D. and Endler, J. A. (ed.) (1989). *Speciation and its consequences*. Sinauer, Sunderland, Mass.

Padian, K. and Chiappe, L. M. (1998*a*). the origin and early evolution of birds. *Biological reviews*, 73, 1–42.

Padian, K. and Chiappe, L. M. (1998*b*). The origin of birds and their flight. *Scientific American*, February 1998, 28–37.

Panchen, A. L. (1992). Classification, evolution and the nature of biology. Cambridge University Press, Cambridge.

Panchen, A. L. and Smithson, T. R. (1987). Character diagnosis, fossils and the origin of tetrapods. *Biological Reviews*, **62**, 341–438.

Patterson, C. (1981). Significance of fossils in determining evolutionary relationships. *Annual Review of Ecology and Systematics*, **12**, 195–223.

Patterson, C. (1982). Classes and cladists or individuals and evolution. *Systematic Zoology*, **31**, 284–6.

Patterson, C. (1988). The impact of evolutionary theories on systematics. In *Prospects in systematics*, Systematics Association Special Volume No. 36, (ed. D. L. Hawksworth), pp. 59–91. Clarendon Press, Oxford.

Patterson, C. and Rosen, D. E. (1977). Review of the ichthydectiform and other Mesozoic teleost fishes and the theory and practice of classifying fossils. *Bulletin of the American Museum of Natural History*, **158**, 81–172.

Patterson, C. and Smith, A. B. (1987). Is periodicity of mass extinctions a taxonomic artefact? *Nature*, **330**, 248–51.

Patterson, C., Williams, D. M., and Humphries, J. (1993). Congruence between molecular and morphological phylogenies. *Annual Review of Ecology and Systematics*, **24**, 153–88.

Paul, C. R. C. (1990). Completeness of the fossil record. In *Palaeobiology: a synthesis*, (ed. D. E. G. Briggs and P. R. Crowther), pp. 298–303. Blackwell Scientific, Oxford.

Paul, C. R. C. and Smith, A. B. (1984). The early radiation and phylogeny of echinoderms. *Biological Reviews*, **59**, 443–81.

Pearson, P. N. (1992). Survivorship analysis of fossil taxa when real time extinction rates vary: the Paleogene planktonic foraminifera. *Paleobiology*, **18**, 115–31.

Platnick, N. (1979). Philosophy and the transformation of cladism. *Systematic Zoology*, **28**, 537–46.

Plot, R. (1677). *The natural history of Oxfordshire, being an essay toward the natural history of England*, Oxford.

Popov, L. Y. (1992). The Cambrian radiation of brachiopods. In *Origin and early evolution of the Metazoa*, (ed. J. H. Lipps and P. W. Signor), pp. 399–423. Plenum, New York.

Raff, R. A. (1996). The shape of life: genes, development, and the evolution of animal form. Chicago University Press, Chicago.

Raff, R. A. and Kauffman, T. C. (1983). *Embryos, genes, and evolution*. Macmillan, New York.

Raup, D. M. (1966). Geometrical analysis of shell coiling: general problems. *Journal of Paleontology*, **40**, 1178–90.

Raup, D. M. (1976). Species diversity in the Paleozoic: a tabulation. *Paleobiology*, **2**, 279–88.

Raup, D. M. (1977). Stochastic models in evolutionary paleontology. In *Patterns of evolution as illustrated by the fossil record*, (ed. A. Hallam), pp. 59–78. Elsevier, Amsterdam.

Raup, D. M. (1986). *The Nemesis affair*. W. W. Norton, New York.

Raup, D. M. (1989). The case for extraterrestrial causes of extinction. *Philosophical Transactions of the Royal Society*, **325B**, 421–35.

Raup, D. M. (1991). *Extinction; bad genes or bad luck?* W. W. Norton, New York.

Raup, D. M. (1995). The role of extinction in evolution. In *Tempo and mode in evolution: genetics and paleontology 50 years after Simpson*, (ed. W. M. Fitch and F. J. Ayala), pp. 109–24. National Academy Press, Washington, DC.

Raup, D. M. and Jablonski, D. (1993). Geography of end-Cretaceous marine bivalve extinctions. *Science*, **260**, 971–3.

Raup, D. M. and Sepkowski, J. J. (1982). Mass extinctions in the marine fossil record. *Science*, **215**, 1501–3.

Raup, D. M. and Sepkowski, J. J. (1984). Periodicity of extinctions in the geologic past. *Proceedings of the National Academy of Sciences, USA*, **81**, 801–5.

Raup, D. M. and Sepkowski, J. J. (1986). Periodic extinction of families and genera. *Science*, **231**, 833–6.

Raup, D. M., Gould, S. J., Schopf, T. J. M., and Simberloff, D. S. (1973). Stochastic models of phylogeny and the evolution of diversity. *Journal of Geology*, **81**, 525–42.

Reidl, R. (1979). *Order in living organisms: a systems analysis of evolution*, (trans. R. P. S. Jefferies). Wiley, Chichester.

Reidl, R. (1983). The role of morphology in the theory of evolution. In *Dimensions of darwinism: themes and counterthemes in twentieth-century evolutionary theory*, (ed. M. Grene), pp. 205–38. Cambridge University Press, Cambridge.

Ridley, M. (1986). *Evolution and classification: the reformation of cladism*. Longman: London.

Ridley, M. (1993). Analysis of the Burgess Shale. *Paleobiology*, **19**, 519–21.

Ridley, M. (1996). *Evolution* (2nd edn). Blackwell Science: Oxford.

Romer, A. S. (1966). *Vertebrate palentology*, 3rd edn. Chicago University Press, Chicago.

Rosen, D. E., Forey, P. L., Gardiner, B. G., and Patterson, C. (1981). Lungfishes, tetrapods, paleontology and plesiomorphy. *Bull. Am. Mus. Nat. Hist.*, **167**, 159–276.

Rozensweig, M. L. and McCord, R. D. (1991). Incumbent replacement: evidence for long-term evolutionary progress. *Paleobiology*, **17**, 202–13.

Rudwick, M. (1964). The inference of function from structure in fossils. *British Journal of Philosophy of Science*, **15**, 27–40.

Sadler, P. M. (1981). Sediment accumulation rates and the completeness of stratigraphic sections. *J. Geology*, **89**, 569–84.

Sadler, P. M. and Strauss, D. J. (1990). Estimation of completeness of stratigraphical sections using empirical data and theoretical models. *Journal of the Geological Society*, **147**, 471–85.

Sarich, V. M. and Wilson, A. C. (1967). Immunological time scale for hominid evolution. *Science*, **158**, 1200–3.

Save-Soderbergh, G. (1933). The dermal bones of the head and the lateral line system in *Osteolepis macrolepidotus* Ag. *Nova Acta Regiae Societatis Scientiarum Upsaliensis*, Series 4, Vol. 9, 1–130.

Schaeffer, B., Hecht, M. K., and Eldredge, N. (1972). Phylogeny and paleontology. *Evolutionary Biology*, **6**, 31–46.

Schindewolf, O. H. (1993). *Basic questions in paleontology; geological time, organic evolution, and biological systematics.* Chicago University Press, Chicago. (First published in 1950 as *Grundfragen der Paläontologie*, Schweizerbart'sche Verlagsbuchhandlung, Erwin Nagele, Stuttgart.)

Schopf, J. W. (1995). Disparate rates, differing fates: tempo and mode of evolution changed from the Precambrian to the Phanerozoic. In *Tempo and mode in evolution: genetics and paleontology 50 years after Simpson*, (ed. W. M. Fitch and F. J. Ayala), pp. 41–61. National Academy Press: Washington, DC.

Schopf, T. J. M., Raup, D. M., Gould, S. J., and Simberloff, D. S. (1975). Genomic versus morphological rates of evolution: influence of morphologic complexity. *Paleobiology*, **1**, 63–70.

Schulze, H.-P. (1987). Dipnoans as sarcopterygians. In *The biology and evolution of lungfishes*, (ed. W. E. Bemis, W. W. Burggren, and N. E. Kemp). *Journal of Morphology Supplement* No.1, 39–74.

Schulze, H.-P. (1994). Comparison of hypotheses on the relationships of sarcopterygians. *Systematic Biology*, **43**, 155–73.

Seilacher, A. (1990). The sand-dollar syndrome: a polyphyletic constructional breakthrough. In *Evolutionary innovations*, (ed. M. H. Nitecki). Chicago University Press, Chicago.

Sepkoski, J. J. (1981). A factor analytic description of the marine fossil record. *Paleobiology*, 7, 36–53.

Sepkoski, J. J. (1982). A compendium of fossil marine families. *Milwaukee Public Museum Contributions in Biology and Geology*, **83**, 1–156.

Sepkoski, J. J. (1984). A kinetic model of Phanerozoic taxonomic diversity. III. Post-Paleozoic families and mass extinctions. *Paleobiology*, **10**, 246–67.

Sepkoski, J. J. (1996*a*). Patterns of Phanerozoic extinction: a perspective from global data bases. In *Global events and event stratigraphy in the Paleozoic*, (ed. O. H. Walliser), pp. 35–51. Springer-Verlag, Berlin.

Sepkoski, J. J. (1996*b*). Competition in macroevolution: the double wedge revisited. In *Evolutionary paleobiology: in honour of James W. Valentine*, (ed. D. Jablonski, D. H. Erwin, and J. H. Lipps), pp. 211–55. Chicago University Press, Chicago.

Sepkoski, J. J. and Hulver, M. L. (1985). An atlas of Phanerozoic clade diversity diagrams. In *Phanerozoic diversity patterns: profiles in macroevolution*, (ed. J. W. Valentine), pp. 11–39. Princeton University Press, Princeton, NJ.

Sepkoski, J. J. and Kendrick, D. C. (1993). Numerical experiments with model monophyletic and paraphyletic taxa. *Paleobiology*, **19**, 168–84.

Sepkoski, J. J. and Koch, C. F. (1996). Evaluating paleontologic data relating to bioevents. In *Global events and event stratigraphy in the Phanerozoic*, (ed. O. H. Walliser), pp. 21–34. Springer-Verlag, Berlin.

Sepkoski, J. J. and Miller, A. I. (1985). Evolutionary faunas and the distribution of Paleozoic benthic communities in space and time. In *Phanerozoic diversity patterns: models in macroevolution*, (ed. J. W.Valentine), pp. 153–90. Princeton University Press, Princeton, NJ.

Sepkoski, J. J., Bambach, R. K., Raup, D. M., and Valentine, J. W. (1981). Phanerozoic marine diversity and the fossil record. *Nature*, **293**, 435–7.

Sheldon, P. R. (1993). Making sense of microevolutionary patterns. In *Evolutionary patterns and processes*, (ed. D. R. Lees and D. Edwards), pp. 19–31. Academic Press, London.

Sheldon, P. R. (1996). *Plus ça change*—a model for stasis and evolution in different environments. *Palaeogeography, Palaeoclimatology, Palaeoecology*, **127**, 209–27.

Shoemaker, E. M. (1984). Large body impacts through geologic time. In *Patterns of change in Earth evolution*, (ed. H. D. Holland and A. F. Trendall), pp. 15–40. Springer-Verlag, Berlin.

Sidor, C. A. and Hopson, J. A. (1998). Ghost lineages and 'mammalness': assessing the temporal pattern of character acquisition in the Synapsida. *Paleobiology*, **24**, 254–73.

Signor, P. W. III (1985). Real and apparent trends in species richness through time. In *Phanerozoic diversity patterns: profiles in macroevolution*, (ed. J. W. Valentine), pp. 129–50. Princeton University Press, Princeton, NJ.

Signor, P. W. III (1990). The geologic history of diversity. *Annual Review of Ecology and Systematics*, **21**, 509–39.

Signor, P. W. III (1992). Taxonomic diversity and faunal turnover in the early Cambrian: did the most severe mass extinction of the Phanerozoic occur in the Botomian Stage? *Fifth North American Paleontological Convention, Abstracts with Programs*, p. 272.

Signor, P. W. III and Lipps, J. H. (1982). Sampling bias, gradual extinction patterns, and catastrophes in the fossil record. In *Geological implications of large asteroids and comets on the Earth*, (ed. L. T. Silver and P. H. Schulz), pp. 291–6. Geological Society of America, Special Paper, 190.

Signor, P. W. III and Vermeij, G. J. (1994). The plankton and the benthos: origin and early history of an evolving relationship. *Paleobiology*, **20**, 297–319.

Simpson, G. G. (1944). *Tempo and mode in evolution*. Columbia University Press, New York.

Simpson, G. G. (1952). How many species? *Evolution*, **6**, 342.

Simpson, G. G. (1953). *The major features of evolution.* Columbia University Press, New York.

Skelton, P. W., Crame, J. A., Morris, N. J., and Harper, E. M. (1990). Adaptive divergence and taxonomic radiation in post-Palaeozoic bivalves. In *Major evolutionary radiations*, (ed. P. D. Taylor and G. P. Larwood), pp. 91–117. Clarendon Press, Oxford.

Smith, A. B. (1984). Classification of the Echinodermata. *Palaeontology*, **27**, 431–59.

Smith, A. B. (1990). Evolutionary diversification of echinoderms during the early Palaeozoic. In *Major evolutionary radiations*, (ed. P. D. Taylor and G. P. Larwood), pp. 265–86. Clarendon Press, Oxford.

Smith, A. B. (1994). *Systematics and the fossil record: documenting evolutionary patterns.* Blackwell Scientific, Oxford.

Sober, E. (1984). *The nature of selection: evolutionary theory in philosophical focus.* MIT Press, Cambridge, Mass.

Sober, E. (1988). *Reconstructing the past: parsimony, evolution and inference.* MIT Press, Cambridge, Mass.

Sokal, P. H. A. and Sneath, R. R. (1973). *Numerical taxonomy: the principles and practice of numerical classification.* W. H. Freemanm San Francisco.

Somit, A. and Peterson, S. A. (1989). The dynamics of evolution: the punctuated equilibrium debate in the natural and social sciences. Cornell University Press, Ithaca, N.Y.

Sprinkle, J. (1992). Radiation of Echinodermata. In *Origin and early evolution of the Metazoa*, (ed. J. H. Lipps and P. W. Signor), pp. 375–98. Plenum, New York.

Stanley, S. M. (1977). Trends, rates, and patterns of evolution in the Bivalvia. In *Patterns of evolution as illustrated by the fossil record*, (ed. A. Hallam), pp. 209–50. Elsevier, Amsterdam.

Stanley, S. M. (1979). *Macroevolution: pattern and process.* W. H. Freeman, San Francisco.

Stanley, S. M. (1985). Rates of evolution. *Paleobiology*, **11**, 13–26.

Stanley, S. M. (1987). *Extinction.* Scientific American Library, New York.

Stanley, S. M. (1990). The general correlation between rate of speciation and rate of extinction: fortuitous causal linkages. In *Causes of evolution: a paleontological perspective*, (ed. R. M. Ross and W. D. Allmon), pp. 103–27. Chicago University Press, Chicago.

Stanley, S. M., Signor, P. W., Lidgard, S., and Karr, A. F. (1981). Natural clades differ from 'random' clades: simulations and analyses. *Paleobiology*, **7**, 115–27.

Stenseth, N. C. and Maynard Smith, J. (1984). Coevolution in ecosystems: Red Queen evolution or stasis? *Evolution*, **38**, 870–80.

Swofford, D. L. (1991). When are phylogeny estimates from molecular and morphological data incongruent? In *Phylogenetic analysis of DNA sequences*, (ed. M. M.Miyamoto and J. Cracraft), pp. 295–333. Oxford University Press, Oxford.

Teichert, C. (1990). The Permian-Triassic boundary revisited. In *Extinction events in Earth history*, (ed. E. G. Kauffman and O. H. Walliser), pp. 199–238. Springer-Verlag, Berlin.

Thomason, J. J. (ed.) (1995). *Functional morphology in vertebrate paleontology.* Cambridge University Press, Cambridge.

Thompson, d'A. W. (1942). *On growth and form*, (2nd edn). Cambridge University Press, Cambridge.

Thomson, K. S. (1966). The evolution of the tetrapod middle ear in the rhipidistian-tetrapod transition. *American Zoologist*, **6**, 379–97.

Thomson, K. S. (1988). *Morphogenesis and evolution*. Oxford University Press, Oxford.

Thomson, K. S. (1993). The origin of the tetrapods. *American Journal of Science*, **293A**, 33–62.

Thulborn, T. (1990). *Dinosaur tracks*. Chapman and Hall, London.

Trueman, A. E. (1922). The use of *Gryphaea* in the correlation of the Lower Lias. *Geological Magazine*, **59**, 256–68.

Uhen, M. D. (1996). An evaluation of clade shape statistics using simulations and extinct families of mammals. *Paleobiology*, **22**, 8–22.

Valentine, J. W. (ed.) (1985). *Phanerozoic diversity patterns*. Princeton University Press, Princeton, N.J.

Van Valen, L. M. (1973). A new evolutionary law. *Evolutionary Theory*, **1**, 1–30.

Van Valen, L. M. (1985). How constant is extinction? *Evolutionary Theory*, **7**, 93–106.

Vermeij, G. J. (1987). *Evolution and escalation: an ecological history of life*. Princeton University Press, Princeton, NJ.

Vrba, E. S. (1980). Evolution, species and fossils: how does life evolve? *South African Journal of Science*, **76**, 61–84.

Vrba, E. S. (1984). Patterns in the fossil record and evolutionary processes. In *Beyond neo-darwinism: an introduction to the new evolutionary paradigm*, (ed. M.-W. Ho and P. T. Saunders), pp. 115–42. Academic Press, London.

Vrba, E. S. (1985). Environment and evolution: alternative causes of the temporal distribution of evolutionary events. *South African Journal of Science*, **81**, 229–36.

Waggoner, B. M. (1998). Phylogenetic hypotheses of the relationships of arthropods to Precambrian and Cambrian problematic taxa. *Systematic Zoology*, **45**, 190–222.

Wagner, P. J. (1995). Stratigraphic tests of cladistic hypotheses. *Paleobiology*, **21**, 153–78.

Wake, M. M. (1992). Morphology, the study of form and function in modern evolutionary biology. *Oxford Surveys in Evolutionary Biology*, **8**, 289–346.

Wake, D. B. and Roth, G. (ed.) (1989). Complex organismal functions: integration and evolution in vertebrates. Wiley, Chichester.

Walliser, O. H. (1996*a*). Global events in the Devonian and Carboniferous. In *Global events and event stratigraphy in the Phanerozoic*, (ed. O. H. Walliser), pp. 225–50. Springer-Verlag, Berlin.

Walliser, O. H. (ed.) (1996*b*). *Global events and event stratigraphy in the Phanerozoic*. Springer-Verlag, Berlin.

Wang, K., Geldsetzer, H. H. J., Goodfellow, W. D., and Krouse, H. R. (1996). Carbon and sulphur isotope anomalies across the Frasnian-Famennian extinction boundary, Alberta, Canada. *Geology*, **24**, 187–91.

Webb, T and Bartlein, P. J. (1992). Global changes during the last 3 million years: climatic controls and biotic responses. *Annual Review of Ecology and Systematics*, **23**, 141–73.

Wei, K.-Y. and Kennett, J. P. (1983). Nonconstant extinctionrates of Neogene planktonic foraminifera. *Nature*, **305**, 218–20.

Weishampel, D. B. (1995). Fossils, function, and phylogeny. In *Functional morphology in vertebrate paleontology*, (ed. J. J. Thomason), pp. 34–54. Cambridge University Press, Cambridge.

Wheeler, W. C., Cartwright, P., and Hayasha, C.-Y. (1993). Arthropod phylogeny: a combined approach. *Cladistics*, **9**, 1–39.

Whetstone, K. N. (1983). Braincase of the Mesozoic birds: 1. New preparation of the London *Archaeopteryx*. *Journal of Vertebrate Paleontology*, **2**, 439–52.

Whitmire, D. P. and Jackson, A. A. (1984). Are periodic mass extinctions driven by a distant solar companion? *Nature*, **308**, 713–15.

Wilde, P. and Berry, W. B. N. (1984). Destabilisation of the oceanic density structure and its significance to marine 'extinction' events. *Palaeogeography, Palaeoclimatology, Palaeoecology*, **48**, 143–62.

Williams, G. C. (1966). *Adaptation and natural selection*. Princeton University Press, Princeton, N.J.

Williams, M. A. J., Dunkerley, D. L., De Dekker, P., Kershaw, A. P., and Stokes, T. (1993). *Quaternary environments*. Edward Arnold, London.

Williamson, P. G. (1981). Paleontological documentation of speciation in Cenozoic molluscs from Turkana Basin. *Nature*, **293**, 437–43.

Willmer, P. (1990). *Invertebrate relationships*. Cambridge University Press, Cambridge.

Wills, M. A. (1998). Cambrian and Recent disparity: the picture from priapulids. *Paleobiology*, **24**, 177–99.

Wills, M. A., Briggs, D. E. G., and Fortey, R. A. (1994). Disparity as an evolutionary index: a comparison of Cambrian and Recent arthropods. *Paleobiology*, **20**, 93–130.

Wilson, M. V. H. (1992). Importance for phylogeny of single and multiple stem group fossil species with examples from freshwater fishes. *Systematic Biology*, **41**, 462–70.

Witmer, L. M. (1995). The extant bracket and the importance of reconstructing soft tissues in fossils. In *Functional morphology in vertebrate paleontology*, (ed. J. J. Thomason), pp. 19–33.

Wray, G. A., Levinton, J. S., and Shapiro, L. H. (1996). Molecular evidence for deep Precambrian divergences among metazoan phyla. *Science*, **274**, 568–73.

Wright, S. (1986). *Evolution: selected papers*, (ed. W. B. Provine). Chicago University Press, Chicago.

Xu, G. (1991). Stratigraphical time-correlation and mass extinction event near Permian-Triassic boundary in South China. *Journal of China University of Geoscience*, **2**, 36–46.

# Index

Page numbers in **bold** refer to illustrations


*ad hoc* assumptions
  epistemological gap 22
  and Occam's razor 17–18, 19, 60
adelphotaxa 54
age of fossils
  ghost lineages **115**
  and phylogenetic reconstruction 114–18
  species longevities 177–8, 252
allopatry, defined 131
ammonites
  Cretaceous, first and last occurrences **193**
  Devonian, heterochrony **220**
  preservational bias, apparent mass extinction **192–3**
  uncoiled **30**
ammonoids
  end-Triassic event 211
  Red Queen hypothesis 174–6, **175**
Amniota
  outgroup (amphibians) 58
  phylogeny **49**, **111**, 121–2
anatomy, *see* functional anatomy
ancestral character state 59, 108–9
ancestry
  hypothetical in cladistics 56, 108–9
  sister groups 67
angiosperms 168–9
  K–T boundary 212
  origins 246
anoxia
  black shales 195
  Frasnian–Famennian event 209
antelopes, generalists vs specialists 153
*Archaeopteryx* 108–9, 122–3, **237**
arthropods
  Burgess Shale 67, 72, 239, 240–1
  Cambrian explosion **244**
  insects, end-Permian event 210
  'inverted-cone' model (Gould) **242**
  *Lagerstätten* 72, 170, 239
  onophyly 240
  origins 239–45
  trackways, Ediacaran fauna 243
Ashgillian (end-Ordovician) event 205–7

bases, nucleic acids, transitions vs transversions 63
biogeographic effects, clade histories 176–7
biogeographical incompleteness 72, 100–3
birds
  cladogram–tree–scenario trio 18–20
  origins 235–9
  stem group cladogram 235, **236**
black shales, association with mass extinctions 195
Botomian–Toyonian event 205
bovids, phylogeny **152**
brachiopods
  example of stasis 141–2
  *see also* molluscs
bryozoans, example of stasis 141, **142**
Burgess Shale arthropods 67, 72, 239–41

$^{13}C$
  end-Permian event 210
  Frasnian–Famennian event 207–8
Cambrian faunal diversity 163, **164**, 248
  arthropod explosion **244**
  guilds 165–7
carbon, radioisotope ecological information 95–6
case studies, higher taxa 227–46
Cenozoic diversity, number of species 162–3
character polarity, embryology 58
character weighting
  circularity 120
  a priori 62–3, 120
Chicxulub Crater, and K–T boundary 199, 212
choana, lungfish–tetrapod relationship 112, **113**
chordates, divergence 243

chronospecies concept 84
clades
clade-rank vs age-rank 116–17
diversity and species turnover, stochasticity 173–4
histories, biogeographic effects 176–7
interactions, taxonomic turnover 179–88
possible patterns of replacement **183**, 184–6
incumbent replacement 186
opportunistic replacement 181–4
*see also* mass extinctions
shape rules 171
as units of selection 43–4
cladistics 51–66, 108
'bush' problem 64, **65**
cladograms 18–20, 110–12
consensus trees 119, **120**
ghost lineages **115**
molecular-based phylogeny 119–21
*see also named taxa*
cladogram–tree–scenario trio 18–20
hypothetical ancestry 56
limits 64–6
monophyly 52–6
parsimony 59–63, 110–12
confidence levels 66
principles of analysis 52
rank 54
sister taxa 54
species level 64–5
synapomorphy 57–9
classification
as bias source 139
cladistic **53**
formal 121–7
non-utility of fossil evidence 109
problems 12, 14
climatic chnage, causing apparent evolutionary change 100–3
coccoliths, Red Queen hypothesis 174–6
coelacanths, cladograms **113**
cohesion concept, species 83
comparative test, functional anatomy 78
competitive processes
difficulties in ecological extrapolation 179–86
diminishing role in replacement models **182**
*Compsognathus* **101**, 235
congruence
parsimony 62
synapomorphy 59
consensus trees, defined 119, **120**
constraints, contemporary disputes 28–34, **33**
correlated progression, higher taxa 225–6
Cretaceous–Tertiary boundary (K–T) 212–13
Chicxulub Crater 199
crocodiles, and Therocephalia, hindlimb structure 75, **76**, 78
crown group 127, 231
Cyanobacteria, stromatolites 159, **160**

*Danaus*, paraspecies 64, **65**
Deccan Traps, volcanism 198, 212
Deep Sea Drilling Project cores
FADs and LADs 91, **92**
phyletic gradualism 144–5, **146**
design test, functional anatomy 75–8
developmental feedback, morphological integration, higher taxa 226
*Diacodexis metsiacus*, variation within species **86**
Diapsida, possible patterns of clade replacement **183**
dichotomies vs multiple splitting 54, **55**
dinosaurs
and birds 235–7
footprints **99**
opportunistic replacement concept 181
possible patterns of clade replacement **183**
Dipnoi, alternative cladograms **113**
directional evolutionary trends, causation 29
discontinuties of records, causes 102–3
disparity
and phenetics 66–9
trilobites **68**
diversity, *see* species diversity
DNA
introns, neutral evolutionary process 37
silent 41–2
*see also* molecular-based phylogeny
*Drosophila*, homeotic mutants 34, **35**

Earth, Milankovitch cycles 150–1
echinoderms
divergence 243, 246
phylogeny **247**
ecological cline, apparent evolutionary change **103**
ecological incompleteness of information 72, 94–100
geochemistry 95–6
palaeoautecology 100
structural sedimentology 95

taphonomy 97–9
trace fossils 96–7
ecological processes, extrapolation, competitive processes, difficulties with concept 179–86
*Ectocion osbornianus*, variation within species **86**
Ediacaran fauna 161, **204**
arthropod trackways 243
mass extinction 198
embryology, character polarity 58
end-Cretaceous event, *see* Cretaceous–Tertiary boundary (K–T)
end-Ordovician event 205–7
end-Permian event **180**, 209–11, **209**
end-Triassic event 211–12
epistemological gap 21–2, 251
*ad hoc* assumptions 22
epistemology of mass extinctions 189–201
escalation of diversity 168
eukaryotes, unicells 159
*Eusthenopteron*, lungfish–tetrapod relationship 112, **113**, **126**, 238
evaporites 210
evolutionary change, and observation 12–13
evolutionary process, and taxonomic patterns 11–18, **13**
evolutionary theory, current mainstream 26–45
contemporary disputes 27–45
constraints 28–34, **33**
non-adaptive evolution 34–9
units of selection 39–45, **44**
five component theories 26–7
hierarchical expansion 45
extinction rate
vs speciation rate 165, 178
*see also* mass extinctions
extraterrestrial impacts 198–9

fish
lungfish–tetrapod relationship 112, **113**
*P. bechei*, status as sister group to teleosts 54, 122, **123**
flood basalt rocks, and mass extinctions 197–8
footprints, dinosaurs **99**
foraminiferans
DSDP cores, FADs and LADs **91**, 92
evolution of *Globorotalia* **90**, 139
species longevities 252
form, *see* functional anatomy
fossils
misinterpreted 2
soft tissues **74**
Frasnian–Famennian boundary 207–9
Upper Kellwasser event 207
functional anatomy 73–80
comparative test 78
design test 75–8
phylogenetic test 79–80

gastropod molluscs, example of stasis and gradualism 142–4
generalists, vs specialists 153
genes, as units of selection 41
genetic drift 34
following isolation 132
genetic variation theory 26
geochemistry, radioisotope ecological information 95–6
geographic isolation, and genetic drift 132
ghost lineages **115**, 116
global climate change 214, **215**
*Globorotalia*, DSDP cores, FADs and LADs **90**, 92, 144–5, **146**
graphic correlations 92
greenhouse effects, global climate change, Phanerozoic **215**
*Gryphaea* 139, **140**
guilds, ecological diversity 165–7
gymnosperms 168–9

heterochrony 219–20
birds 237
Devonian ammonites **220**
hierarchical expansion of evolutionary theory 45
hierarchical (phenetic) relationships 49–51
higher taxa
abruptness of appearance 253
case studies of origins
arthropods 239–45
mammals 227–34
other higher taxa 246
other vertebrate groups 235–9
causation of evolutionary trends
chance 222
natural selection 223
orthogenesis 222–3
species selection 223–4
defined 217
evolutionary steps 218–21
generalizations 246–9

maintenance of morphological integration 224, **225**
correlated progression 225–6
developmental feedback 226
origin 217–50
quantum evolution 218, **219**
as units of diversity change 157–9
hindlimb structure, Therocephalia 75, **76**, 78
homoplasy 62
defined 59
horses
clade-rank vs age-rank 116–17
orthogenetic pattern of evolution **30**
species selection 224
Hox (homeotic) genes 34, **35**, 221

ichnology, trace fossils 96
incompleteness of information 72–104
as bias source 139
biogeographical 72, 100–3
ecological 72, 94–100
organismic 71, 72–85
stratigraphic 71, 85–93
incumbent replacement, clades 186
India, Deccan Traps 198, 212
inheritance theory 26
iridium, sources 196–7, 198
iridium anomalies, and Cretaceous–Tertiary boundary (K–T) 188–9, 208, 212
isotopes, *see* radioisotopes

*Lagerstätten* 72, 170, 239
Late Devonian (Frasnian–Famennian) event 207–9
Lower Cambrian (Botomian–Toyonian) event 205

Maastrichtian chalk community 213
macroevolutionary theory 27
decoupling from microevolution 138
inferences from stasis 137–8
macromutational (saltational) theories 31, 221, 231
magnetic reversal, method of estimating completeness of information **91**, 93
mammal-like reptiles
phylogenetic tree **228**
possible patterns of clade replacement **183**
mammals
clade-rank vs age-rank 116–17
cladogram **229**
hypothetical ancestors, inference of acquisition of characteristics 231
integration of structure and future prospects **233**
interrelationships 113–15
metabolic characteristics 232, **233**
origins 227–34
phylogenetic tree **228**
species selection 234
*Mammuthus*, phyletic gradualism 144, **145**
Manicouagan Crater 211
marine provinciality, and global diversity 165
marine regressions, correlation with mass extinctions 194–5
mass extinctions 188–216, 253
abiotic correlations 194–9
black shales 195
element abundances 196–7
radioisotope ratios 195–6
sea-level changes 194–5
volcanism 197–8
big seven bioevents 202–13
extinction rate **191**
biotic correlations 200–1
causation 214–15
defined 189
effect of preservational bias **192–3**
epistemology 189–201
taxonomic units 189–90
time-course 190–4
extraterrestrial impacts 198–9
global Phanerozoic pattern 201–2
pattern **203**
periodicity 188–9, 202, **203**
Raup's kill curve **203**
and tectonic activity 198
*Mediospirifer audaculus* 141–2, **143**
*Melanopsis* 143–4
*Metrarabdotus* 141, **142**
Milankovitch cycles 150–1, 214
modern faunal diversity 163, **164**
guilds 165–7
molecular analysis
vs morphology
primate relationships 121, **122**
reptile relationships 121, **122**
protein substitutions vs divergence times 36
molecular developmental genetics 254–5
molecular-based phylogeny 118–21, 254
cladograms 119–21
evolution rates 37

molluscs
brachiopods
Ashgillian extinction **206**
dominance in Palaeozoic 179–81, 186
example of stasis 141–2, **143**
completeness of record 93
completeness of shells **101**
gastropod
example of stasis and gradualism 142–4
species selection, planktotrophic species 153–4
inoceramid bivalve
diversity in higher latitudes 177
dominance in Permian 179, **180**, 186
lophospirid, cladograms **115**, 116
Palaeozoic, dominance 179–81
punctuated equilibrium, explanations 148–50
shell, parameters of form 32, **33**
monophyletic groups
crown group 127
plesions 124
stem group 127
and taxonomic turnover 158–9
monophyly
cladistics 52–6
defined 52
morphological integration, maintenance 224, **225**
morphological reversals (oscillations) 142–3, 151
morphospace 66

natural selection, causation of evolution of higher taxa 223
Nemesis hypothesis 202
neo-Darwinism, *see* evolutionary theory, current mainstream
neoteny theory 220
neutron activation analysis 196–7
non-adaptive evolution, contemporary disputes 34–9
nucleic acids, transitions vs transversions 63

objectivity, defined 6
Occam's razor
and *ad hoc* assumptions 17–18, 19, 60
defined 60
onshore–offshore rule 176–7
ontogeny, von Baer's rule 58
opportunistic replacement concept 181–4
Ordovician mass extinctions 205–7
organism-level selection, vs species selection 153–4
organismic incompleteness 72, 72–85
functional anatomy 73–80
orthogenesis, causation of evolution of higher taxa 222–3
Osteolepiformes, alternative cladograms **113**
outgroup comparisons 57–8
oxygen, radioisotope ecological information 95–6

palaeoautecology 100
palaeomagnetic method of estimating completeness of information **91**, 93
palaeospecies, defined 84
Palaeozoic faunal diversity 163, **164**
dominant molluscs, changes 179–81
guilds 165–7
Pangaea, formation 210
panspermia theory 219
paradigm test 75–7
parapatry, defined 131, 138
paraspecies, defined 64, 65
parsimony (cladistics) 59–63
cladograms 110–12
defined 5
maximum-congruence 60–2
minimum-evolution 60, **61**
missing forms 114
*Parvancorina* 243, **244**
pattern vs process circularity 11–23, **13**, 85
Permian–Triassic boundary
diversity changes in molluscs **180**
marine invertebrate group extinctions **209**
mass extinctions 253
Phanerozoic
extinction events **196**
extinction rate, families and genera **191**
global climate change, ice-house vs greenhouse effects 214, **215**
mass extinctions 201–2
species diversity 161, **162**
vs time **162**, 163
phenetics 49–51
and disparity 66–9
sister groups 54, 67, 80
phenon, defined 84
*Pholidophorus bechei*, status as sister group to teleosts 54, 122, **123**
phyletic gradualism
*Mammuthus* 144, **145**
*Melanopsis* 143–4

phylogenetic fuse 248, **249**
phylogenetic gradualism, defined 135–6
phylogenetic tree
  extant phylogenetic bracket **79**
  tetrapods **5**
phylogeny
  formal classification 121–7
  molecular-based 118–21
  patterns 11–23
  phylogenetic analysis 107–18
  phylogenetic test, functional anatomy 79–80
plants, land, four successive floras 168–70, **169**
plesion
  in classification of teleosts **125**
  defined 124
pollen grains **102**
polychotomies, unresolved 54–6
polyploidy events 221
population structure, and species selection 153
preadaptation 238
Precambrian–Cambrian event 204–5
present-day diversity, *see* modern faunal diversity
preservational bias, apparent mass extinction **192–3**
priapulids, origins 246
primates
  clade-rank vs age-rank 116–17
  plesiadapid, patterns of diversity changes in Palaeocene **185**
  relationships, molecular analysis vs morphology 121
process
  analysing 24–47
  pattern vs process circularity 11–23
prokaryotes 159
proteins, molecular substitutions vs divergence times **36**
protostomes 243
pteridophyta 168–9
punctuated equilibrium 135–8, 148
  explanations 148–50
  problems 138–9, 148–50

quantum evolution 218, **219**

radioisotopes
  ecological information 95–6, 150
  ratio shifts
    association with mass extinctions 195–6
    past environments 254
recognition concept, species 83
Red Queen hypothesis 174–6
reptile relationships, vs morphology, molecular analysis 121
Rhyniophyta 168–9
rodents, patterns of diversity changes in Palaeocene **185**

$^{34}$S, end-Permian event 210
saltational theories 31
sauropods, and maximum size attainable 4
scenario, defined **20**
sea-level changes, correlation with mass extinctions 194–5
selection
  Darwinian logic 39–41
  units of selection, contemporary disputes 39–45, **44**
self-organization 31
selfish DNA hypothesis 42, 45
Siberian Traps, volcanism 20, 198
Silurian land plant, modelling selection 24, **25**
Simpson, G.G. 14, 16
sister groups 54, 67, 80, 122
size, maximum 4
sorting theory 26–7
speciation rate, vs extinction rate 165, 178
speciation theory 27, 129–55
  debate 138–55
    implications for species selection 152–5
    phyletic gradualism 141–5, **146**
    punctuation 144, **146**
    stasis 139–4, **146**
    theoretical interpretation 147–51
    variability of pattern 146–7
  neontology
    allopatry, parapatry and sympatry 131, **132**
    sources of variation 129–33
  palaeontology
    frequency of speciation 134–5
    time-course of speciation 135–8
  turnover pulse hypothesis 146
species
  chronospecies concept 84
  cohesion concept 83
  recognition concept 83
  recognition of fossil species 81–5
  rival concepts of defining characters 82–3
  as units of selection 42–3, **44**

species diversity 159–70
changes within taxa 171–8
defined 66
Earth's biota 158–70
escalation of diversity 168
Phanerozoic, correction factors 161, **162**
Red Queen hypothesis 174–6
taxonomic turnover 156–87
vs time, spindle diagrams 171, **172**
species longevities 252
taxonomic turnover rates 177–8
species selection
causation of evolution of higher taxa 223–4
vs organism-level selection 153–4
planktotrophic species 153–4
and population structure 153
species turnover, *see* taxonomic turnover
specific mate-recognition system (SMRS) 84
spindle diagrams, species diversity vs time 171, **172**
squamates, cladogram–tree–scenario trio **20**
stasis 137–8
co-ordinated 147
examples
beetles 141
brachiopods 141–2
bryozoans 141
oyster 139–40
explanations 151
theoretical interpretation 147–55
stem group 127, 231
'Strangelove Ocean' 195, 212
stratigraphy
and adequacy of fossil record 89
age of fossils, and phylogenetic reconstruction 114–18
depositional rates **87**
DSDP cores, FADs and LADs **91**, 92
DSDP FADs and LADs **91**
ghost lineages **115**
incompleteness of information 72, 85–93
models for soil structure **91**, 92
mammal phylogenetic tree **228**
palaeomagnetic method of estimating completeness of information **91**, 93
resolution 88–9
time interval vs percentage completion **88**
stromatolites 159, **160**
strontium, radioisotope ratios, association with mass extinctions 195–6
sulphur, radioisotope ecological information 96
sympatry, defined 131
synapomorphy
cladistics 57–9
congruence test 59
defined 52
Synapsida
origins 227
possible patterns of clade replacement **183**

taphonomy 97–9
ecological incompleteness of information 97–9
taxon, origination and extinction 157
taxonomic patterns
analysis 49–70
and evolutionary process 11–18, **13**
non-utility of fossil evidence 110
sister taxa 54, 67, 80
*see also* higher taxa
taxonomic turnover 156–87, 252–3
clade interactions 179–88
and competitive processes concept 179–86
diversity changes within taxa 171–8
diversity of Earth's biota 158–70
higher taxa as units of diversity change 157–9
numbers and rate of increase within higher taxa 157–8
rates, species longevities 177–8
units 156–9
*see also* clades; cladistics
teleosts
cladogram **123**
classification, plesions **125**
Tetrapoda
alternative cladograms **113**
changing diversity **169**, 170
crown and stem groups 127
end-Triassic event 211
*Eusthenopteron*, lungfish–tetrapod relationship 112, **113**, **126**, 238
interrelationships 112, **113**, 124–7
origins 238–9
phylogenetic tree **5**, **126**
Therapsida, phylogenetic tree **229**
Therocephalia
hindlimb, structure 75, **76**, 78, 81
phylogenetic tree **229**
extant phylogenetic bracket **79**
sister groups 80
time
geological vs ecological 22
*see also* stasis

Tommotian fauna **204**, 205
trace fossils, ichnology 96
tree, defined **20**
trilobites, disparity **68**
Turkana, Lake, East Africa, punctuated equilibrium of molluscs **148**
turnover pulse hypothesis, speciation theory 146

unit character, defined 63
units of selection 39–45, **44**
units of taxonomic turnover 156–9
Upper Kellwasser event 207

vertebrate groups
  origins 235–9
  tetrapods, phylogenetic tree **5**
  *see also* mammals; Tetrapoda; *specific examples*
volcanism
  Cretaceous–Tertiary boundary (K–T) 212–13
  Deccan Traps 198
  element abundances 196–7
  and K–T boundary 198
  and mass extinctions 198
von Baer's rule, ontogeny 58